DICTIONARY OF BUILDING

DICTIONARY OF BUILDING

Randall McMullan

NICHOLS PUBLISHING: NEW YORK

First published in the United States of America by
NICHOLS PUBLISHING
Post Office Box 96
New York, NY 10024

Nichols Publishing is an imprint of GP Publishing Inc.

Library of Congress Cataloguing-in-Publication Data

McMullan, Randall.
Dictionary of building.

1. Building—Dictionaries. 2. Building—Great Britain—Dictionaries. I. Title.
TH9.M36 1988 690'.03'21 88-12598
ISBN 0-89397-319-X

Printed in Great Britain by
Richard Clay Ltd, Bungay, Suffolk

Contents

Preface

The words in the dictionary have been chosen from the written material used in current building practice. Modern specifications, technical standards, reports and product literature cover a surprisingly wide range of disciplines, many of them new. The age-old activities of building have meanwhile left a rich store of words.

New terms have displaced some older words: it is hoped that their omission from this dictionary has been by design rather than by accident or oversight. Many classical architectural terms have been excluded: only those commonly employed in modern construction or maintenance appear. Words that describe specialized aspects of trades and professions are also restricted to those in general use. Trades with long histories, such as masonry and carpentry, have accumulated a rich variety of terms which deserve complete books of their own.

The geographical spread of building terms within the English-speaking world is also complex. The use of a word can vary between areas such as the North and South of Britain, or the East Coast and West Coast of North America. Meanings can vary or even be reversed within relatively small areas, with most local users happily ignorant of other meanings. Entries in this dictionary are therefore cautious in indicating the geographical origin of words, though they do list different meanings where a word has more than one. The context will usually indicate which meaning is involved and at least some of the other possibilities for ambiguity will be known. Despite some differences between countries, there is more agreement about words than one might expect: an encouraging trend for an industry with multinational links and projects.

London
1988

How to use the Dictionary

Headings to entries appear in alphabetical order, disregarding breaks between words. In other words, entries will appear in the order:

fir
fire alarm
fireback
fire extinguishing equipment
fireplace
fire resistance
firing

and not:

fir
fire alarm
fire extinguishing equipment
fire resistance
fireback
fireplace
firing

Words appearing in small capitals (e.g. BRICK) indicate that there is an entry on that subject elsewhere in the dictionary. The instruction *See* tells the reader that information on the topic he or she has looked up is to be found under another heading. The instructions *See also* and *Compare* at the end of an entry tell the reader of other entries in the dictionary that may be of relevance.

The tags (*UK*) or (*USA*) after a heading, are as the Preface explains, only rough guides as to usage: (*UK*) should be taken to mean 'most of the British Isles', and (*USA*) 'most of North America'.

The following abbreviations are also used:

abbr. abbreviation
adj. advective
pl. plural
vb. verb

A

A. *abbr.* AMPERE.

ABS. *abbr.* Acrylonitrile-butadiene-styrene. A PLASTIC material with high toughness and low toxicity. It is used for cold water services and waste pipes but is not suitable for prolonged use at high temperatures.

absolute humidity. A measure of the actual quantity of water VAPOUR in the air. *See also* HUMIDITY.

absolute pressure. The measurement of gas PRESSURE on a scale that starts at absolute zero of pressure (vacuum). Absolute pressure equals atmospheric pressure plus GAUGE PRESSURE.

absolute temperature. TEMPERATURE measured in degrees KELVIN on the scale that starts at the absolute zero of temperature (-273.16°C). It is also known as the 'thermodynamic' or 'kelvin' temperature. *Compare* CELSIUS TEMPERATURE, FAHRENHEIT TEMPERATURE.

absolute viscosity. A measure of VISCOSITY or the amount of resistance given by a fluid to the tendency to flow. It can be calculated as the tangential stress on the fluid divided by the velocity gradient.

absorber. A building material or device that is designed to increase SOUND ABSORPTION and control the ACOUSTICS of a room. *See also* CAVITY ABSORBER, PANEL ABSORBER.

absorption. In general, the process that occurs when a material takes in substances or energy from its outside surfaces. (1) The water retained by brick, concrete or other material expressed as a percentage of its original dry weight. (2) *See* SOUND ABSORPTION.

absorption coefficient. *See* SOUND ABSORPTION COEFFICIENT.

absorption refrigerator. A refrigerator, such as a gas-operated type, that works without a compressor. The REFRIGERATION CYCLE uses heat to vaporize and circulate the ammonia REFRIGERANT which is later absorbed into a secondary cycle of water. *Compare* COMPRESSION REFRIGERATOR.

absorptivity. (1) The fraction of radiant heat energy absorbed by a surface compared to that absorbed by a technically perfect absorber (known as a 'black body'). The value varies with the temperature (wavelength) of the heat source. Building surfaces are usually rated for solar radiation. *Compare* EMISSIVITY. (2) The ability of a material to absorb sound. *See* SOUND ABSORPTION COEFFICIENT.

abstracting. The collection of items and quantities involved in a building contract in preparation for a BILL OF QUANTITIES. The descriptions and quantities are arranged in the recognized sequence in works sections and reduced to the appropriate units of measurement.

abutment. In general, an end-to-end meeting of different constructions. (1) A pier, wall or other mass that resists the horizontal thrust from an ARCH. (2) The junction of a roof surface and a wall.

AC. *abbr.* ALTERNATING CURRENT.

accelerator. In general, an ADMIXTURE that speeds the setting and hardening of cements, plasters and resins. (1) A chemical added to cement used in concrete or mortar. Calcium chloride can have unwanted effects and its use is now restricted. (2) An inorganic salt, such as potassium sulphate, that is added to anhydrous calcium sulphate to make wall PLASTER. (3) A pump used to help the circulation of water though a heating system, especially one with small-bore pipes.

access eye (cleaning eye, *USA*: clean-out). An opening in a DRAIN or duct through which a cleaning rod can be inserted. The opening is often located at a bend and covered with an access plate.

accounts. (1) In book-keeping, a list of debits and credits relating to a particular activity of a business. (2) A collection of regular business clients who pay money owed on a deferred basis, such as monthly.

ACH. *abbr.* Air changes per hour. *See* AIR CHANGE.

ACI. *abbr.* American Concrete Institute.

acid. A class of corrosive chemical compounds containing hydrogen that can be released as hydrogen IONS. Acids turn LITMUS paper red, give pH readings less than 7, and can be neutralized by alkalis (bases). *Compare* BASE. *See also* PH VALUE.

acidic. *adj.* The property of a substance that is ACID or contains acids.

acidity of water. In some natural waters, an ACIDIC quality that is caused by dissolved carbon dioxide and organic substances such as peat. Acidic waters are corrosive and tend to dissolve lead. *See also* PH VALUE, PLUMBO-SOLVENCY.

acoustic board. *See* ACOUSTIC MATERIALS.

acoustic clip. A type of FLOOR CLIP used to support a floating wooden floor on top of a screed. The clip includes a resilient layer to reduce the transmission of sound.

acoustic construction. Any form of construction designed to reduce the transmission of sound in a building. Methods include the use of dense materials, discontinuous construction and absorption.

acoustic materials. Special types of plaster, boards and tiles used on walls or ceilings in order to control sound INSULATION and REVERBERATION. The surfaces of such materials are porous and have a high SOUND ABSORPTION COEFFICIENT.

acoustics. (1) In general, the science of sound. (2) The study of sound quality in enclosed spaces such as a room or an auditorium.

acoustic tile. *See* ACOUSTIC MATERIALS.

acre. A non-metric unit of land area. 1 acre = 43,560 square feet or approximately 0.4 HECTARES.

Acrow prop. Trade name for a tubular metal PROP that can be adjusted for height. It is commonly use as a temporary support for window and door openings. *See also* NEEDLE.

acrylics. A family of POLYMERS used to make PLASTIC materials which can take many forms. Acrylic materials include Perspex, acrylic paints and transparent roofing materials.

acrylonitrile-butadiene-styrene. *See* ABS.

activated carbon. A porous form of carbon that is highly adsorptive. It is used to remove impurities from gases and liquids.

activator. A small amount of substance that is added to another material to start a reaction or process. *See* TWO-PACK.

active solar energy. Energy from the sun's RADIATION that is directly put to use, such as by collection in a solar roof panel. *Compare* PASSIVE SOLAR CONSTRUCTION.

activity. An action or process that takes time during a project. In a NETWORK ANALYSIS diagram activities can be represented either by arrows or by the junctions (nodes) between arrows.

activity-on-node (AON). *See* PRECEDENCE DIAGRAM.

acuity. The ability of the eye to distinguish between details that are very close together. Visual acuity increases as lighting levels increase.

AD. *abbr.* Above datum.

adaption. The visual process that occurs as the eyes take time to adjust to different lighting conditions.

adaptor. A connector used for joining electrical fittings or pipes of different sizes or designs. For example, adaptors are used for connecting several electrical devices to one outlet, or for joining copper pipe to steel tube.

additive. *See* ADMIXTURE.

additive colour. The reproduction of COLOURS by adding various proportions of the three additive primary colours—red, green and blue. This relatively rare form of colour reproduction is used by stage lighting and television screens. *Compare* MONOCHROMATIC LIGHT, SUBTRACTIVE COLOUR.

adhesion. (1) The joining together of materials by bonds between their surfaces, aided by adhesives such as CEMENT or GLUE. (2) The bond between a film of paint and the ground coat beneath.

adhesive. A material, such as a glue, that encourages ADHESION or bonding between surfaces.

adj. *abbr.* Adjustable.

admittance. A measurement of heat transmission through a construction which, unlike a U-VALUE, takes account of the thermal capacity of the materials. A heavyweight structure has a higher admittance (or 'Y-value') than a lightweight structure.

admixture. A small quantity of substance added to concrete, mortar or plaster to change the properties such as the workability of the mix or the speed of hardening. *See also* ACCELERATOR, RETARDER.

adobe. BRICKS or BLOCKS, reinforced with chopped straw, that are baked in the sun rather than fired in a kiln. They are used in hot dry areas such as Central America and Australia.

adze (*USA*: adz). An axe-like tool used to surface timbers roughly. The cutting blade is set at right angles to the handle.

aerated concrete (foamed concrete, gas concrete). A lightweight cellular concrete, usually made into blocks, which has low strength but good thermal INSULATION. The cellular structure is produced by a small quantity of additive, such as aluminium powder, which reacts chemically to release bubbles of gas.

aerobic bacteria. Bacteria that require air to survive. In the SECONDARY TREATMENT of sewage, aerobic bacteria are present in the FILTER BED. *Compare* ANAEROBIC BACTERIA.

AFNOR. *abbr.* Association Française de Normalisation. *See* STANDARDS ORGANIZATIONS.

A-frame. A structural frame made by two inclined members meeting at a point. These legs are tied by a horizontal member, giving a shape similar to the letter A.

African mahogany. *See* MAHOGANY.

afrormosia. A HARDWOOD tree of tropical Africa. The wood is used as a substitute for TEAK.

after-flush. The small amount of water remaining in a cistern after a WC pan is flushed. This water trickles down to remake the water seal in the pan.

agate. A form of quartz stone which usually contains bands of colour.

ageing. Changes in the properties of certain materials that occur over a period of time. The ageing of metals may involve heat treatments and give increased strength and hardness. The term 'age-hardening' can also mean an increased brittleness with age.

agenda. A list of matters to be discussed and attended to at a meeting.

agglomerate. In general, the formation of smaller parts into a large object.

aggregate. Crushed stone, slag, gravel, sand or other mineral material which form the bulk of concrete, plaster, asphalt or tarmacadam. FINE AGGREGATE is defined as that which can pass through a screen with holes 5 mm square, the remainder is COARSE AGGREGATE. *See also* GRADED AGGREGATE, LIGHTWEIGHT AGGREGATE, SIEVE ANALYSIS.

aging. (*USA*) *See* AGEING.

Agrément Board. *See* BRITISH BOARD OF AGREMENT.

Agrément certificate. A certificate associated with a product or process that has been approved by the BRITISH BOARD OF AGREMENT.

agricultural drain. *See* FIELD DRAIN.

AHU. *abbr.* AIR-HANDLING UNIT.

AIA. *abbr.* American Institute of Architects. *See also* ARCHITECT.

AIEE. *abbr.* American Institute of Electrical Engineers.

air-admittance valve (vacuum breaker). A valve that prevents the formation of a vacuum in a pipe system, such as a SOIL STACK, where the vacuum might break the water seal in a TRAP.

airborne sound. In the study of sound insulation, airborne sound is that sound which travels through the air before reaching the partition being assessed. Voices heard through a floor below are airborne sound while footsteps are usually IMPACT SOUND. The distinction is important when choosing methods of SOUND INSULATION.

air brick (ventilating brick). A brick made with perforations through which air can pass into a building. It is commonly used to ventilate the space beneath a suspended timber floor. *See also* CAVITY SLEEVE.

air brush. A small paint spray gun used for decorating and illustrating. It can produce delicate brush-like effects.

air change. A complete change of air for a room, which is equal to the volume of the room. The number of air changes per hour (ACH) can be used to specify ventilation rates and typical figures for a house range 0.5 to 2 ACH.

air cock. A small valve used to release air trapped in a hot water radiator. *See also* AIR LOCK.

air conditioning. A system of full control over the internal air supply to give specified comfort or other conditions. It is a form of mechanical ventilation that has the extra ability to cool and heat the air, and to change HUMIDITY.

air curtain. A strong downward or upward flow of warm air produced by mechanical units set above entrances such as loading bays that need to be kept open. The air pattern is designed to limit the amount of cold air entering from outside.

air-dry. *See* AIR-SEASONED.

air duct. A rectangular or circular channel that carries air in a ventilation system.

air entrainment. The general process of adding air to materials. Air-entrained concrete is produced by a chemical additive in the mix which produces a stable dispersion of small air bubbles. The higher air content of the concrete improves the workability of the mix and gives better frost resistance to the hardened concrete.

air grating. A slotted plate that covers an AIR DUCT where it enters a room. *Compare* REGISTER.

air-handling unit. A central item of equipment that supplies conditioned air to a VENTILATION or AIR CONDITIONING system.

air house. *See* AIR-SUPPORTED STRUCTURE.

air lock. (1) An entrance LOBBY designed to prevent the loss of air from a building. A simple system has two doors which are never both open at the same time. (2) (*USA*) *See* WEATHERSTRIP. (3) An obstruction in a water supply or heating system caused by a bubble of air trapped in a pipe or fitting.

air-seasoned. *adj.* Describing timber that is left in the air and sheltered from the rain until its MOISTURE CONTENT is in equilibrium with the surrounding air. The measured moisture content is typically around 20 per cent, with variations that depend on the type of wood and local climate. *Compare* KILN-DRIED.

air shaft. *See* LIGHT WELL.

air slaking. The absorption of moisture from the air by the chemical reaction of CEMENT or QUICKLIME.

air speed. The velocity of air as measured while passing though a ventilation duct or outlet.

air-supported structure (air house, pneumatic structure). A structure that is kept erect by air pressure. The flexible fabric roof may be supported by higher air pressure inside the building, in which case air lock and constant fan supplies are needed. An alternative is to support the membrane by tubular ribs stiffened by air.

air temperature. The average temperature of air measured with an ordinary DRY-BULB THERMOMETER. *Compare* RADIANT TEMPERATURE, ENVIRONMENTAL TEMPERATURE.

air test. A method for detecting leaks in drains and gas pipes. The length of the system under test is plugged at all openings and pressurized with air to a specified maximum. A U-gauge connected to one of the plugs is sensitive to small leakages from the system. *Compare* HYDRAULIC TEST, SMOKE TEST.

air-to-air resistance. The total THERMAL RESISTANCE of all thermal components occurring through a structure from the inside surface layer to the outside surface layer. *See also* U-VALUE.

air-to-air transmission coefficient. *See* U-VALUE.

air washer. Part of an AIR CONDITIONING system where air is passed through water sprays or over wet plates. The air is cleaned and, if subsequently heated, has a lower HUMIDITY.

AITC. *abbr.* American Institute of Timber Construction.

Al. Chemical symbol for ALUMINIUM.

ALA. *abbr.* American Lumber Association.

alabaster. A white, semi-translucent form of MINERAL, usually based on GYPSUM.

alarm call point. A place where an alarm system, such as a fire alarm, can be manually operated.

alburnum. The botanical name for SAPWOOD of DECIDUOUS TREES.

alkali. A soluble form of the chemical compounds called bases. Strong alkalis are corrosive and can neutralize ACIDS. Alkalis will turn litmus paper blue and give a pH reading greater than 7. The calcium compounds (lime) in concrete, mortar and plaster make them alkaline. *See also* PH VALUE.

alkali resistance. A property of materials, such as some paints, that reduces the disruptive effects of coming into contact with alkalis. New concrete, brickwork and plaster contains alkaline calcium compounds (lime) which are alkaline and can break down the resins in paints.

alkali silica reaction (ASR). A reaction between the alkali in some cements and the silica in certain aggregates used in concrete. The reaction causes expansion and disruption of the concrete when it is penetrated by moisture. *See also* CONCRETE CANCER.

alkyd. A synthetic resin that is based on POLYESTER and used as the BINDER in some paints.

all-in aggregate. Aggregate for concrete that includes both COARSE AGGREGATE and FINE AGGREGATE.

all-in contract. *See* PACKAGE DEAL.

allotropy. The ability of a chemical element to exist in two or more different physical forms. Coal and diamond, for example, are two allotropes of carbon.

alloy. A metal whose properties have been modified by adding another material, usually a metal. For example: BRONZE is an alloy of copper and tin; STEEL is iron that contains small amounts of carbon.

alluvial. *adj.* Pertaining to ALLUVIUM.

alluvium. Mud, soil and other materials deposited by the flow of water over a flood plain or river bed.

alternating current (AC). A pulsing form of electric CURRENT that continually reverses its direction of flow. These changes usually follow the repetitive form of a sine wave. Mains supplies of electricity give alternating current while batteries give DIRECT CURRENT. *See also* FREQUENCY, ROOT MEAN SQUARE, TRANSFORMER.

alternator. *See* DYNAMO.

altitude. (1) The height of a location above sea level. (2) The sighting angle an object compared to the horizontal.

alum. A chemical compound of aluminium, potassium and sulphate. It is used to control the setting of plaster.

alumina. Aluminium oxide, Al_2O_3. A chemical found naturally as a primary ingredient in all CLAYS. ALUMINIUM is refined from clays with a high alumina content, such as BAUXITE.

aluminium (*USA*: aluminum). A pure metal element (chemical symbol Al) of low density and a white silvery colour. ALLOYS of aluminum, which are used in construction, have a higher strength than aluminium. Aluminium is a good conductor of heat and electricity and has a high resistance to corrosion.

aluminium foil. A thin sheet form of aluminium, less than 0.15 mm thick, which is often bonded to another material such as PLASTERBOARD. It is used as an insulator against radiant heat and as a VAPOUR BARRIER.

aluminium paint. A paint based on aluminium powder which is used for the protection of metals.

aluminous cement. *See* HIGH-ALUMINA CEMENT.

aluminum. (*USA*) *See* ALUMINIUM.

ambient temperature. The temperature of the air immediately surrounding an object.

ambrosia beetle. One of several species of BEETLE whose life cycle is destructive to wood.

American bidding. A form of TENDER in which payment may include goods or services instead of money.

American bond. *See* ENGLISH GARDEN WALL BOND.

amino resin. A synthetic RESIN that is used in paint for metal surfaces.

ammeter. An instrument that measures electric CURRENT in AMPERES.

ammonia (NH_3). A volatile and toxic gas which is used as a REFRIGERANT.

amp. *abbr.* AMPERE.

ampere (amp, A). The unit of electrical CURRENT. One ampere is equivalent to a flow of one coulomb per second, where the COULOMB is the unit of static electricity. *See also* AMMETER.

ampere-hour. An alternative to the COULOMB as a measure of the quantity of electricity. The unit is used for rating the capacity of a rechargeable battery.

amplitude. For vibrations in general, amplitude is the maximum distance in either direction from the rest position. The term is used in connection with sound waves and with alternating current electricity.

amyl acetate ($C_7H_{14}O_2$). A solvent used to dissolve certain paints such as cellulose lacquers. It has a strong banana-like smell.

anaerobic bacteria. Bacteria that require no air to survive. In SEWAGE treatment, anaerobic bacteria start the breakdown of raw sewage. *Compare* AEROBIC BACTERIA.

anaglyptic wallpaper. A heavy type of WALLPAPER that is embossed with a pattern.

anchor. In general, any means of obtaining a fixing on to a building so as to support or restrain another part of the structure. Methods include the use of anchor plugs, clips, blocks, and plates. *See also* EXPANSION BOLT.

anemometer. An instrument that measures the movement of air, such as that in ventilation ducts. *See also* HOT-WIRE ANEMOMETER.

aneroid barometer. A simple form of BAROMETER which measures variations in atmospheric pressure by detecting changes on the sides of a sealed thin metal box. It is used for approximate measurements of ALTITUDE.

angiosperms. Flowering plants whose seeds are enclosed within an ovary. Many species of HARDWOOD trees are angiosperms. *Compare* GYMNOSPERMS.

angle bead (plaster bead, plaster stop). A strip of metal or plastic used to reinforce the edges of plaster and plasterboard.

angle block (glue block). A triangular piece of wood used to stiffen a joint, such as between the TREAD and RISER in a timber stair.

angle bracket (gallows bracket). A bracket that extends from a wall in the shape of an inverted letter L and is often used to support shelving.

angle closer. A brick cut to a shape that completes and continues the BOND at the corner of two wall surfaces.

angle grinder. An electric hand tool that uses a rotating abrasive disk to cut metal and masonry.

angle iron (angle section). A standard length of rolled steel with a cross-section shaped like a letter L.

angle of elevation. The angle, measured in a vertical plane, between the horizontal and the observed point.

angle of incidence. The angle at which sound, light or solar radiation strikes a surface. The angle is measured between the incident rays and the line at right angles to the surface (the 'normal').

angle of repose. The steepest angle at which a heap of loose material, such as soil, will come to rest, or will remain unsupported.

angle section. *See* ANGLE IRON.

angle tie (dragon tie). A horizontal timber beam across the corner of a roof. It ties the two walls and may also support the DRAGON BEAM.

anhydrite. An ANHYDROUS form of CALCIUM SULPHATE from which anhydrite plaster is made. This is not the same as ANHYDROUS GYPSUM PLASTER.

anhydrous. *adj.* A chemical term used to describe a substance without WATER OF CRYSTALLIZATION.

anhydrous gypsum plaster. A form of PLASTER made from GYPSUM that has been heated to remove all WATER OF CRYSTALLIZATION. A setting ACCELERATOR is added to make a hard usable plaster such as Keene's cement and Sirapite. *Compare* ANHYDRITE.

anion. In ELECTROLYSIS, an ION that is attracted to the ANODE.

annealing. A process of HEAT TREATMENT that is used to toughen glass or metals. The sample is heated and then gradually cooled: the formation of a new crystalline structure reduces stresses and brittleness.

annual ring (year ring). One of the circular growth rings seen when a tree trunk is cut through.

annular nail. A type of NAIL with circular ridges around the shaft. The rings make the nail less likely to rise out of a material than an ordinary nail.

annular bit. *See* HOLE SAW.

anobium puntatum. *See* FURNITURE BEETLE.

anode. (1) In ELECTROLYSIS, the positive conductor or ELECTRODE. (2) The positive terminal of an electric cell or battery to which current flows. *Compare* CATHODE.

anodic protection. An electrical method of protecting iron against corrosion by connecting an external source of direct current. It is similar in principle to CATHODIC PROTECTION.

anodizing. The use of ELECTROLYSIS to produce a protective or decorative coating on a metal such as aluminium. The metal object itself is connected so as to act as the ANODE.

ANSI. *abbr.* American National Standards Institute. *See* STANDARDS ORGANIZATIONS.

anti-actinic glass. *See* HEAT-ABSORBING GLASS.

anti-capillary groove. *See* CAPILLARY GROOVE.

anti-condensation paint. A bulky paint that absorbs moisture and may provide a little thermal insulation. It can hide the effects of small amounts of surface CONDENSATION but, despite its name, it is not a cure for condensation.

anti-corrosive paint. A paint containing PIGMENTS that inhibit CORROSION of metal surfaces. *See also* INHIBITING PIGMENT.

antimony oxide (Sb_2O_3). A dense white PIGMENT for paints which gives good hiding power.

anti-siphon pipe (puff pipe). A small ventilation pipe in a plumbing system that admits air downstream from the trap in a WC, sink or similar device. It is designed to prevent the water seal in the trap being sucked away by the reduced pressure that occurs when water is flushed down the nearby SOIL STACK. *See also* SINGLE-STACK SYSTEM. *Compare* AIR-ADMITTANCE VALVE.

anti-siphon trap (deep-seal trap). A type of drainage TRAP below a sink or washbasin. It contains a large volume of water which makes it difficult for the water seal to be broken. It can be a replacement for an ANTI-SIPHON PIPE.

anti-skinning paint. A paint containing an additive that helps prevent a surface skin forming during storage.

AON. *abbr.* Activity-on-node. *See* PRECEDENCE DIAGRAM.

APA. *abbr.* American Plywood Association.

apartment. (*USA*) One of a number of self-contained dwellings within a single building. A FLAT. *Compare* CONDOMINIUM.

apex. The topmost stone of a gable, vault or arch. *See also* KEYSTONE.

apostilb. An alternative unit of surface brightness or LUMINANCE. It is measured by the LUMINOUS FLUX emitted per unit area. 1 apostilb = 1 lumen per square metre.

appd. *abbr.* Approved (used on drawings).

apprentice. A person learning a craft or TRADE. A traditional apprenticeship was a legal contract (INDENTURE) between the apprentice and the employer.

approximate quantities. A quick method for giving an initial estimate of the cost of a project. The major elements are costed using an overall rate for their construction and an average cost of minor items is included. *Compare* BILL OF QUANTITIES.

apron eaves piece. A single piece of FLASHING material that is inserted to protect the base of an inclined roof.

apron flashing. A single piece of FLASHING material used where a vertical surface, such as a chimney, penetrates a sloping roof. *Compare* SOAKER. *See* ROOF diagram.

apron wall. (*USA*) A SPANDREL between the windows in a multi-storey building.

aquifer. A layer of porous rock or soil, like limestone or gravel, that supplies good quantities of water. An aquifer can also act like a pipe and transmit water over a distance.

arbitration. The settlement of a building contract dispute by an independent arbitrator rather than by legal action in the courts. Both sides agree to abide by the decision of the arbitrator, who may be named in the CONTRACT.

arc. (1) A discharge of electric CURRENT across an air gap. *See* ARC WELDING. (2) Part of the circumference of a circle.

arcade. A passageway that is roofed by a system of arches and often lined with shops.

arch. A curved structure that supports a heavy load such as a roof or bridge. The downward thrust is transferred by compression within the structure and the sideways thrust from the arch is supported by ABUTMENTS. Arches can be constructed from most construction materials but the principle is particularly important for stones and bricks which can support one another within an arch. *See also* AXED ARCH, GAUGED ARCH.

arch brick (voussoir). A wedge-shaped brick used for building an ARCH or other curved structure. *See also* KEYSTONE.

arch dam. A type of DAM built in a curve against the water. The structure uses the principle of the ARCH to transfer the thrust of the water to the sides of the valley.

architect. A person who designs buildings, prepares drawings and contracts, and supervises construction. In many countries, such as the United Kingdom, the law requires that an architect must be appropriately qualified and registered. *See also* AIA, RIBA.

architectural ironmongery. *See* HARDWARE.

architrave. A TRIM, often of wood, that covers the join between the wall and the frame of a door or window.

architrave block (foot block, skirting block). A junction block at the bottom edge of a door surround that acts as junction between the vertical ARCHITRAVE trim and the horizontal SKIRTING BOARD.

arch stone (voussoir). A wedge-shaped stone used for building an ARCH or other curved structure.

ARCUK. *abbr.* Architects' Registration Council of the United Kingdom.

arc welding. A method of joining metals by means of a hot electric arc produced between the metal and an electrode. The melting and refusion of the metal, together with a filler metal, gives a strong join that is used in heavy structural work.

armature. The revolving part of an electric motor or DYNAMO on which the coils are wound.

armoured cable. Electrical CABLE that has been wound with layers of steel strip or wire to give increased mechanical protection for use beneath streets and in tunnels or mines.

armour-plate glass. A type of FLOAT GLASS that has been heat-treated to give increased resistance to impact and to temperature changes.

arris. The sharp edge to part of a building such as a brick, a plaster wall or skirting board.

arris gutter (Yankee gutter). A V-shaped wooden gutter.

arris rail. A three-sided fence rail with a triangular cross-section. Several such rails are set horizontally into the fence posts and vertical boards are attached to them. *See also* CLOSE-BOARDED FENCING.

arrow diagram. A form of CRITICAL PATH ANALYSIS in which the activities are represented by arrows linked in a network which represents their sequence and duration. *Compare* PRECEDENCE DIAGRAM.

artesian well. A WELL or borehole that delivers a constant supply of water without pumping. The pressure occurs because the head of the well is below the WATER TABLE.

Artex. A brand name for a fibrous type of plaster that gives a TEXTURED FINISH.

articles of association. (*UK*) The regulations of a registered company.

artificial light. Light that is obtained from a lamp. *Compare* NATURAL LIGHT.

artificial sky. A room with specially placed lamps, reflectors and diffusers that can be controlled to imitate the distribution of light from the sky. *See also* HELIDON, STANDARD SKY.

artificial stone (cast stone, reconstructed stone). An imitation stone material made from a mixture of crushed natural stone and CEMENT.

ASA. *abbr.* American Standards Association. *See* STANDARDS ORGANIZATIONS.

asbestos. A silica-based mineral with thin tough fibres and a high fire resistance. *See* ASBESTOS MATERIALS.

asbestos cement. A composite material of cement and asbestos fibres which has been used to make wall sheets, roof materials, flue pipes and water pipes. *See* ASBESTOS MATERIALS.

asbestos materials. ASBESTOS has been widely used in building, usually in composite materials such as concrete, because of the reinforcement properties of its fibres and its high fire resistance. However, it is now known that the fine dust or fibres produced when working with asbestos causes lung disease. The use of asbestos and the removal of existing asbestos in buildings is therefore subject to severe safety restrictions and requires specialist treatment. Where possible, manufacturers of asbestos materials now use substitute fibres such as plastic, glass and textiles.

A-scale. The most common WEIGHTING NETWORK in a SOUND LEVEL METER. It is a standardized electronic setting which emphasizes the middle frequencies of sound in a similar manner to human hearing. The symbols dB(A) indicate that a sound is measured in DECIBELS using the A-scale.

ASCE. *abbr.* American Society of Civil Engineers.

ash. (1) A HARDWOOD tree of the genus *Fraxinus*. The pale-coloured wood is used for VENEER and for handles of tools such as hammers. (2) The carbon-based products left in a fireplace or furnace after the process of burning.

ashlar (ashler). (1) Walls or facings made of squared building stone laid in courses with thin joints. *Compare* RUBBLE, RUBBLE ASHLAR. (2) (*USA*) Walls made of burnt clay blocks that are larger than bricks. *See* TERRA COTTA.

ashlering. Short vertical timber STUDS used in an attic room to form the low partition between the pitched roof and floor.

ashpit. An area beneath a FIREPLACE or burner designed to collect and hold ash before it is removed.

ASHRAE. *abbr.* American Society of Heating, Refrigerating and Air-conditioning Engineers.

ASHVE. *abbr.* American Society of Heating and Ventilating Engineers.

asphalt. (1) A composite material made from mineral AGGREGATE bonded together by BITUMEN. Some natural asphalts which occur in rocks and at Lake Asphalt, Trinidad are used as sources of bitumen. *See also* ASPHALT MATERIALS, COLD ASPHALT, MASTIC ASPHALT, ROCK ASPHALT. (2) (*USA*) Pure BITUMEN.

asphalt materials. ASPHALT has properties of plastic flow, adhesion and waterproofing which are valuable in building and road construction. Asphalt roofing materials combine layers of asphalt with a felt or matting. *See also* ROOFING FELT.

asphalt roofing. *See* BUILT-UP ROOFING.

asphalt shingle. (*USA*) A roof SHINGLE made from a fibrous material, such as fibreglass, that is soaked in ASPHALT or BITUMEN and then coated in aggregate.

ASR. *abbr.* ALKALI SILICA REACTION.

assembly drawing. A type of drawing that displays all the parts of a construction and indicates the joins between them.

ASTM. *abbr.* American Society for Testing and Materials. *See* STANDARDS ORGANIZATIONS.

astragal. A raised MOULDING with a T-shaped cross-section.

atmosphere. (1) The layer of gases that surrounds the earth. (2) The pressure exerted by the earth's atmosphere under certain standard conditions.

atom. The smallest quantity of a chemical element that can take part in a chemical change.

atomization. The process in a spray gun which breaks up fluids, such as paints, into very fine droplets.

atrium. In modern buildings, a large entrance hall or LOBBY that rises for the height of several floors which open on to the space. The atrium is often extensively glazed and landscaped inside.

attendance. Help given to a sub-contractor by the main CONTRACTOR. It may include use of PLANT on site.

attenuation. A gradual reduction in sound energy, during transmission through a wall for example.

attic. The area between a roof and ceiling.

attic wall. A low partition built in a roof to provide extra support for the RAFTERS.

audio frequency. A sound wave of a frequency that can be detected by the human ear. It covers a range of approximately 20–20,000 Hz.

audiology. The scientific study of hearing.

auditorium. The area of a large hall or theatre in which the audience sits.

auger. A tool shaped like a corkscrew, used for boring holes in timber or the ground.

austenite. A form of STEEL structure that occurs at temperatures above 900°C when the iron crystals dissolve all the carbon present. Austenitic steels, which contain nickel and manganese, have an austenite structure at normal temperatures. *Compare* MARTENSITE.

autoclave. A large strong chamber vessel used to cure materials, such as CALCIUM SILICATE BRICKS, with high pressure steam.

autogenous healing. The process in which small breaks in concrete seal themselves. It is usually promoted by keeping the concrete surfaces damp and in contact.

automatic flushing cistern. A FLUSHING CISTERN that fills continuously and flushes automatically when the water level reaches a certain height. It may be used to flush urinals.

automatic level. A type of surveyor's LEVEL which uses suspended prisms in the telescope to avoid the need for the frequent adjustments needed by a DUMPY LEVEL or TILTING LEVEL.

AWG. *abbr.* (*USA*) American Wire Gauge.

awl (scratch awl, scribe). A steel-pointed tool used by carpenters for marking wood or piercing thin materials.

awning. A retractable fabric screen that projects above a window to protect it from sunlight. A common form of awning is made from a folding frame covered with canvas.

axe. (1) A long-handled tool with a flat blade used for cutting down trees and for roughly dressing timber. (2) (*USA*) A BRICKLAYER'S HAMMER.

axed arch. An ARCH for which the bricks are roughly cut. *Compare* GAUGED ARCH.

axed brick. A brick cut with an AXE or BOLSTER. *Compare* GAUGED BRICK. Axed brickwork is laid with thicker mortar joints than gauged brickwork.

axial load. A load on a structure where the line of force passes through its centre of area (centre of gravity).

axman. (*USA*) *See* CHAINMAN.

axonometric projection. A form of drawing PROJECTION that gives a three-dimensional view of the building. It is similar to the ISOMETRIC PROJECTION except that the horizontal lines are drawn at 45° to the horizontal. *Compare* ORTHOGRAPHIC PROJECTION.

azimuth. The angle measured clockwise in the horizontal plane from a reference line, such as that running north.

B

backactor. A popular type of multi-purpose EXCAVATOR used to excavate pits and trenches. It is based on a tractor power unit with an excavating bucket (backhoe) at the rear and a loading scoop in the front. *See also* JCB.

back boiler. A boiler built into the back of an open fire or stove to provide hot water or CENTRAL HEATING.

back drop. A vertical pipe in a MANHOLE which drops from a higher level than the main drain. Connecting a branch by this method avoids excessive excavation and fall.

back filling. (1) (back fill) Material such as earth that is returned after excavations for foundations and drains have been completed. (2) (*USA*) Brick nogging. The brickwork in a wall that is hidden by the FACING BRICKS.

back-flap hinge. A door HINGE that is attached to the face of a door because the edge of the door is too thin for a BUTT HINGE.

back flow. A flow of liquid in a pipe that moves in the opposite direction to the designed flow. The backing up of sewage in a blocked drain is an example.

background heating. A heating system designed to be supplemented by another heat source. By itself background heating does not normally provide a high enough temperature for full thermal comfort.

back gutter (chimney gutter). A metal gutter behind a chimney at the junction with a pitched roof.

backhoe. (1) (*USA*) A multi-purpose excavating machine with a bucket at the back and a loading scoop at the front. (2) The excavating bucket at the rear of a BACKACTOR or JCB.

backing. The brickwork in a wall that is hidden by the FACING BRICKS.

backing insulation. Insulating material that is shielded from damage or high temperatures by a layer of more resistant material, such as the FIRE BRICKS in a furnace.

back-inlet gulley. A drainage GULLEY with an inlet pipe connected below the top opening or grating.

back lintel. A type of LINTEL that carries a load across the backing of a wall but not the face.

back putty. The part of the putty around a window that remains between the glass and the rebate of the frame. It may also include the BED PUTTY.

back saw. *See* TENON SAW.

back shore. An outer part of a RAKING SHORE.

back sight (*USA*: plus sight). A sighting made in surveying after an instrument has been moved. The observation is taken back to the previous sighting. *Compare* FORE SIGHT.

back siphonage. A form of liquid BACK FLOW into a supply system caused by unequal pressures or 'siphonage' developing in the pipes. Water authority regulations are designed to prevent the back siphonage of polluted household water into the main water supply.

backup. (*USA*) *See* BACK FILLING.

back vent. (*USA*) *See* LOOP VENT.

baffle. In general, a device used to deflect a liquid or gas. (1) A section fitted between cladding panels to prevent the direct entry of rain. (2) An acoustic reflecting surface that modifies the distribution of sound.

bag plug. An inflatable bag used as a drain plug when testing a drain for leaks. *See also* AIR TEST, HYDRAULIC TEST.

Bakelite. A THERMOSETTING type of plastic from the family of PHENOL-FORMALDEHYDE RESINS. Bakelite, brown in colour, was one of the earliest synthetic resins and may still be used for electrical fittings.

baking. *See* STOVING.

balanced base. A type of foundation used to divert pressure from existing structures beneath the ground. A connecting beam between the bases of two columns allows one of the columns to transfer forces to the other.

balanced construction. A form of construction, such as that in PLYWOOD, which resists warping and other changes by balancing forces in the materials. On either side of the central line there are layers of material that match in position, thickness, grain and other characteristics. Adjoining layers of plywood have grains running in opposite directions.

balanced flue. A ventilation system for a heating appliance, such as a gas heater, that is independent of the room air. The construction of the outside terminal includes both the inlet for fresh air and the outlet for exhaust gases. The flows in the inlet and outlet balance one another without mixing. *See also* FLUE.

balanced step. *See* DANCING STEP.

balanced weight. A weight inside the casing of a SASH WINDOW. It is usually one of a pair which balance the weight of the window and make it easier to slide.

balcony. A platform that projects from an upper floor of a building, usually with access by window or door.

Bales's catch. A type of BALL CATCH.

balk. *See* BAULK.

ballast. (1) Ungraded gravel which can consist of sand, grit and stones or clinker. It is used to ballast or hold down the SLEEPERS of a railway track. (2) A mixture of COARSE AGGREGATE and FINE AGGREGATE used to make concrete. (3) An electrical coil or device used to restrict or 'choke' the flow of current through a DISCHARGE LAMP. It also provides a voltage pulse to start the discharge.

ball catch. A type of door fastening in which a spring-loaded ball projects from the door and engages with a hole on the door frame.

ball cock. *See* BALL VALVE.

balloon (bonnet). A globe-shaped wire cage that covers the open end of a VENT PIPE or outlet to a GUTTER.

balloon frame. A type of TIMBER-FRAME CONSTRUCTION which may be used for multi-storied houses. The STUDS are continuous from the ground to roof level and carry the JOISTS for intermediate floor levels. *Compare* PLATFORM FRAME. *See also* RIBBON BOARD.

ball valve (ball cock, float valve). A valve on the supply pipe to a CISTERN, such as for a WC. The valve is connected by lever to a floating ball which causes the supply to be turned off as the water level rises and turned on again when the water level falls.

baluster (banister, bannister). A pillar or post in the BALUSTRADE of a bridge or of stairs. *See* STAIRS diagram.

balustrade. A row of BALUSTERS that support handrails, such as at the edge of a stair or bridge. *See* STAIRS diagram.

band-and-gudgeon hinge. *See* BAND-AND-HOOK HINGE.

band-and-hook hinge (band-and-gudgeon hinge). A heavy type of T-HINGE with a wrought iron strap that is bent into a circle at one end. This loop drops on to a pin (gudgeon) and forms the hinge.

band saw. A power SAW that consists of an endless steel loop of teeth. Small versions are used for cutting difficult shapes and sawmill versions are used for the CONVERSION of timber.

banister. *See* BALUSTER.

banker. A board or platform on which concrete, mortar or plaster is mixed by hand. The mixture is contained by edge boards on three sides of the platform.

bannister. *See* BALUSTER.

bar. A straight length of STEEL or other metal as used for the REINFORCEMENT of concrete. The cross-section is usually round or many-sided.

barbed nail. *See* ANNULAR NAIL.

bar-bending machine. A machine used to bend the steel bars used in REINFORCED CONCRETE.

bar chart. A display of the different activities in a programme of work using the same horizontal time scale. The bars represent the relative starting times and durations of the tasks. In project management bar charts are also known as GANTT CHARTS. *See also* NETWORK ANALYSIS, PROGRESS CHART.

barefaced tenon. A type of TENON with a shoulder that projects from one side only.

barge board (gable board, verge board). A board attached to the sloping edge of a GABLE that covers and protects the ends of roof timbers. A pair of barge boards usually make the shape of an upside-down letter V.

barge course. A COPING of bricks or tiles running along the top edge of a roof GABLE.

barium plaster. A dense type of PLASTER made from BARYTES aggregate bound together with GYPSUM PLASTER. It is used for protection against X-rays.

bark. Dead corky tissues that form the protective outer layer of a tree trunk. *See also* WANE.

barometer. An instrument that measures BAROMETRIC PRESSURE. *See also* ANEROID BAROMETER.

barometric pressure. The pressure produced on a surface by the gases in the atmosphere. Changes in the pressure can be used to estimate altitude above sea level and to predict weather. *See also* ISOBAR.

barrel. The main part of a pipe where the bore diameter and the wall thickness is constant.

barrel bolt. A door fastening made from a metal rod that is slid by hand along a slotted guide.

barrel vault. A semi-circular roof in the form of a continuous ARCH. Although traditionally built of brick or masonry, this type of roof is now often made from concrete or lightweight materials.

barrow run. A path on a building site used for wheeling a loaded WHEEL BARROW. It is often made from SCAFFOLD BOARDS laid on the ground.

barytes. A barium ore ($BaSO_4$) used as a heavy AGGREGATE. *See also* BARIUM PLASTER.

basalt. A dark igneous type of rock. *See also* IGNEOUS ROCKS.

base. (1) The lowest part of a wall, column or similar part of a construction. (2) The concentrated body of a paint that needs to be diluted before use. The base may consist of the PIGMENT or the MEDIUM. (3) A substance that reacts with an ACID to give a SALT and water only. The compounds in rocks, concrete, mortar and plaster tend to make them basic. An ALKALI is a base that can be dissolved in water.

baseboard. (*USA*) *See* SKIRTING BOARD.

base course. (1) The lowest visible layer of bricks or stonework in a masonry wall. (2) In road construction, the surface layers other than the WEARING COURSE.

base exchange. A small-scale method of WATER SOFTENING. The compounds responsible for hardness are converted, via exchange of IONS, into other compounds. The water is diffused through a ZEOLITE compound which must regularly be regenerated by flushing with a sodium chloride solution. *Compare* SODA-LIME PROCESS.

base line. In land surveying, a line that is accurately measured and used as the basis for subsequent calculations of length and area. *See also* TRIANGULATION.

basement. The lowest storey in a structure that is partly or wholly beneath ground level.

base plate. A steel plate used to distribute the load beneath a steel stanchion or item of heavy equipment. *See also* PADSTONE.

basil. The angle at which a cutting blade is ground sharp.

basilar membrane. The group of nerve endings that detect sound vibrations within the human inner ear.

basin. *See* WASHBASIN.

bastard ashlar. The facing stonework of a RUBBLE wall that is dressed and built like ASHLAR.

bat. (1) A CLOSER brick that is cut short in order to complete a BOND. (2) (batt) A slab of material, such as MINERAL WOOL or EXPANDED PLASTIC, that can be placed in a wall to improve the THERMAL INSULATION. (3) *See* LEAD WEDGE.

batch. In general, the quantity of material used or produced at one time. The quantity of CONCRETE, MORTAR or PLASTER obtained from a single mixing operation.

batch box (gauge box). A measuring frame that contains and controls the quantity of aggregate and sand used in a mix of CONCRETE. The box usually has four sides and no bottom.

batching plant. The equipment used to measure, by weight or by volume, the ingredients of a concrete MIX.

bath (bathtub, tub). A bathroom SANITARY APPLIANCE in the form of a large container in which a person can be immersed in water. Baths may be made of plastic, enamelled steel or iron.

baton. *See* BATTEN.

batten (baton). (1) A length of square-sawn softwood with a section that typically measures 50–100 mm by 100–200 mm. (2) Horizontal strips of timber to which are attached coverings such as tiles, slates and plasterboard.

battenboard. A type of COREBOARD.

batten door. *See* BOARDED DOOR.

battening. Strips of wood attached to a wall as COMMON GROUNDS.

batter. *See* RAKE.

batterboard. (*USA*) *See* PROFILE.

batter brace. (*USA*) A diagonal BRACE.

batter peg. A peg driven into the ground to set out the limits of an earth slope.

battery lavatories. (*USA*) An installation of multiple WASHBASINS, as used in public washrooms.

batting (tooling). A regular pattern of chisel marks on the surface of stone. *Compare* BOASTING.

baulk (balk). Softwood timber that is squared with dimensions usually greater than 100 by 125 mm.

bauxite. Hydrated ALUMINA. The main ore from which ALUMINIUM is extracted.

bay. In general, a uniform division of space in a building such as the area between four columns.

bayonet cap (BC). A type of connector for electric LAMPS found in some domestic lamps. Two small pins are used to engage slots in the lampholder.

bay window. A window structure that projects beyond the general line of a building. The window rests on its own sill wall and foundations.

BBA. *abbr.* BRITISH BOARD OF AGREMENT.

BBC. *abbr.* (*USA*) Basic Building Code.

BC. *abbr.* BAYONET CAP.

BCIS. *abbr.* (*UK*) Building Cost Information Service.

bd. *abbr.* BOARD.

BDA. *abbr.* (*UK*) Brick Development Association.

bead (beading). (1) A length of moulding used to hide a join or to retain glass in a window frame. Beading is often semi-circular in cross-section. *See also* COCK BEAD, QUIRK BEAD. (2) A built-up line of paint at the bottom of a painted area which is caused by excessive flow.

beading tool. *See* BEAD PLANE.

bead plane (beading tool). A type of PLANE used for forming wood into the shape of a BEAD.

beam. A structural member designed to resist forces that tend to bend it. Simple beams are horizontal and transfer their load to supports at each end. Solid wooden floor JOISTS and steel RSJs are examples of beams.

beam and slab floor. A type of concrete floor in which the slab is supported on REINFORCED CONCRETE beams.

beam box. The FORMWORK for casting a concrete beam.

beam casing. An external covering that increases the FIRE RESISTANCE of a steel beam. Claddings of concrete and plaster are often used.

beam compass. *See* TRAMMEL.

beam fill (beam filling). Brickwork or masonry filling between the members of a floor or roof. It stiffens the structure and improves fire resistance.

beam hanger. *See* STIRRUP STRAP.

bearer. A horizontal timber that distributes a load or supports another timber such as a JOIST.

bearing. (1) The structural area where a BEAM is supported. (2) In surveying, the horizontal angle between a survey line and a reference direction such as true north. *See also* QUADRANT BEARING, TRUE BEARING, WHOLE CIRCLE BEARING.

bearing capacity. The BEARING STRESS that a subsoil can withstand before it starts to compress. It is measured by the load divided by the bearing surface.

bearing pile. A PILE that carries a downward weight rather than some form of thrust at an angle. *See also* END-BEARING PILE, FRICTION PILE. *Compare* SHEET PILE.

bearing plate. *See* BASE PLATE.

bearing pressure. *See* BEARING STRESS.

bearing stress (bearing pressure). The load on a BEARING divided by the area of the bearing surface.

BEC. *abbr.* (*UK*) Building Employers' Confederation.

bed. The bottom surface of brick, stone or tile that is laid on to the MORTAR.

bedding. A continuous layer of material that supports part of a structure. Examples of bedding include the mortar beneath bricks and stones, the glazing putty beneath window glass and the concrete beneath a drain.

bed joint. (1) A horizontal joint in brickwork or stonework. *Compare* CROSS JOINT. *See also* FACE JOINT. (2) The sloping joints in an ARCH.

bed putty. The strip of PUTTY in a window frame on which the glass is seated.

bedrock. In general, a hard rock that lies beneath gravel or surface soil.

beech. A HARDWOOD tree of the genus *Fagus*. The light-brown wood is traditionally used for making furniture.

beeswax. A wax produced by the honeybee and traditionally used in stains and polishes for wood.

beetle. (1) A heavy wooden MALLET often used for positioning paving slabs without damage. (2) An insect whose life cycle damages timber. *See* DEATH-WATCH BEETLE, FURNITURE BEETLE, HOUSE LONGHORN BEETLE, LYCTUS POWDER-POST BEETLE.

bel. A measure of SOUND LEVEL. 1 bel = 10 DECIBELS.

Belfast sink (butler's sink). A deep-sided kitchen SINK made of stoneware.

Belfast truss. A wooden TRUSS that consists of a curved upper chord and a horizontal bottom TIE called the string piece. Small diagonal timbers run between the members and brace the structure.

bell cast. A bottom edge, such as that used on RENDERING, that is expanded outward like the edge of bell.

belt sander. A SANDING MACHINE that uses a continuous belt of GLASSPAPER to smooth timber surfaces. *Compare* ORBITAL SANDER.

benching. (1) The concrete base of a MANHOLE. It has a shallow V-shaped cross-section which slopes into the drain channel at the bottom. *See also* SLIPPER. (2) A horizontal ledge above a ditch. *See* BERM.

bench mark. A fixed point of known level that is used as a local reference for surveying and levelling. *See also* DATUM, ORDNANCE BENCH MARK, TEMPORARY BENCH MARK.

bench plane. A type of PLANE that is used mainly at the work bench.

bench sander. A fixed, non-portable type of SANDING MACHINE that smooths timber with strong GLASSPAPER or cloth.

bench stop. An adjustable peg that protrudes through a work bench to hold wood still while it is planed.

bending iron. A curved steel bar used by plumbers for straightening or widening lead pipe.

bending moment. A measure of the tendency of a BEAM to bend. The bending moment is considered at one particular section of the beam and is equal to the algebraic sum of all MOMENTS OF FORCE acting to the left (or the right) of that section. *See also* MOMENT OF RESISTANCE.

bending spring. A length of tight steel spring inserted inside copper or lead tube to keep a circular cross-section while the tube is spent.

bentonite. A clay, like FULLER'S EARTH, that contains SILICA and ALUMINA and expands significantly with increased water content. Bentonite mud is used to support the sides of trenches and as a lubricant for drilling or PILING.

benzine. A volatile compound of carbon and hydrogen used as a SOLVENT in paints.

berm. (1) A narrow ledge on an earth bank or cutting. (2) Earth that is banked against the walls of a house.

Bernoulli's theorem. An effect of the principle of conservation of energy within a moving liquid or gas. The total energy possessed by a particle of moving water, for example, remains constant when the different types of energies change their proportions. The potential energy plus pressure energy plus kinetic energy plus friction losses is constant.

best reed. Reed used for THATCHING roofs.

bevel. (1) A slanted surface used for joining materials. *Compare* CHAMFER. (2) A blade tool used for setting and marking angles.

bevelled closer. A brick cut lengthwise on its edge from the middle of one end to a far corner. *See* CLOSER.

bevel siding. *See* CLAPBOARD.

BFB. *abbr.* Broad flanged beam (used in structural steelwork).

bibcock. *See* BIB TAP.

bib tap (bibcock). A water TAP fed by a horizontal pipe from the wall instead of from below like a PILLAR TAP. Bib taps are commonly installed above BELFAST SINKS.

bid. (*USA*) A TENDER or offer from a contractor to carry out specified work for a stated sum of money.

bidet. A bathroom SANITARY APPLIANCE in the form of a low-level washbasin fitted with a mixing valve. A person sits across a bidet and uses it to wash the lower parts of the body.

bifurcated. *adj.* Describing a tool or piece of work that is divided into two pointed forks.

bilection. *See* BOLECTION.

billet. (1) A thick piece of wood with at least one side left unsawn. (2) A small bar of steel.

billing. The writing out of a BILL OF QUANTITIES.

bill of quantities (BQ). A complete list of the materials and labour needed for a building or civil engineering project. Items in the bill are traditionally grouped together by the process of ABSTRACTING. The document is often used as a basis for a contractor's TENDER.

bi-metallic strip. A length of two different metals bonded together and often used to make a THERMOSTAT. The metals have different expansion properties which cause the strip to bend as the temperature changes.

binder. (1) A material used to provide adhesion and support for a mixture of other filling materials. For example, PORTLAND CEMENT is a binder for the AGGREGATE in CONCRETE. (2) In paint, the non-volatile part of the MEDIUM that contains the PIGMENT. LINSEED OIL was a traditional binder and is now replaced by artificial RESINS. (3) A small-diameter steel rod used to hold the main steel together in REINFORCED CONCRETE.

binding rafter. *See* PURLOIN.

biological filter. In general, a filter for water supply or sewage treatment that works by the action of micro-organisms rather than by mechanical action. *See* GRAVITY FILTER.

birch. A pale-coloured type of HARDWOOD of many varieties. It is used for furniture and veneers.

birdseye. A small circular decorative pattern in the grain of wood caused by a depression in the ANNUAL RINGS during the growth of the tree.

birdsmouth joint (plate cut, *USA*: foot cut). A notch cut into the end of a timber such as a RAFTER to allow the timber to fit over a cross member such as a WALL PLATE. *See* TIMBER JOINTS diagram.

biscuiting. The production of a glazed bricks by a second firing after the bricks are coated in glaze.

bit. (1) The detachable cutting mechanism of a DRILL or BRACE. *See also* CENTRE BIT, CHUCK. (2) The heated head of a soldering iron.

bitty. A paint defect. *See* NIB.

bitumen. A black resinous mixture of hydrocarbons that are soluble in carbon disulphide. Bitumen is found as naturally-occurring ROCK ASPHALT or LAKE ASPHALT and it is also refined from petroleum. It softens with increase in temperature and its adhesive and waterproof qualities make it a useful BINDER in ASPHALT and other BITUMEN MATERIALS. *See also* CUTBACK BITUMEN, STRAIGHT-RUN BITUMEN. *Compare* TAR.

bitumen macadam. *See* COATED MACADAM.

bitumen materials. For use in construction, BITUMEN is often combined with other materials to give convenient forms and better wearing. Sheets of woven fibres, such as glass or textiles, are impregnated with bitumen to form various types of bitumen felt, ROOFING FELT and sarking felt. ASPHALT is also a composite material based on BITUMEN.

bituminous paint. A paint that contains a

high proportion of BITUMEN and is used for its waterproof qualities.

black Japan. A type of BITUMINOUS PAINT used as a protective varnish for ironmongery.

blank. A cut piece of timber or metal before it is shaped into an article.

blanket (quilt). A flexible sheet or strip of THERMAL INSULATION material such as fibre glass. The form is convenient for inclusion inside wall and roof constructions. *See also* BAT.

blank wall. A wall without an opening.

blast cleaning. *See* SHOT BLASTING.

blast-furnace cement. A type of PORTLAND CEMENT containing powdered BLAST-FURNACE SLAG which provides POZZOLANIC properties.

blast-furnace slag. SLAG that is produced during the smelting of iron ore and used as AGGREGATE for concrete. *See also* PORTLAND BLAST-FURNACE CEMENT.

bleaching. The removal of colour from paint or woodwork by exposure, by natural ageing or by application of chemicals.

bleeder tile. (*USA*) A drain pipe that runs through a basement wall and leads water from the outside soil to drains inside the building.

bleeding. In PAINT and other decorative surfaces, a defect caused by the penetration of colour or glue through the surface layer.

blender. A soft, blunt paint brush used to blend colours and brush marks.

blind. An adjustable screen that covers a window to give privacy or to reduce daylight and solar radiation. *See also* ROLLER BLIND, VENETIAN BLIND.

blind floor. (*USA*) *See* SUB-FLOOR.

blind header. (1) A half brick or HEADER that cannot be seen from the face of a wall. (2) The visible heads of half bricks that are used to give the decorative effect of a HEADER brick.

blinding. (1) A layer of lean concrete used to seal soil before reinforcement is laid on it. (2) Sand or fine gravel used to fill the spaces in a HARDCORE bed.

blind mortise. A MORTISE that receives a STUB TENON.

blind nailing. *See* SECRET NAILING.

blistering. (1) Bubbles in a paint surface caused by the vaporization of moisture behind the paint film. It is often caused by painting over damp surfaces. (2) (*USA*: blub) Swelling in a plaster surface caused by continued SLAKING of lime in the plaster.

block. (1) A regular-shaped precast building unit, solid or hollow, that is usually laid in MORTAR. Blocks are commonly made from concrete, glass or hardwood. *See* BLOCKWORK, STRUCTURAL CLAY TILE. *Compare* BRICK. (2) *See* ANGLE BLOCK.

blockboard. A composite wooden board made from core strips up to 30 mm wide that are glued between sheets of PLYWOOD. *See also* COREBOARD, LAMINBOARD.

block-bonding. A join between two different areas of brickwork that BONDS several courses at a time instead of staggering them. The method is usually used for the join between new and existing brickwork.

block coin. *See* BLOCK QUOIN.

blocking course. The courses of bricks or stones laid over a stone CORNICE to hold it in position.

block plan. A plan showing the outline of a building in relation to its surrounding boundaries, roads and other buildings. *See also* LOCATION PLAN.

block quoin (block coin). A pattern of projecting or contrasting bricks that are laid to form blocks of pattern at the corner of a building. *See also* QUOIN.

blockwork. Wall construction using precast building BLOCKS laid in mortar. Lightweight blocks of AERATED CONCRETE are used on the inner leaves of modern masonry construction and give increased thermal insulation. *See* WALL diagram.

bloom. (1) In painting, a dull film that develops on a gloss surface. It can be caused by defective paint or by application in humid conditions. (2) A discoloration of metal surfaces.

blow. *See* BOIL.

blowing (pitting, popping). Small pits in the surface of plaster caused by expansion of unslaked material behind the surface.

blow lamp. A hand-held burner with a hot flame that can be directed against surfaces. It is used to melt SOLDER for pipe joints and to burn old paint from surfaces. *See also* HOT-AIR STRIPPER.

blown bitumen. *See* STRAIGHT-RUN BITUMEN.

blown joint. A SOLDER joint made between two lead pipes in which one end is heated and widened while the other end is tapered.

blub. (*USA*) *See* BLISTERING.

blue bricks. *See* STAFFORDSHIRE BLUES.

blue stain (sap stain). A blue discoloration of SAPWOOD caused by a fungus which does not reduce the strength of the timber.

blushing. A milky clouding within a LACQUER, often caused by moisture.

BM. *abbr.* BENCH MARK.

bm. *abbr.* BEAM.

board. (1) In softwood, timber that is square sawn and usually less than 50 mm thick and 100 mm or more wide. (2) In hardwood, timber less than 50 mm thick and of varying width. (3) A manufactured sheet of rigid building material such as wood CHIPBOARD and PLASTERBOARD.

boarded door (batten door, ledged door). A simple door without a frame. It is made of vertical BOARDS attached to horizontal LEDGES, which are diagonally braced if necessary. The boards may be MATCHBOARD. *See also* LEDGED AND BRACED DOOR. *See* DOOR diagram.

boarding (close boarding). A continuous surface of BOARDS laid over RAFTERS, JOISTS or STUDS. Tiles, insulation or other cladding is laid over the boarding.

boarding joist. *See* COMMON JOIST.

boasted ashlar. A form of ASHLAR stone that is roughly finished by BOASTING.

boaster (bolster, *USA*: brick set). A broad CHISEL (40–80 mm wide) used by masons to dress the surface of stone.

boasting. The dressing or finishing of a stone surface with parallel strokes from a BOASTER. The pattern does not cover the complete face. *Compare* BATTING.

BOCA. *abbr.* (*USA*) Building Officials and Code Administrators.

body. The general stiffness of a PAINT or other coating.

body coat. An UNDERCOAT.

boil (blow). A flow of fine soil, such as sand or silt, into the bottom of an excavation that is under pressure of water or air.

boiler. In general, the equipment used to heat water. Domestic boilers may use gas, electricity or solid fuel to provide hot water for taps and for central heating. Industrial boilers often produce steam. *See* HEATING diagram.

bole. The trunk of a tree.

bolection (bilection). A MOULDING used to cover the join between surfaces at different levels such as between the panel and frame of a door. The profile of the moulding projects beyond the joint.

bollard. (1) A short strong post set into a quay and used for securing ships. (2) A short post or sign used to separate areas on a roadway.

bolster. (1) A timber cap over a post that gives an increased bearing area. (2) A wide bricklayer's CHISEL. *See* BOASTER. *See* MASONRY TOOLS diagram.

bolt. (1) A cylindrical length of strong metal used to fix two objects together. The bolt has a head at one and a thread for a NUT at the other end. *See also* HOOK BOLT, WAISTED BOLT. (2) A shaped metal bar used to fasten a door or window. *See also* INDICATOR BOLT.

bond. (1) The adhesion between the surfaces of two materials caused by a bonding agent such as a glue or adhesive. (2) A method of laying bricks, blocks, stones, tiles and slates in patterns so that vertical joints are never one above the other. The bonding pattern adds to the strength supplied by the mortar. Common bonds for walls include STRETCHER BOND, HEADER BOND, ENGLISH BOND, FLEMISH BOND. *See* BRICKWORK diagram. (3) Adhesion between neighbouring atoms which depends on the arrangement of their outer electrons. *See* COVALENT BOND, ELECTROVALENT BOND. (4) Part of the payment to a contractor that is covered by sureties or insurance if the work is not satisfactorily completed.

bond course. A course of HEADERS.

bonder. *See* BOND STONE.

bonding agent. A substance that is applied to a smooth surface and acts as a KEY to improve adhesion for subsequent coatings.

bonding conductor. A short conductor that earths the metal frame of electrical equipment.

bond stone (bonder, through bonder, throughstone). A long stone laid as a HEADER in a wall so that it extends through the wall to form a bond with another wall.

boning (boning rod). A method of setting out a regular slope on a site. Two T-shaped rods are set at correct levels and a third rod is moved and lined up between them.

bonnet. (1) *See* BALLOON. (2) A roof over a BAY WINDOW.

bonnet tile (bonnet hip). A type of HIP TILE with a flared top that is bedded in mortar on to the next tile.

bonus payment. *See* INCENTIVE SYSTEM.

boom. *See* JIB.

boot lintel. A type of LINTEL that bridges an opening in a CAVITY WALL. A projecting ledge carries the outer brickwork and has a cross-section that resembles a boot.

bore. (1) The internal diameter of a pipe or cylinder. (2) *See* BOREHOLE. (3) A wave that advances upstream during the incoming tide in an estuary such as the Severn estuary in Britain.

bored pile. A PILE formed by pouring concrete into a hole containing some reinforcement. Often used for extending foundations through soft soil on to bearing ground. *Compare* DRIVEN PILE.

borehole. A hole made in the ground to investigate the nature of the subsoil. *See also* SITE INVESTIGATION.

borer. A BEETLE or other insect that tunnels into wood. *See also* FLIGHT HOLE.

borrowed light. (1) Natural light that is brought from one room to another. (2) A window set in an internal partition.

boss. (1) A decorative rounded shape that projects, usually downward, from a ceiling or a NEWEL POST. (2) A wooden cone for widening the end of a lead pipe. (3) A projection round an opening, such as on a pipe, that receives a connection.

bottle-nose drip. A round-edged form of DRIP on a FLEXIBLE METAL roof.

bottle trap. A common form of TRAP on the waste pipe just below a sink or washbasin. The WATER SEAL is kept in a round container with central inlet pipe below the water level and the outlet at the side.

bottoming. (1) Large stones used for road formation or for the BALLAST on a railway line. (2) The use of a spade to ensure that the bottom of an excavation is smooth and level.

bottom rail. The horizontal member that forms the bottom part of a door or casement. *See also* RAIL.

boundary temperature. The temperature at the junction between two layers of materials within the construction of a wall, roof or other element. Boundary temperatures can be calculated from a knowledge of the THERMAL RESISTANCE of the layers. A TEMPERATURE GRADIENT can then be drawn between each boundary temperature.

boundary wall. A wall of a building that also forms the boundary of the property. *See also* PARTY WALL.

boundary wall gutter. A roof GUTTER with one side flat and the other side recessed beneath the edge of a pitched roof.

Bourdon gauge. A simple form of gauge for measuring gas pressure. It uses a hollow metal tube that tends to uncoil as pressure increases. *Compare* MANOMETER.

bow. A distortion caused by the bending of timber or other structural components.

Bow's notation. In STRUCTURAL DESIGN, a standard method for labelling the spaces between the forces acting at a point. *See also* POLYGON OF FORCES.

bow window. A BAY WINDOW with a continuous curve.

box beam (box girder, hollow-web girder). A hollow GIRDER with a square or rectangular section. Large steel box girders have been assembled as bridge structures. Timber box beams may use plywood panels to join the timber CHORDS.

boxed eaves (closed eaves). EAVES of a roof that are covered in with a FASCIA BOARD and a SOFFIT.

boxed frame. *See* CASED FRAME.

boxed mullion. A hollow MULLION that contains the counterweights of a SASH window.

box gutter (parallel gutter, parapet gutter, trough gutter). A roof GUTTER of rectangular cross section used in valleys or parapets. The gutter may be made of wood and lined with flexible metal or BITUMEN MATERIALS.

boxing. (1) *See* CASED FRAME. (2) (*USA*) *See* BOARDING. (3) *See* BOXING SHUTTER.

boxing shutter (folding shutter). An internal window SHUTTER that folds into a window recess called the 'boxing'.

boxwood. A variety of HARDWOODS, not always from the box tree, which are used for making tools.

Boyle's law. One of the GAS LAWS. The principle that, for a fixed mass of gas, the volume changes in inverse proportion to the pressure. For example, if the pressure is doubled then the volume is halved. *Compare* CHARLES'S LAW, DALTON'S LAW.

BQ. *abbr.* BILL OF QUANTITIES.

brace. (1) A structural component, such as a diagonal rod, that stiffens a frame or structure. (2) A carpenters' tool that revolves a BIT for drilling holes in wood. *See also* RATCHET BRACE.

braced frame. A type of timber frame construction in which widely-spaced edge posts are framed with horizontal binders to carry the floor load. The intermediate studs between the edge posts carry no floor load, unlike a BALLOON FRAME.

bracket. (1) A projection from a wall that supports a horizontal member or platform such as a shelf. (2) A short wooden support beneath a wide tread in a stair.

brad (floor brad). A NAIL with a constant thickness but tapering width. The brad has a short head on one side only and is used for fixing floorboards.

bradawl. A short pointed hand tool used for making holes in wood for nails and screws. *See* WOODWORKING TOOLS diagram.

brad punch (*USA*: brad setter). A type of NAIL PUNCH.

branch. A subsidiary connection to a pipe or cable.

brander. A LATH nailed beneath a wide floor joist to which the other plaster laths are nailed. This gives a better key for the ceiling plaster beneath the joist.

brash. (1) Small pieces of rock or other material. (2) Timber that breaks up easily.

brass. An ALLOY of copper and zinc which is resistant to corrosion. It was traditionally used for taps and other fittings. *See also* MUNTZ METAL.

brazing. The joining of red-hot metal surfaces with a capillary joint of HARD SOLDER using a flux such as borax.

BRE. *abbr.* (*UK*) Building Research Establishment.

breaking joint. *See* STAGGERING.

breakwater (mole). A massive wall projecting into the sea to protect a harbour from waves. *See also* TETRAPOD.

breast. (1) A wall that projects into a room and usually contains a FIREPLACE. (2) The wall beneath the SILL of a window. (3) The RISER of a stair.

breast wall. A low earth-retaining wall.

breather paper. A building paper that repels liquid water and is used as additional waterproofing behind CLADDING. The paper 'breathes' by allowing water vapour to escape from inside the structure. *See* WALL diagram.

breeze block. A BUILDING BLOCK originally made from coke. The material is no longer available but the term is sometimes used for blocks made from CLINKER.

BRE protractor. *See* DAYLIGHT FACTOR PROTRACTOR.

bressummer. A long LINTEL, originally made of timber, that spans a wide opening such as a shop window.

brick. A preformed building unit in the shape of a rectangular block which is usually laid in COURSES with MORTAR to form walls and other constructions. The common clay brick is moulded or cut into shape and then baked in a KILN. Variations in materials and manufacture give other forms of brick such as CALCIUM SILICATE (sand-lime) and ADOBE (sun-baked). Bricks need to be conveniently placed by hand and various 'standard' sizes have existed. In the United Kingdom, the usual size is 215 by 102.5 by 65 mm, to which is added a 10 mm allowance for mortar. *See also* MODULAR BRICK. Bricks may also be classified by varieties (COMMON BRICK, FACING BRICK, ENGINEERING BRICK); by qualities such as internal, ordinary, special; and by type such as solid, perforated, hollow. Some local names of bricks now apply to a general type of brick such as FLETTONS, LONDON STOCKS, SOUTHWATER REDS and STAFFORDSHIRE BLUES. *See also* BOND.

brick-and-a-half wall. (*UK*) A wall that has a thickness of one HEADER plus one STRETCHER.

brick axe. *See* BRICKLAYER'S HAMMER.

bricklayer's hammer (brick axe). A hand hammer with a sharp wedge shape balancing the hammer head. It is used for dressing bricks.

brick masonry. (*USA*) *See* BRICKWORK.

brick nogging. *See* BACK FILLING.

brick set. (*USA*) *See* BOASTER.

brick veneer. An outer skin of decorative brickwork used to cover a timber-framed house. The skin is usually one half-brick thick and is non-structural.

brickwork. The use of BRICKS laid in a BOND for a wall or other construction. *See* BRICKWORK diagram.

bridge bearing. The support that transfers the weight of a bridge on to the bridge pier. The bearings are usually spherical or cylindrical and may allow for expansion.

bridge board. *See* CUT STRING.

bridging. (1) The spanning of an area with COMMON JOISTS. (2) FLOOR STRUTTING.

briefing. A meeting at which information and instructions are given to a contractor.

bright. *adj.* Describing timber that is freshly-sawn and free from discoloration.

brindled brick. A brick that has variations in colour but is otherwise usable.

brine. A solution of sodium chloride (common salt) in water.

Brinell hardness test. A test for HARDNESS in which a standardized hard steel ball is pressed into a surface for a set time and the diameter of the indentation is measured. The Brinell hardness number is equal to the load (in kilograms) divided by the spherical area of impression (in square millimeters). *Compare* ROCKWELL HARDNESS TEST.

brise-soleil. A system of sun-screens, such as horizontal or vertical slats, that are fixed to the outside of a building.

British Board of Agrément (BBA). An independent body set up the United Kingdom government to test and assess new building materials and techniques which are submitted by manufacturers on a voluntary basis. An Agrément Certificate indicates that a particular product has been passed. The Board is a member of the European Union of Agrément.

British Standard (BS). A publication of the BRITISH STANDARDS INSTITUTION. It details acceptable standards for the manufacture, design and installation of materials and equipment such as those used in construction. *Compare* CODE OF PRACTICE.

British Standards Institution (BSI). (*UK*) The national organization that establishes acceptable standards for the manufacture, design and installation of materials and equipment such as those used in construction. They publish BRITISH STANDARDS (BS) and CODES OF PRACTICE (CP). *See also* STANDARDS ORGANIZATIONS.

British thermal unit (BTU). A non-metric unit of heat sometimes found in specifications for services. 1 BTU = 1055 joules. 100,000 BTU = 1 therm. *See also* JOULE.

British Zonal System. A method of classifying light fittings or LUMINAIRES. Ten standard light distributions have BZ numbers from 1 (mainly downward) to 10 (mainly upward). *Compare* LIGHT OUTPUT RATIO.

brittleness. (1) A general property of materials that crack easily. A lack of TOUGHNESS or flexibility. *See also* CHARPY TEST. (2) A defect in a paint surface that cracks when stretched.

bronze. An ALLOY traditionally made of copper and tin. The tin may be replaced by other metals such as aluminium. *See also* GUNMETAL.

bronzing. A variety of treatments used to give a golden brown colour to surfaces.

brooming. The process of scratching a FLOATING COAT of PLASTER to form a key for another layer of plaster.

browning. *See* RETARDED HEMIHYDRATE PLASTER.

browning coat. The middle coat of PLASTER in three-coat work. It is usually RETARDED HEMIHYDRATE PLASTER.

browning plaster. *See* RETARDED HEMIHYDRATE PLASTER.

brushability. A property of a paint that makes it easy to be brushed on to a surface without leaving defects.

BS. *abbr.* BRITISH STANDARD.

BSCP. *abbr.* British Standard Code of Practice. *See* BRITISH STANDARDS INSTITUTION.

BTEC. *abbr.* Business and Technician Education Council.

Bthu. *abbr.* BRITISH THERMAL UNIT.

BTU. *abbr.* BRITISH THERMAL UNIT.

bubbling. A defect in paints or other films that contain volatile SOLVENTS. Bubbles of solvent vapour disrupt the surface but may disappear before the paint dries.

bucket sink (housemaid's sink). A type of SINK installed near floor level so that it can be easily used to fill and empty buckets.

buckle. A type of SPAR used in thatching.

buckling. The deformation of a structural member bending out of line because of an excessive compressive load.

buff. The abrasive polishing of a surface with buffing wheels, as for a TERRAZZO floor.

buggy. (*USA*) A cart that carries concrete from a mixer to the forms.

builder. In general, a person or organization who contracts to construct a building. *See also* CONTRACTOR.

builder's level. *See* LEVEL.

builder's square. A large timber frame in the form of a right-angled triangle that is used in SETTING OUT buildings.

building block. *See* BLOCK.

building board. Various forms of manufactured sheet material used in construction. Examples include BLOCKBOARD, CHIPBOARD, PLASTERBOARD and PLYWOOD.

building brick. (*USA*) *See* COMMON BRICK.

building codes. (*USA*) Local building laws that are usually based on the UNIFORM BUILDING CODE.

building control officer. (*UK*) *See* BUILDING INSPECTOR.

building inspector (building control officer, district surveyor). A qualified local authority official with responsibility for the safety aspects of the design and construction of buildings in a particular area. The inspector approves plans and inspects stages of construction to ensure that the process follows appropriate BUILDING LAWS. The inspector has the power to stop illegal or unsafe work. *See also* BUILDING CODES, BUILDING REGULATIONS.

building laws. A legal requirement for minimum standards of materials and methods used in construction. *See also* BUILDING CODES, BUILDING INSPECTOR, BUILDING REGULATIONS.

building line. (1) The line made by the external face of a building. (2) A line set by the local planning authority to keep streets and buildings uniform. Construction in front of the building line is not usually allowed.

building paper. A strong, reinforced waterproof paper used to protect concrete from soil, to separate different building materials, and for linings in walls and roofs. The BITUMEN content of the paper may be replaced by plastic sheeting. *See also* SHEATHING PAPER.

Building Regulations (Building Standards). The national BUILDING LAWS of the United Kingdom.

Building Research Establishment (BRE). (*UK*) A government-sponsored research organization that also publishes important digests, leaflets and papers on all aspects of construction.

building services engineer. An engineer trained in one or more of the areas of heating, ventilation, lighting, electricity, gas, water supply and drainage for buildings.

Building Standards. *See* BUILDING REGULATIONS.

building surveyor. A person who is trained in technical, economic and legal aspects of building. The building surveyor gives advice on alterations and renovations to buildings in use. *See also* SURVEYOR.

building system (industrialized building). An organized method of construction that uses prefabricated parts and less labour during construction.

built-in. *adj.* Describing a building component that is fixed by mortaring into a brick or block wall. Wall ties and joists are commonly built in to a wall.

built-up roofing. A form of ROOF CLADDING formed from consecutive layers of ROOFING FELT laid on ASPHALT. The method is often used for flat roofs. *See also* MINERAL-SURFACED BITUMEN FELT.

bulb angle. A length of steel angle SECTION that is enlarged to form a bulb at one end.

bulb of pressure. The region of compressed soil below a loaded foundation or footing. Lines of equal stress in the soil give a typical bulbous shape.

bulk density. The DENSITY (mass per unit volume) of a material such as soil or aggregate, including any voids and water that it contains. *Compare* DRY DENSITY.

bulkhead. A box-shaped cover or roof, often above a stairwell or water tank.

bulkhead fitting. A sealed light fitting mounted directly on a wall or ceiling. It is often used externally.

bulking. The increase in volume of sand, soil or aggregate caused by moisture content. The measurement of this effect is important when BATCHING materials because slightly damp sand can occupy 40 per cent more volume than dry sand.

bulk modulus. *See* ELASTIC MODULUS.

bulldog clip. *See* FLOOR CLIP.

bulldozer. A caterpillar tractor with a wide adjustable blade mounted across the front. It is used to push loose materials such as excavated earth.

bullnose. A gently-rounded corner such as found on certain bricks or steps. *See also* ARRIS, NOSING. *See* STAIRS diagram.

bullseye. A small circular window or opening.

bund. A continuous low wall, often of earth, built around an oil tank or similar container and designed to retain the contents in case of leakage. *See also* BERM.

bungalow. (1) A low house with a wide verandah. (2) (*UK*) Any single-storied house.

bungalow siding. *See* WEATHERBOARDING.

burl. *See* BURR.

burlap. *See* HESSIAN.

burning off. The removal of old paint surfaces by softening with the heat of a BLOW LAMP or HOT-AIR STRIPPER, followed by immediate scraping.

burnt lime. *See* QUICKLIME.

burr. (1) (burl) A curly figure in the grain of woods such as walnut. (2) A rough edge left on metal after cutting.

bus bar. A bare electrical conductor often in the form of a thick copper or aluminium bar to which other leads are clamped.

bush. A metal lining within a hole.

bush hammer. A percussive tool used to remove the outer skin of concrete or stone surfaces to provide a final texture.

bushing. The joining of two pipes by a short length of threaded pipe inserted into them.

butadiene. A chemical component found in several forms of synthetic rubber.

butane. A member of the alkane (paraffin) series of HYDROCARBON compounds, C_4H_{10}. In liquefied form, it is used as a fuel. *See also* LIQUEFIED PETROLEUM GAS.

butler's sink. *See* BELFAST SINK.

butter. *vb.* To use a trowel to place MORTAR on a brick or block.

butterfly fastener. *See* BUTTERFLY TIE.

butterfly roof. A reversed form of pitched roof in which two surfaces slope from the outside down to a valley.

butterfly tie (butterfly fastener). A WALL TIE made of galvanized steel wire twisted in a figure of eight shape. The twist helps prevent water running across the tie. *See* WALL diagram.

butt hinge. A common type of door HINGE that is sunk into both the door frame and the

HANGING STILE so that two surfaces of the hinge fold together. *See* DOOR diagram.

butt joint (square joint, straight joint). A straight join between two materials or components that meet without an overlap. *Compare* SCARF JOINT. *See* TIMBER JOINTS diagram.

buttress. A structure built at right angles against a wall to help resist outward forces from the wall.

butyl rubber. A form of synthetic rubber.

BWF. *abbr.* British Woodwork Federation.

byatt. A horizontal beam used as support for timber decking and walkways in excavations.

BZ number. *See* BRITISH ZONAL SYSTEM.

C

C. *abbr.* COULOMB.

cabinet scraper. *See* SCRAPER.

cabin hook. A simple hooked bar that drops in to a SCREW EYE and is used to fasten a cupboard door or window.

cable. (1) A flexible length of electrical conductor, often made of multiple wire strands, surrounded by insulation. *See also* FLEX, TWIN CABLE. (2) A length of steel rope used in TENSION structures such as suspension bridges and PRE-STRESSED CONCRETE.

cable trunking. A system of preformed covering for electrical CABLES.

CAD. *abbr.* Computer-aided design. A system of construction drawing that uses a computer to eliminate repetitive drawing chores and to increase versatility. Drawings are composed and edited on the screen of a visual display unit (VDU). Standard parts or 'entities' of drawings can be stored and re-used with different scales and viewpoints. A single drawing stored by the computer can generate 'hard copy' on a variety of printers or PLOTTERS. *See also* DIGITIZER, MOUSE, WIRE FRAME.

cadastral survey. A land survey used to prepare plans that show and define legal boundaries between properties. *Compare* GEODESY, TOPOGRAPHICAL SURVEY.

cadmium plating. A protective coating given to some hardware items such as screws and nails. *See also* ELECTROPLATING.

caisson. A large watertight chamber or ring-wall used to prevent water and soft ground from entering excavations. The bottom may be open or closed and the caisson may be left as part of the structure.

calcine. *vb.* To heat a substance, such as LIME, to make it easy to crush into powder.

calcium carbonate ($CaCO_3$). The main chemical compound of chalk, LIMESTONE, MARBLE and TUFA.

calcium chloride ($CaCl_2$). A compound that has been used in concrete mixes as an ACCELERATOR.

calcium hydroxide ($Ca(OH)_2$). *See* SLAKED LIME.

Calcium oxide (quicklime, CaO). The main compound in LIME.

calcium silicate brick (sand-lime brick). A type of BRICK made from a blend of sand or flint and slaked lime that is moulded and then cured in an AUTOCLAVE. The bricks are smooth and regular in shape with a similar range of crushing strengths to clay bricks.

calcium sulphate ($CaSO_4$). The main chemical compound in GYPSUM. *See also* ANHYDRITE.

calculated brickwork. A structure made up from brickwork whose load-bearing performance is calculated rather than assumed.

calendering. The use of rollers to produce continuous films or sheets of materials such as PVC.

callow. *adj.* Describing bricks and tiles that are underburnt in the kiln.

callus. A thickened mass of wood that grows over an injury to the tree.

calorie. A metric unit of heat which is now replaced by the SI UNIT of heat: the JOULE. 1 calorie = 4.187 joules.

calorific value. A measure of the total heat energy content or PRIMARY ENERGY of a raw fuel such as oil or coal. The values, which are expressed in joules/kilogram or joules/cubic metre, are used to compare fuels and to predict consumption. *See also* CALORIMETER.

calorifier. A type of HEAT EXCHANGER that transfers heat from coils of hot water or steam pipes to cooler water surrounding the pipes. *Compare* INDIRECT CYLINDER.

calorimeter. Equipment used to measure the heat energy given off during the complete combustion of a substance. *See also* CALORIFIC VALUE.

CAM. *abbr.* Computer-aided management.

cam. A wheel with a non-circular edge which, when it turns in contact with another component, imparts to it a rocking motion.

camber (hog). (1) The upward curve of a road surface that helps water to drain away and vehicles to corner. (2) An upward arching (hog) given to a beam to help counteract sag.

camber board. A TEMPLATE used for forming a CAMBER.

camber slip. A piece of wood with a CAMBER in its upper surface that is used for CENTERING a flat brick arch.

cambium. The layer of active cells, just beneath the bark of a tree trunk, that provides the yearly growth. *See also* WOOD.

came. An H-shaped strip of metal such as lead that joins pieces of glass in a LEADED LIGHT or stained-glass windows. *See also* ELECTRO-COPPER GLAZING.

candela. The SI unit for LUMINOUS INTENSITY. The effect of 1 candela is approximately the same as the earlier unit of 1 candlepower. *See also* LUMEN.

candle power. *See* CANDELA.

canopy. A roof structure that protects an outside area such as the entrance to a building.

canopy switch. (*USA*) A small switch attached to a light fitting such as a reading lamp.

cant. *vb.* (1) To turn or tilt. (2) To cut the WANE from a log.

cantilever. A projecting structure, such as a beam, that is fixed at one end only.

cant strip. (*USA*) A triangular strip of timber installed where a flat roof covering meets a wall. The shape of the strip helps to shed water. *See also* TILE FILLET, UPSTAND.

cap. (1) A plate fitted to the top of a post such as a NEWEL. (2) A cover that screws over the end of a pipe to close it. (3) The part of an electric LAMP that attaches to the holder.

capillary action (capillary attraction). The tendency of water and other liquids to move through very small holes or gaps. The movement, which can be vertical, is caused by unbalanced molecular forces. *See also* CAPILLARY GROOVE, RISING DAMP.

capillary groove (anti-capillary groove). A groove or space between two components that is made large enough to prevent CAPILLARY ACTION. The opening frames of wooden windows, for example, need a capillary groove around them.

capillary joint (*USA*: sweated joint). A joint made between two pipes by inserting one into another and letting solder fill the narrow gap between them by CAPILLARY ACTION. It is smaller than a COMPRESSION JOINT.

cap iron. A metal sheath that covers the cutting PLANE IRON of a woodworking PLANE. It breaks up the wood shavings and prevents tearing.

capital. The head of a COLUMN where the load is transferred from the beam or other structure above. Traditional stonework capitals are decorated with carving. *See* NECKING.

capping. (1) A metal strip or section that covers a gap between sections of flexible roofing or glazing. (2) The top of a low wall or part of a wall that does not reach the ceiling. (3) *See* DADO.

capstone. *See* COPING.

carbonation. A chemical reaction with natural carbon dioxide in the air. The reaction causes the slow hardening of LIME MORTAR as it is converted to CALCIUM CARBONATE. Carbonation disrupts the composition of concrete and can corrode the reinforcement. *Compare* ALKALI SILICA REACTION.

carbon black. A pure type of carbon in the form of very fine powder. *See also* LAMP BLACK.

carbon dioxide (CO_2). An odourless gas that is present in the atmosphere and does not allow burning. *See also* CARBONATION, CARBON DIOXIDE FIRE EXTINGUISHER.

carbon dioxide fire extinguisher. A device or method of putting out fires that depends on the properties of CARBON DIOXIDE gas, which does not allow burning and is heavier than air. The carbon dioxide is stored as a liquid under pressure and becomes a gas when released.

carbon steel. Since all steel is an alloy of IRON and carbon, the term 'carbon steel' is generally used to mean steel whose properties depend mainly on carbon and not on other alloying elements such as silicon and manganese. *See also* LOW CARBON STEEL, MILD STEEL, HIGH CARBON STEEL. *Compare* STAINLESS STEEL.

carborundum. A very hard compound of silicon and carbon that is used as an abrasive. *See* SILICON CARBIDE.

carcase (carcass). The basic load-bearing structure of a building such as walls and roof (that is, without the windows, doors, plaster and other finishes).

carpenter. A person who erects structural timberwork. *See* CARPENTRY, JOINERY.

carpentry. The structural timber work or CARCASE of a building such as floors, wall and roof frames. In the USA the term includes JOINERY.

carpet strip (saddle-back board). A strip of metal or timber that secures the edge of a carpet to the floor at a door opening.

carport. A simple shelter for a car which is usually built as a roof without sides.

carriage. A long slanted timber that is placed as extra support between two STRINGS of a stair.

carriage bolt. (*USA*) *See* COACH BOLT.

carrying capacity. The maximum electrical current in AMPERES that a CABLE or FUSE can carry without excessive heating or VOLTAGE DROP.

Cartesian coordinates. *See* COORDINATES.

cartographer. *See* LAND SURVEYOR.

cartridge fuse (*USA*: enclosed fuse). A FUSE that for safety is contained within a tube of incombustible material. The complete fuse must be replaced if it blows.

case. A surface layer that is harder than the surrounding material.

cased beam (cased column). A structural steel beam or column that has been surrounded with a fire resistant material such as concrete or plaster. The casing increases the relatively low FIRE RESISTANCE of unprotected steel.

cased column. *See* CASED BEAM.

cased frame (boxed frame). The covered, hollow window frame of a SASH WINDOW that contains the system of pulleys and sash weights. *See* WINDOW diagram.

case hardening. (1) The surface hardening of a material, usually steel, by a heat treatment that introduces carbon or nitrogen into the surface. (2) An effect during the natural seasoning of timber where the outer skin dries and hardens more quickly than the inner layers. The stress may cause defects such as warping.

casein glue. A wood glue made from milk protein. It is water-soluble before it hardens and has useful gap-filling properties.

casement. *See* CASEMENT WINDOW.

casement door (French door, French window). A fully-glazed door. It is usually one of a pair that hinge at the outer edges and meet in the centre.

casement fastener. A locking device that holds a CASEMENT WINDOW closed.

casement stay. A device, such as a metal arm, that holds a CASEMENT WINDOW open.

casement window. A window, part of which is hinged to open. The hinges for the main opening are attached to the 'casement' which is generally vertical, like a door. *See* WINDOW diagram.

cash flow. The changing pattern of finance available to a company or a particular project over a set period. A simple cash forecast uses the estimated costs and incomes to calculate the cumulative profits or losses.

casing. (1) A timber lining around the opening of a window or door. (2) A continuous boxed enclosure that hides pipes or cables on wall or ceiling. (3) *See* FORMWORK.

castella beam (castellated beam). A beam made by cutting a ROLLED STEEL JOIST in a zigzag shape along the web and rejoining the parts at their crests.

casting resin. A synthetic resin, such as a PHENOL-FORMALDEHYDE RESIN or EPOXY RESIN, that can be poured into moulds and shaped into solid articles.

cast-in-situ. A description of concrete or plaster that is poured in place, such as for a pile or lintel. *Compare* PRECAST CONCRETE.

cast iron. A form of IRON which contains a higher proportion of carbon (usually 2–4 per cent) than STEEL. Cast or 'pig iron' can be obtained directly form the blast furnace and is easily poured or 'cast' into moulds. Cast-iron products are BRITTLE but have good resistance to corrosion.

cast stone. *See* ARTIFICIAL STONE.

casual heat gains. Those HEAT GAINS that are supplied to a building as a by-product of using it. The main sources are body heat, lighting, cooking and washing. *Compare* FUEL.

catalyst. A substance that alters the rate of a chemical reaction but remains unchanged at the end of the reaction.

catchment area. The area that drains into a particular watercourse such as a river or surface water sewer, or into a reservoir. *See also* RUN OFF.

catchment tray. A tray or vertical guttering that collects water behind the OPEN JOINT of a RAIN SCREEN CLADDING.

catch pit. A trap placed in a drainage system that is designed to collect solid matter and prevent it blocking the drain.

catenary. The natural curve formed by a uniformly-loaded cable hung between two points, such as that in seen a power line or suspension bridge.

cathedral glass. A translucent glass with a texture on one surface.

cathode. (1) In ELECTROLYSIS, the negative ELECTRODE. (2) On an electric CELL or battery the negative terminal from which current flows. *Compare* ANODE.

cathodic protection. The principle of protecting a metal against CORROSION by making it the CATHODE in the corrosion process. The metal to be protected is connected to an external source of direct current or to a block of 'sacrificial' metal. *See* GALVANIC SERIES.

catladder (duckboard). A ladder, or board with cross bars, laid on a sloping surface such as a roof to provide access.

Catnic. A trade name for a steel LINTEL.

caulking. (1) The general process of filling a seam between two pipes or materials with a sealant to make the join water-tight or air-tight. (2) Splitting and bending the ends of a metal bar to give it a better grip in mortar.

caustic soda. Sodium hydroxide (NaOH), a strong ALKALI. *See also* PICKLING.

cavil (kevil). A type of AXE used to cut stone.

cavitation. The formation and collapse of cavities in a liquid when parts of a pump move faster than the liquid in the pump. The oxygen released from water can corrode metal.

cavity absorber (Helmholtz resonator). A form of acoustic ABSORBER that absorbs sound over a narrow range of frequencies. It consists of an enclosure that can be tuned to absorb specific frequencies.

cavity barrier. *See* FIRE STOP.

cavity flashing. *See* CAVITY TRAY.

cavity insulation. Various forms of THERMAL INSULATION, such as boards, particles and foam, that are installed in the gap of a CAVITY WALL.

cavity sleeve. A hollow lining that extends the passage of an AIR BRICK through a CAVITY WALL.

cavity tie. *See* WALL TIE.

cavity tray (cavity flashing). A sloped DAMP-PROOF COURSE that bridges the gap in a CAVITY WALL and drains water away from the inner leaf of the wall.

cavity wall (*USA*: multi-unit wall). A method of wall construction that uses two separate skins of bricks or blocks with a continuous gap between them. The 'leafs' are structurally tied across the cavity with WALL TIES. The wall cavity was introduced to minimize rain penetration of brick walls but it is also used to contain THERMAL INSULATION. *Compare* FACING WALL, VENEERED WALL. *See* WALL diagram.

c/c. *abbr.* Centre to centre (used on drawings).

CCA. *abbr.* (*UK*) Cement and Concrete Association.

cedar. Wood from various SOFTWOOD conifer trees.

ceiling. The lining on the overhead surface of a room. It may be made of plaster, panels or boards.

ceiling floor (ceiling joist). A floor or JOIST that carries a CEILING below it but has no floor above.

ceiling joist. *See* CEILING FLOOR.

ceiling rose. An electrical junction on the ceiling to which LUMINAIRES are attached or suspended.

cell. (1) A small cavity in a material. (2) A device that produces electrical current by the chemical action between certain materials. A 'battery' is a combination of several cells. Primary cells like those in a radio battery can not be reused, whereas secondary cells like those in a car battery can be recharged. *See also* SIMPLE CELL.

cellar rot. *See* WET ROT.

cellular block (cellular brick). In general, a type of BLOCK or BRICK with one or more large cavities that occupy up to 50 per cent of the volume. The cavities increase the lightness and thermal insulation of the block, which may be used to build external walls and partitions.

cellular materials. A material, such as lightweight concrete or expanded plastic, that contains many small cavities dispersed throughout its bulk. The cells may all be separate and closed or they may be open and interconnecting, as in mineral wool.

cellulose. A carbohydrate that forms the cell walls of plants and timber. It is used in the manufacture of products such as paper and paint. *See also* LIGNIN.

cellulose acetate. A THERMOPLASTIC material derived from cellulose.

cellulose nitrate. *See* NITROCELLULOSE.

cellulose paint (cellulose lacquer). A flexible tough PAINT, such as motor car paint, made from CELLULOSE compounds. This form of paint dries by evaporation of the SOLVENTS in the paint rather than by the chemical processes that harden other paints.

Celsius temperature. A scale of TEMPERATURE with units in 'degrees', based on the melting point of ice as 0°C and the boiling point of water as 100°C. The less correct term 'centigrade' is also widespread. *Compare* ABSOLUTE TEMPERATURE, FAHRENHEIT TEMPERATURE.

cement. (1) In general, a BINDER or GLUE. (2) A powder that reacts chemically with water and provides adhesion and support for the AGGREGATE in CONCRETE. In building the term 'cement' usually means PORTLAND CEMENT. *See also* HIGH-ALUMINA CEMENT, POZZOLAN.

cementation. The injection of cement GROUT into rocks or gravel to make the ground stronger and more watertight.

cement fillet. A triangular strip of MORTAR used to seal a junction such as that between a wall and roof. Flexible metal SOAKERS are recommended as a better form of seal. *Compare* TILE FILLET.

cement fondu. *See* HIGH-ALUMINA CEMENT.

cement grout. *See* GROUT.

cementite. A hard brittle compound of iron and carbon (Fe_3C) which causes the brittleness of CAST IRON.

cement mortar. A common form of MORTAR made from one part of CEMENT to four or more parts of sand with water. Lime or plasticizer is also usually added.

cement paint. A form of exterior MASONRY PAINT that contains cement to provide durability.

cement rendering. A coating made from cement, sand and water that is used to waterproof brickwork. When used on walls, LIME or some other plasticizer is needed to prevent shrinkage of the render when it dries.

cement screed. A layer of cement and sand mortar laid on a floor to provide a level surface for a floor finish such as tiles.

cement slurry. A liquid mixture of cement and water that can be injected into gaps or as a KEY coat on a wall or floor.

centering (centring, centres). A temporary framework, usually wooden, that is used to support an arch or dome during construction.

centigrade. *See* CELSIUS TEMPERATURE.

central heating. A form of SPACE HEATING in which the entire building is heated from a small number of sources with a unified system of control. Sources of heat include hot water pipes, WARM AIR HEATING and electric storage heaters. *See* HEATING diagram.

centre bit (Forstner bit). A type of BIT used to drill relatively large-diameter holes in wood. It has a central point and two side cutters or NICKERS.

centre-hung window. A type of window that swings open on pivots fixed halfway down each side of the frame.

centre nailing. A method of fixing SLATES through a nail hole that is near the middle of the slate and just above the line of slates below. *Compare* HEAD NAILING.

centre of gravity. *See* CENTRE OF MASS.

centre of mass (centre of gravity). The point in a body at which it will balance and from which the total weight of the body appears to act.

centre of pressure. The point on a surface subject to fluid pressure at which the resultant force due to the pressure acts on that surface. For example, the centre of pressure on a rectangular surface immersed in water is at two-thirds of its depth below the surface.

centres. *See* CENTERING.

centrifugal fan. A high-speed fan in which air enters near the centre of the blades and is thrown outward by the blades.

centring. *See* CENTERING.

centroid. The balancing point or centre of area of a section. It is also the CENTRE OF MASS if the section has mass.

ceramics. Articles, such as BRICKS and TILES, made by firing clay or other minerals at high temperature until they are hard and VITREOUS. *See also* EARTHENWARE, PORCELAIN.

ceramic tile. A clay tile that has a GLAZE on the surface with a high water resistance. Ceramic tiles are commonly used to protect kitchen and bathroom surfaces.

cermet. A hard COMPOSITE material that is made of CERAMIC and metal and used for cutting tools.

cesspit (cesspool). An underground chamber used for holding SEWAGE from buildings that have no foul drains or sewage treatment available. A cesspit needs pumping out at intervals.

chain. A device for measuring long distances in land surveys. It consists of rigid wire links, usually 300 mm long, jointed together with handles at either end.

chain link. A type of fencing material made from a diamond-shaped mesh of wire.

chainman. A junior member of a survey team who usually carries the CHAIN.

chain saw. A power-driven SAW on which the cutting picks are attached to a travelling chain. It is often used for the hand cutting of logs.

chain survey. A land survey that measures only lengths. Angles may be calculated later using triangles.

chair rail. A form of DADO rail that protects a wall from the backs of chairs.

chalk. A soft SEDIMENTARY rock that mainly consists of CALCIUM CARBONATE.

chalking. A defect in a painted surface caused by the breakup of the BINDER. A dusty film of PIGMENT remains and can be rubbed off.

chamfer. A corner that is symmetrically cut off at 45°. By comparison, a BEVEL takes an unequal amount from each face.

channel. (1) A drainage pipe that is open along the top. (2) A rolled metal section with a U-shaped cross-section.

channel tile (tegula). The flat bottom tile in a system of ITALIAN TILES.

characteristic value (characteristic strength). A value used in structural design that allows for the statistical variation in the strength of batches of steel and concrete. The characteristic value is usually taken to be the value below which not more than 5 per cent of test results fall. *See also* DESIGN STRENGTH, LIMIT STATE DESIGN, MARGIN.

charge hand. A person who is in charge of a group of workers and reports to the foreman or forewoman.

Charles's law. One of the GAS LAWS. The principle that a fixed mass of any gas, held at constant pressure, expands by the same amount when subject to the same temperature rise. *Compare* BOYLE'S LAW, DALTON'S LAW.

Charpy test. A test for the BRITTLENESS of metals measured by the energy needed to break a small specimen with a hammer.

chartered engineer. A person who is a full member of one of the chartered engineering institutions. Admission to membership usually requires both approved academic and professional experience in engineering.

chase. A groove made in the surface of a wall, ceiling or floor. It is used to hide pipes and cables and to receive FLASHINGS.

chattel. An item of movable personal property, such as furniture.

check. (1) A surface crack or split in timber that runs along the grain of converted timber. *Compare* SHAKE, SPLIT. (2) Cracks that extend through a paint coating and cause it to break up. (3) A structure that confines the flow of water in an irrigation system.

check fillet. A raised edge of asphalt on a

flat roof that is used to control the flow of rainwater.

check lock. A device on a LOCK that prevents a key from unlocking the door.

check rail. *See* MEETING RAIL.

check throat. A form of CAPILLARY GROOVE beneath a window or door SILL. It prevents water being drawn into a narrow space.

check valve. A one-way valve that protects a plumbing system from backflow of water.

cheek. A general name for the side surface of some shapes, such as a DORMER window.

cheek nailing. A method of nailing slates by means of a hole in one side of the slate and a notch in the other side.

cheesy. *adj.* Describing a film of paint or varnish that, when dry, is still soft and weak.

chemical closet (chemical toilet). A WC that is not connected to a drain but stores its contents with the help of a liquid deodorant.

chestnut. A HARDWOOD tree of the genus *Castanea*. The light-brown wood is traditionally used for making gates and fences.

chestnut paling. A lightweight form of fencing made from chestnut poles that are set about 100 mm apart and tied together by several lines of galvanized wire.

Chézy formula. A formula that relates the velocity of flow in a drain or open channel to the size and gradient of the pipe. *Compare* DARCY'S FORMULA.

chilling. A deterioration of paint or varnish that has been stored at low temperatures.

chimney. A structure, traditionally built into a wall with brick, that contains a hollow duct or FLUE to carry smoke and fumes from a fire.

chimney back. The wall behind a FIREPLACE.

chimney cap. *See* CHIMNEY POT.

chimney breast. A structure that projects into a room and contains the FIREPLACE and FLUE.

chimney cowl. *See* COWL.

chimney gutter. *See* BACK GUTTER.

chimney pot (chimney cap, chimney hood). The projecting outlet at the top of a CHIMNEY, often made of earthenware, which is designed to lead smoke clear of the chimney and help the draught. *See* ROOF diagram.

chimney stack. The brickwork part of a CHIMNEY that projects above a roof and may contain more than one FLUE. *See* ROOF diagram.

china clay. *See* KAOLIN.

China wood oil. *See* TUNG OIL.

chipboard. An artificial building board made from timber waste that is chipped and compressed together with resin binders. *See also* FIBREBOARD.

chip breaker. *See* CAP IRON.

chippings. *See* SPAR CHIPPINGS.

chisel. In general, a tool used to shape wooden or masonry surfaces with a cutting edge on the end of metal blade. The handle my be repeatedly knocked with a hammer or mallet.

chlorinated polyvinyl chloride. *See* CPVC.

chlorinated rubber paint. A PAINT made from rubber and resins that dries by evaporation of the SOLVENTS. Such paints often have a good resistance to moisture, chemicals and fire.

chlorination. One of several methods of DISINFECTION used in the treatment of water supplies to ensure that a safe quality is maintained in the distribution system. Different chemical forms of chlorine are injected into the water at a controlled rate. *Compare* OZONE TREATMENT.

choke. *See* BALLAST.

chord (flange). The top or bottom parts of a beam or girder. It carries most of the bending forces and is often strengthened. *Compare* WEB.

CHP. *abbr.* COMBINED HEAT AND POWER.

chroma. The intensity or SATURATION of a colour.

chromating. A chemical treatment of metals and alloys by painting or dipping to give increased resistance to corrosion.

chromium plating. A hard, shiny protective coating of chromium metal that is bonded to another metal by ELECTROLYSIS.

CHS. *abbr.* Circular hollow section (used in structural steelwork).

chuck. A device that grips and holds the BIT on a lathe or drill.

CI. *abbr.* CAST IRON.

CIB. *abbr.* Conseil International du Bâtiment.

CIBSE. *abbr.* (*UK*) Chartered Institution of Building Services Engineers.

CIE. *abbr.* Commission Internationale de l'Eclairage. The international body for standards concerned with light.

CIE colour coordinates. *See* COLOUR COORDINATES.

CIE sky. A STANDARD SKY in which the LUMINANCE steadily increases above the horizon so that it is three times brighter at the zenith than at the horizon. *Compare* UNIFORM SKY.

cill. An alternative spelling for SILL.

ciment fondu. HIGH-ALUMINA CEMENT.

cinder block. (*USA*) *See* CLINKER BLOCK.

CIOB. *abbr.* (*UK*) Chartered Institute of Building.

circuit. (1) A connected system of electric cables and devices that provides a complete path for electric CURRENT. (2) A system of pipes and fittings through which hot water circulates for heating or hot water supply.

circuit breaker. A safety device in an electrical CIRCUIT that automatically breaks the circuit if excess current flows. It usually operates by electromagnetism and can be reset when the fault is put right. *See also* EARTH, FUSE.

circular saw. A mechanical SAW with teeth set on the edge of a rotating steel disk. The blade is often fixed into a work bench. *See also* RADIAL ARM SAW.

circulation. (1) The movement of ventilating air within a room. (2) The movement of water within a CIRCUIT of pipes that provides heat or hot water.

CIRIA. *abbr.* (*UK*) Construction Industry Research and Information Association.

CI/SfB. A system for classifying information in the building industry. In one version of the system, building industry subjects are grouped into four main 'tables' which are further subdivided by keywords and codes.

cissing. A defect where PAINT is reluctant to adhere to a hard surface and form a continuous film. 'Cissing down' is the rubbing of a surface with fine sandpaper to prevent cissing. A form of severe cissing is known as CRAWLING.

cistern. (1) A large watertight container used to store water supplies for a building. Cisterns are often installed in the roof and used to feed taps and the heating system. A cistern is usually made of galvanized steel or plastic with a loosely fitting cover. *See* HEATING diagram. *See also* TANK. (2) *See* FLUSHING CISTERN. (3) Any large chamber, often underground, used for the storage of water.

CITB. *abbr.* (*UK*) Construction Industry Training Board.

city planning. *See* TOWN PLANNING.

civil engineering. The design and construction of works that affect the physical environment and the stability of buildings. Civil engineering projects include: foundations, roads, bridges, tunnels, water supplies, drainage, sewerage and harbours. *See also* STRUCTURAL ENGINEERING.

ckd. *abbr.* Checked (used on drawings).

cladding. An outer skin that protects a building from the weather. Brick or concrete panels may be used but modern claddings are usually lightweight panels attached to a separate load-bearing frame. *See also* CURTAIN WALL, RAIN SCREEN CLADDING, WEATHERBOARDING. *See* WALL diagram.

clamp (cramp). (1) A tool used to squeeze pieces of wood together while they are glued. (2) A metal strap used to fix a door or window lining into a wall.

clamping plate. A timber CONNECTOR.

clamp nail. (*USA*) A corrugated fastener used to secure mitre joints, such as in picture frames.

clapboard. A form of WEATHERBOARDING with tapered boards that overlap but have no rebates or tongue and grooves.

clarification. A stage in the treatment of water supplies which uses a process such as COAGULATION to remove very fine particles that do not settle naturally.

clashing strip. A sealing strip fixed on the edge of a FLUSH DOOR.

clasp nail. A strong CUT NAIL with a square cross-section. It is difficult to extract and is useful for fixing components like doors to masonry.

claw. A bar with a split end that is used to draw nails out from wood by sliding the 'claw' beneath the nail head.

claw hammer. A HAMMER with a split end or CLAW on the head. It is a common type of hammer for general carpentry. *See* WOODWORKING TOOLS diagram.

clay. A fine-grained soil that consists mainly of SILICA and ALUMINA. It is a COHESIVE SOIL that expands with moisture and shrinks with drying. *See also* FROST HEAVE.

clay tile. (1) A ROOFING TILE made from burnt clay. (2) A QUARRY TILE.

clayware. *See* VITRIFIED CLAYWARE.

cleaner's sink. A sink at normal height but with taps set at high level to allow the filling of buckets. *Compare* BELFAST SINK.

cleaning eye. *See* ACCESS EYE.

cleaning hinge (easy-clean hinge, offset hinge). A type of hinge that holds a CASEMENT window clear of the frame so that the outside of the glass can be cleaned from the inside.

cleanout. (*USA*) An ACCESS EYE.

clear sky radiation. The loss of heat from a body by RADIATION that occurs when a night sky is dark and cloudless. This effect can cause the temperature of an object, such as a roof, to fall below the surrounding air temperature.

clear span. The distance between the faces of the two supports on which a beam rests. This is usually longer than the EFFECTIVE SPAN.

clearstorey, clearstory. *See* CLERESTORY.

clear timber (free stuff). Timber that is free from visible defects.

cleat. A strip of material, such as a wooden BATTEN, used to locate and fix objects to a wall or other partition. A shelf may rest on top of cleats.

cleavage. A direction in a material, such as timber or metal, along which splits or breakage are most likely to occur.

cleft timber. Timber that has been split.

clenching (clinching, clench nailing). Bending over the point of a nail that protrudes through a piece of wood.

clerestory (clearstorey, clearstory). A row of windows high in a wall near the roof.

clerk of works. An experienced person on a building site who works with the architect and keeps records to ensure that work is carried out exactly as specified.

climbing crane. A form of crane that is housed within the structure and is raised as the structure is built. Upon completion the crane is dismantled and lowered from the building.

clinching. *See* CLENCHING.

clinker. (1) The ash and other remains of coal burnt in a furnace. Clinker of the right quality can be used for HARDCORE and as an aggregate for precast concrete BLOCKS. (2) The fused compounds of clay and chalk produced in a kiln that manufactures PORTLAND CEMENT.

clinker block (*USA*: cinder block). A preformed concrete BLOCK made from CLINKER aggregate. Modern lightweight blocks may also be made of AERATED CONCRETE.

clinometer. An surveying instrument that measures the angle of dip of an inclined plane.

clo. A measurement of the thermal insulation provided by clothing. The clo-scale is used to predict thermal comfort conditions and ranges from 0 clo for nakedness, through 2.0 for heavy winter clothing, to 4.0 maximum.

close-boarded fencing. A fence that is formed by vertical BOARDING nailed to two or more horizontal rails. The boards may be butt jointed or feather edged (*see* BUTT JOINT, FEATHER EDGE). *See also* GRAVEL BOARD.

close boarding. *See* BOARDING.

close-contact adhesive. A type of ADHESIVE or glue that will not work unless the gap between the surfaces is small. *Compare* GAP-FILLING GLUE.

close-coupled cistern. A type of LOW LEVEL FLUSHING CISTERN in which the cistern is directly attached to the WC pan.

close-couple roof (couple roof). A form of ROOF that has the COMMON RAFTERS joined at WALL PLATE level by TIES. It is suitable for spans up to about 4 metres. *See also* SIMPLE ROOF.

close-cut hip. A HIP on a roof where the tiles or slates are cut at an angle so that they almost make a BUTT JOINT on the line of the hip.

close-cut valley. A VALLEY on a roof where the tiles or slates are cut at an angle so that they almost make a BUTT JOINT on the line of the valley.

closed cell. A cellular material, such as some EXPANDED PLASTICS, in which the adjacent cells are not connected. This type of structure helps make thermal insulating materials waterproof. *Compare* OPEN CELL.

closed eaves. *See* BOXED EAVES.

closed string (close string, housed string). On a stair, an outer STRING that has both edges parallel. *Compare* CUT STRING. *See* STAIRS diagram.

closed traverse. A form of TRAVERSE in surveying that finishes at its starting point.

closed valley. (*USA*) *See* SECRET GUTTER.

close-grained. *adj.* Describing wood that has a fine GRAIN.

closer (bat). A brick or block that is cut short in order to complete a BOND. *See* BEVELLED CLOSER, KING CLOSER, QUEEN CLOSER.

close string. *See* CLOSED STRING.

closet. (1) *See* WC. (2) (*USA*). A cupboard or small room.

clout nail (felt nail). A NAIL with a large round flat head. It is often used for fixing roofing felt and plasterboard.

CLS. Canadian Lumber Standards.

club hammer (lump hammer, mash hammer). A heavy hammer, with a relatively short handle. It is used for working with brick and stone. *See* MASONRY TOOLS diagram.

clunch. A hard form of CHALK from which old English cottages and other buildings were built.

coach bolt (*USA*: carriage bolt, lagbolt). A round-headed BOLT with a square section beneath the head which grips into wood as the NUT is tightened on to the bolt.

coach screw (lagscrew). A large pointed SCREW that is driven into wood by turning the square head with a spanner.

coagulation. A process used for removing fine suspended particles during water treatment and sometimes sewage treatment. The addition of a chemical such as ALUM produces a precipitate that coagulates with the fine particles and allows them to be removed.

coal tar. A thick black oily liquid of HYDROCARBONS obtained during the production of coal gas. PITCH is left as a residue after coal tar has been purified. *See also* REFINED TAR.

coarse aggregate. (1) Larger sized AGGREGATE which is mixed with FINE AGGREGATE and CEMENT to made CONCRETE. Coarse aggregate for concrete may be defined as that which does not pass through a screen with holes 5 mm square. (2) Coarse aggregate for ASPHALT is that which does not pass through a screen with holes 3 mm square.

coarse filter. An initial filtration stage sometimes used in the treatment of public water supplies. The filter is usually mechanical in action and reduces the load on the fine filters that follow. *See also* FILTER.

coarse stuff. A form of MORTAR made from sand and LIME. It can be kept for several weeks and used for base coats of lime PLASTER.

coat. (1) One thickness of a surface material, such as plaster or asphalt, that is applied in layers. (2) A film of paint or varnish that is applied in one application and allowed to dry before another application. *See also* PRIMER, UNDERCOAT.

coated macadam. A type of MACADAM road material formed from graded aggregate that is coated with BITUMEN or ROAD TAR.

cob. A traditional building block made of unfired clay, or other earth, that is reinforced with straw. *Compare* ADOBE.

cobble (cobblestone). A rounded stone used for rough or decorative paving.

cock. A valve or other device in a pipe that controls the rate at which liquid or gas flows through the pipe. *Compare* STOPCOCK. *See also* BALL VALVE.

cock bead. A length of BEAD that stands above the surface of the wood.

cocking piece. *See* SPROCKET.

Code of Practice (CP). A publication that details acceptable standards of practice for particular construction operations such as those used in TRADES. The BRITISH STANDARDS INSTITUTION, for example, publishes codes of practice (BS:CP). *See also* BUILDING CODE.

coefficient of elasticity. *See* ELASTIC MODULUS.

coefficient of performance (COP). A measure of the efficiency of a HEAT PUMP. The coefficient can be expressed as the ratio of the heat output from the pump compared to the energy need to operate the pump, but it should also take into account the extra energy needed to defrost coils and supply supplementary heating if necessary. Practical values of COP are often between 2 and 3.

coefficient of thermal conductivity (k-value). A measure of the rate at which heat is conducted through a particular material under specified conditions. The SI unit is in W/m°C. Materials for THERMAL INSULATION are chosen to have low values of thermal conductivity. The coefficient is for an isolated material rather than a composite structure *See* U-VALUE.

coefficient of thermal expansion. A measure of the change in length that occurs in a standard length of a particular material for each degree rise (or fall) in temperature. Similar coefficients measure the change in area and change in volume.

coefficient of utilization. *See* UTILIZATION FACTOR.

coffer. A recessed panel in a ceiling.

cofferdam. A temporary DAM built to exclude water during construction. It is not usually as deep as a CAISSON. *See also* GABION.

cohesion. The general process of attraction between molecules that holds a solid or liquid together.

cohesive soil. A soil, such as CLAY, that sticks together and gives increased SHEAR STRENGTH.

coin. *See* QUOIN.

cold asphalt. A layer of BITUMEN and fine aggregate used as a WEARING COURSE on a roadway. It can be spread and compacted without heating.

cold bridge (thermal bridge). A portion of a structure, such as a LINTEL above a window, that has a lower thermal insulation than the surrounding structure.

cold chisel. A hardened steel CHISEL that is strong enough to cut brick, stone or cold metal. *See* MASONRY TOOLS diagram.

cold-rolled section. A structural steel girder or other profile produced by COLD ROLLING. They require less steel than comparable ROLLED-STEEL JOISTS or UNIVERSAL BEAMS.

cold rolling. The bending or flattening of a metal strip by passing it between rollers without heating. Structural sections are often made by this method. *See also* COLD-ROLLED SECTION. *Compare* HOT ROLLING.

cold roof. In general, a roof where the layer of THERMAL INSULATION is installed near the ceiling rather than near the outside. The air in the roof space is therefore unheated by the building and will be cold in winter. Cold roofs are prone to CONDENSATION and need protection by VAPOUR BARRIERS and ventilation. *Compare* WARM ROOF.

cold water supply. The system of pipes that distributes cold water through a building from a RISING MAIN or cold water storage CISTERN.

cold working. The process of changing the shape of metal without heating it. *Compare* HOT WORKING.

collapse. (1) The general failure of a structure while loaded. (2) An irregular shrinkage in some HARDWOODS such as eucalyptus. *See also* RECONDITIONING.

collar. (1) A section of pipe that is widened or narrowed, usually to help make a join. (2) A circular ridge of asphalt or other roofing material around a vertical pipe where it passes through a roof.

collar beam (span piece, top beam). A horizontal TIE member between a pair of rafters in a roof. It is attached approximately halfway up the length of the rafters.

collector. A device designed to absorb RADIATION from the sun and convert it to heat energy.

collimation line. The line of sight through the centre of a surveying instrument such as a LEVEL or THEODOLITE. The line passes through the intersection of the CROSS HAIRS.

colloid (colloidal suspension). A substance consisting of very fine particles permanently suspended in liquid. *See also* PIGMENT.

colonial bond. *See* ENGLISH GARDEN WALL BOND.

colophony. *See* ROSIN.

colour. (1) The sensation in human vision when different wavelengths of light reach the eye. (2) The property of a surface or PIGMENT that produces the effect of colour in the eye. *See also* ADDITIVE COLOUR, HUE, MONOCHROMATIC LIGHT, SUBTRACTIVE COLOUR.

colour coordinates. A method, standardized by the CIE, that describes any COLOUR as a mixture of three primary colours. *Compare* MUNSELL SYSTEM.

colour rendering. The ability of a light source to reveal the colour appearance of surfaces. The colour rendering index compares the effect of the source with the effect of a reference source such as daylight and uses a value of 100 for an ideal source. *See also* LAMP.

colour temperature. A method of specifying the COLOUR of some light sources by using the temperature, in degrees kelvin, of a theoretical perfect radiator when it gives off a matching colour. The colour temperature does not relate to the actual temperature of the source. Values of colour temperature around 3000 K have a higher content of red light and are considered to be 'warm' white. Higher values of colour temperature, around 6000 K for example, contain more blue light and are interpreted as 'cool' white like daylight. *See also* COLOUR RENDERING, LAMP.

colour wash. The decoration of large areas with relatively simple paints such as DISTEMPER and LIMEWASH.

column. A vertical structural member that normally carries an AXIAL LOAD in compression. It may be made of concrete, stone, brick or metal. *Compare* PILLAR, STANCHION, STRUT.

comb. (1) A toothed metal plate used for spreading adhesive or for scratching the surface of plaster to make a KEY for the next layer. (2) A tool with steel or rubber teeth used for the GRAINING process in decorating.

combed joint (laminated joint). An angle joint between pieces of wood with parallel slotted ends. *Compare* DOVETAIL.

comb hammer (scutch, skutch). A bricklayers' tool used to cut bricks. Instead of a striking face the hammer has PEENS into which steel combs may be fitted.

combination boiler. A gas boiler for central heating that also acts as an INSTANTANEOUS WATER HEATER for the supply of hot water. It does not need a supply tank and storage cylinder for the hot water and therefore reduces the amount of pipework.

combination cylinder and tank. A type of HOT WATER CYLINDER that includes a cold water feed tank. A separate feed tank in the roof tank is unnecessary.

combination pliers. *See* FOOTPRINTS.

combined heat and power (CHP). A system that makes use of waste heat from electricity generation for other purposes such as DISTRICT HEATING.

combined lighting. A lighting system that combines artificial light and natural light. *See* PERMANENT SUPPLEMENTARY ARTIFICIAL LIGHTING OF INTERIORS.

combined system. A drainage system that uses one set of pipes to carry both SURFACE WATER and FOUL WATER. *Compare* SEPARATE SYSTEM.

combustible. *adj.* Capable of burning. *See also* COMBUSTION.

combustion. The chemical combination of a substance with the oxygen of the air to produce heat and light; burning.

comfort temperature. An evaluation of temperature that gives good agreement with the human sensation of thermal comfort. There are several methods of measurement that take into account the air temperature, surface temperature of surroundings, humidity and air movement. *See also* DRY RESULTANT TEMPERATURE, EFFECTIVE TEMPERATURE, ENVIRONMENTAL TEMPERATURE, RADIANT TEMPERATURE.

commode step. A step, usually at the foot of a staircase, that has an outward curve to the RISER.

common bond. (*USA*) *See* ENGLISH GARDEN WALL BOND.

common brick. The local variety of BRICK suitable for general building work but not necessarily used as a FACING BRICK.

common furniture beetle. *See* FURNITURE BEETLE.

common ground (rough ground). A strip of wood that is attached to a wall and acts as a base for plasterboard, skirting board or other lining.

common joist (floor joist, boarding joist). One of a series of parallel timber beams to which the floor is nailed. *See also* JOIST.

common rafter (intermediate rafter). A RAFTER that runs between the WALL PLATE and the RIDGE of the roof and may have intermediate support from horizontal PURLINS. *Compare* PRINCIPAL RAFTER, TRUSSED RAFTER.

common wall. (*USA*) A PARTY WALL.

communicating pipe. The pipe between the Water Authority's MAIN and the consumer's valve or boundary.

compaction. The compression of a soil by mechanical means such as rolling or vibration. It is not the same as CONSOLIDATION.

compartmentation. The division of a large building into areas that, under conditions of fire, are separated by elements with a high FIRE RESISTANCE. The purpose of compartmentation is to prevent the fire spreading beyond the compartment where it started.

compartment floor. *See* COMPARTMENTATION.

compartment wall. A wall built with a high FIRE RESISTANCE as part of a system of COMPARTMENTATION.

compass brick. *See* RADIAL BRICK.

compass saw. *See* PAD SAW.

compatibility. The general ability of a material, such as paint, to be installed in close contact with another material without harmful effects to either.

compo. (1) MORTAR made from cement and lime. (2) An ALLOY of lead used to make gas supply pipes with some flexibility.

composite. A combination of materials that give a new type of material with useful properties. CONCRETE is a composite of cement and aggregate but the term composite is sometimes reserved for more modern materials such as GLASS REINFORCED CONCRETE, GLASS REINFORCED PLASTIC and CERMET.

composite board. A building board made up from more than one type of material, such as the bonding of aluminium foil to plasterboard to make a VAPOUR BARRIER.

composite construction. A combination of different materials used together in one part of a construction, such as brickwork combined with concrete.

composition block. A hard-wearing floor block made from a mixture of cement, wood particles, gypsum, calcium carbonate, colouring and oils. The blocks are laid in a bed of cement and sand.

composition flooring. *See* JOINTLESS FLOORING.

composition roofing. (*USA*) Roofing constructed with BITUMEN felt materials. *See* BUILT-UP ROOFING.

compressed straw slab (strawboard). A building BOARD made from compressed straw that is faced with paper on each side.

compression. The effect of a force when it pushes and tends to shorten a structure. The opposite of TENSION.

compression cycle. *See* COMPRESSION REFRIGERATOR.

compression glazing. A system of GLAZING in which the glass is squeezed between GASKETS rather than held by PUTTY.

compression joint. A connection made between copper piping by means of screw fittings that squash soft metal rings (GLANDS) against the pipe walls to make the seal. *See also* MANIPULATIVE JOINT.

compression refrigerator. A refrigerator, such as the common household type, that uses a compressor to change the REFRIGERANT from vapour to liquid. The compressor may also act as a circulation pump for the REFRIGERATION CYCLE. *Compare* ABSORPTION REFRIGERATOR.

compression test. (1) The controlled crushing of a sample to determine the CRUSHING STRENGTH of the material. *See also* CUBE TEST, MORTAR-CUBE TEST. (2) A laboratory test for the strength of soils. *See* TRIAXIAL COMPRESSION TEST.

compressive stress (compressive strength). The force per unit area needed to compress a material. Metric units in practical use include N/mm^2 and MN/m^2.

computer-aided design. *See* CAD.

concentrated load. *See* POINT LOAD.

concrete. A common building material made from specified proportions of CEMENT, sand and stone AGGREGATE, and water. The concrete hardens by a hydration reaction with the water to produce a material with high COMPRESSIVE STRENGTH. Concrete is often produced on the building site using local aggregates and may be combined with other materials such as steel. *See* GLASS-REINFORCED CONCRETE, REINFORCED CONCRETE. *Compare* MORTAR, which does not contain COARSE AGGREGATE.

concrete block. One of several types of uniform building block made from LIGHTWEIGHT CONCRETE. *See* BLOCK, BLOCKWORK.

concrete cancer. A general term for concrete that fails its purpose because of expansion and disruption when it is penetrated by moisture. One cause of the expansion is a reaction between the alkali in cement and the silica in aggregates. *See also* CARBONATION.

concrete insert. A PLUG of fibre or metal set into concrete and used to receive a SCREW.

concrete nail. (*USA*) A type of MASONRY NAIL.

condensate. The liquid water produced from moist air by the process of CONDENSATION.

condensation. (1) The general change of state of a substance from a vapour to liquid. (2) The formation of liquid water whenever moist air is cooled to the temperature (DEW POINT) at which it becomes saturated. *See* INTERSTITIAL CONDENSATION, SURFACE CONDENSATION. (3) Dampness in buildings and the associated problems caused by condensation of water VAPOUR in the air. (4) A type of POLYMERIZATION reaction that has a simple by-product such as water. The production of nylon, for example, involves a condensation reaction.

condensation groove (water channel). A channel or other detail of a window that is designed to collect CONDENSATION water from the glass and carry it to the outside.

condenser. (1) A container in which a substance changes from a vapour to a liquid, accompanied by the emission of latent heat. (2) The part of a refrigerator or heat pump that gives off heat. *Compare* EVAPORATOR.

condominium. An APARTMENT building in which each dwelling is owned by individuals and the common areas are jointly owned. *See also* FLAT.

conductance. *See* THERMAL CONDUCTANCE.

conduction. (1) The transfer of heat energy through a material from a region of higher temperature to a region of lower temperature by means of interactions between neighbouring molecules or free electrons within the material. *Compare* CONVECTION, RADIATION. *See also* COEFFICIENT OF THERMAL CONDUCTIVITY. (2) The flow of electric current through a body by the movement of FREE ELECTRONS or IONS.

conductivity. (1) A measure of a material's ability to transfer heat by thermal CONDUCTION. *See* COEFFICIENT OF THERMAL CONDUCTIVITY. (2) A measure of a material's ability to carry an electric CURRENT.

conductor. (1) A material with a high CONDUCTIVITY for the passage of heat or electricity. Metals tend to be good thermal and electrical conductors. *Compare* INSULATOR. (2) A wire or cable used in the distribution of electricity. (3) (*USA*) *See* DOWNPIPE.

conduit. (1) A length of protective tube or trough which contains insulated electric CABLES. The conduit may be embedded in a wall or other part of a building, and new cables pulled through the conduit. (2) A pipe or open channel for the supply of water.

conduit box. *See* JUNCTION BOX.

conduit bushing. A short insulating sleeve that attaches to the opening of a CONDUIT to protect the cables.

cone penetration test. A test for predicting the resistance of some soils to penetration, such as for the driving of piles. The depth of indentation is measured when a standard-sized cone is pressed into the ground under a known load.

conifer. A family of trees and shrubs that usually have cones and evergreen leaves. Conifers include PINES and FIRS.

Coniophora cerebella, Coniophora puteana. *See* WET ROT.

connector (timber connector). A mechanical FASTENER used to join sections of timber frameworks such as roof TRUSSES. Forms of connector include toothed rings held by bolts or plates with nail-like teeth.

consolidation. The gradual compression of a soil, mainly clay, as water and air are driven out of it. It is not the same as COMPACTION.

construction joint. A joint between two areas of concrete that have been placed and set on different days.

consultant. A professionally qualified person, such as an architect, engineer or surveyor, who acts on behalf of the client in some aspect of the building process.

consumer control unit (*USA*: cutout box). An electrical junction box that contains the main isolation switch and FUSES. The consumer unit is usually installed just past the electricity authority's cable and meters.

contiguous piling. A continuous series of reinforced concrete PILES that lock together forming a continuous wall to keep water out of excavation works.

contingency sum. A sum of money included in a CONTRACT to allow for unforeseen work during the construction.

contour (contour line). A line on a map that connects points at the same height above a set level or DATUM such as high water sea level.

contract. An agreement between a client and a contractor to complete specified building works for a certain sum of money. *See also* CONTINGENCY SUM, COST-PLUS CONTRACT, DAMAGES, FIXED-PRICE CONTRACT, JOINT CONTRACTS TRIBUNAL, SUB-CONTRACT, TENDER.

contract documents. The documents that define the basis of a CONTRACT and that are signed by both parties. The contract documents often include drawings, specification and the BILL OF QUANTITIES.

contraction joint (shrinkage joint). A MOVEMENT JOINT in concrete or masonry that allows for shrinkage caused by drying and temperature changes.

contract manager. An experienced member of the construction team who takes responsibility for the completion of the CONTRACT and deals with the client's representative on site.

contractor. A person, or organization such as building firm, who signs a CONTRACT to complete a specified work. *See also* SUB-CONTRACT.

contraflexure (point of contraflexure, inflexion point). A point in a beam where there is a change in direction of bending from either HOGGING or SAGGING. The BENDING MOMENT is zero at this point.

contrast. Differences in brightness or in colours seen by the eye at one time because they are in the same VISUAL FIELD.

control gear. The electric devices included in the circuit of a gas DISCHARGE LAMP like a fluorescent lamp. The control gear may include a BALLAST coil and capacitor. *See also* QUICK START CIRCUIT.

controlling dimension. A dimension between reference planes such as floor levels.

convection. The transfer of heat energy through a liquid or gas by the bodily movement of particles. Natural convection occurs when part of a sample becomes warm and less dense; the cooler heavier air then displaces the warm air, which rises. *Compare* CONDUCTION, RADIATION.

convection current. The pattern of movement caused by CONVECTION within a liquid or gas. The currents may be driven by natural convection caused by uneven heating or by mechanical convection. *See also* STACK EFFECT.

convector (convector heater). A form of heater that distributes heat by the process of CONVECTION. Heat may be transferred from the heating surfaces by natural convection or with the aid of a fan.

conversion. (1) The process of producing TIMBER from logs. *See also* FLITCH. (2) A slow change in the chemical or mechanical properties of a material that is usually caused by exposure conditions. HIGH ALUMINA CEMENT, for example, loses strength in a warm humid environment. (3) The process of changing a building to a different use.

converted timber. *See* SQUARE-SAWN TIMBER.

cooling tower. A large tower in which flowing water transfers heat from steam CONDENSERS into the air. This heat is usually associated with large thermal power stations in isolated locations. *See also* COMBINED HEAT AND POWER, DISTRICT HEATING.

coordinates. A system of measurements that define the position of a point on paper or in space. The commonly used Cartesian coordinates use the distances from two fixed straight lines called axes. *See also* POLAR COORDINATES.

COP. *abbr.* COEFFICIENT OF PERFORMANCE.

coping. (1) A protective cap of brick, stone or concrete that runs along the top of a wall and usually has an overhang. *See also* PARALLEL COPING, SADDLE-BACK COPING, WEATHERING. (2) The process of splitting stones by driving wedges into a line of drilled holes.

co-polymer. A POLYMER formed by the combination of more than one type of starting MONOMER molecule.

copper. A pure metal element (chemical symbol Cu) with a reddish colour and a good resistance to corrosion. Its high MALLEABILITY makes it easy to work and copper is used for pipes and some roofing. Copper is also a good conductor of heat and electricity and its DUCTILITY makes it suitable for electrical CABLES.

copper plating. A protective surface coating of COPPER which is applied to steel and other articles by the process of ELECTROPLATING.

copper roofing. The use of copper sheets for FLEXIBLE METAL roofing.

corbel. A projection of stone, brick or timber that juts out from a wall to support a load such as a roof beam. *See also* OVERSAILING COURSE.

corbelling. Brick or stonework in which each course slightly juts out over the course beneath to form a CORBEL.

core. (1) A cylindrical sample of material such as rock, soil or concrete. (2) A metal rod that runs beneath a handrail. (3) The inner layers of COMPOSITE BOARD, FLUSH DOORS and similar constructions.

coreboard (battenboard, *USA*: lumber core). A general name for a composite wooden board made from strips of wood glued together. *See also* BLOCKBOARD, LAMINBOARD.

core rail. A horizontal steel bar that sometimes supports a timber HANDRAIL.

cork. The BARK of the cork oak tree. In building it is generally used in the form of loose-fill insulation or CORKBOARD.

corkboard (cork tile). A board made from compressed granulated CORK. It is generally used as floor or wall decoration that also provides some sound and thermal insulation.

cork tile. *See* CORKBOARD.

cornice. A length of MOULDING at the top

of a wall. It usually covers the join between a wall or ceiling, or projects from an external wall. *See also* COVE, DENTIL.

correlated colour temperature. *See* COLOUR TEMPERATURE.

corrosion. The gradual destruction of a material by chemical effects produced by its environment. Some corrosion is by direct combination with gases in the atmosphere such as oxygen and sulphur dioxide. Most corrosion occurs by a form of ELECTROLYSIS where differences between metals generate the required electric current by a SIMPLE CELL action. *See also* CATHODIC PROTECTION.

corrugated fastener (joint fastener, mitre brad, wiggle nail). A fastener in the form of a wavy piece of metal that is driven, like a NAIL, across the join between two timber boards.

corrugated sheet. A sheet of light cladding material, such as metal or plastic, with a corrugated profile that increases the rigidity and useful length of the sheet.

cost benefit analysis. An economic evaluation of a project that includes the costs and benefits to people in addition to the construction costs. *See also* LIFE-CYCLE COST, TEROTECHNOLOGY.

cost control. A system that determines the actual cost of work as it proceeds and compares this to the planned cost so that action can be taken if necessary.

cost plan. A breakdown of the estimated costs of a building project. The plan may be used to provide target costs during the design stage of a project.

cost-plus contract. A type of CONTRACT under which the CONTRACTOR is paid the actual cost of materials and labour for a job, plus an agreed percentage to cover overheads and profit. Such a contract is usually restricted to small or emergency works.

coulomb (C). The unit used to measure electric charge or static electricity. *Compare* AMPERE, AMPERE-HOUR, CURRENT.

counter batten. (1) Roof BATTENS that are fixed on top of the boarding or felt and run parallel to the RAFTERS. The battens for the tiles or slates are then nailed to the counter battens, leaving a drainage gap for any water that gets into the roof. (2) A stiffening batten nailed across the back of joined boards.

counter flap hinge. A form of hinge that, when fitted on the flap of counter, allows the flap to be folded back on itself.

counter flashing (cover flashing). A length of metal FLASHING that projects from a chimney or parapet wall and is turned down over the top of the roof covering or flashing.

countersinking. The process or effect of making a conical hole in the surface of timber or other material. It allows the 'countersunk' head of a screw to be flush with the surface.

couple roof. *See* CLOSE-COUPLE ROOF.

coupling. A collar type of fitting used to join two lengths of pipe or scaffolding.

course. (1) A layer of bricks, blocks, stones, tiles etc. that is laid parallel to the layer beneath and usually makes a BOND. (2) The direction and length of a survey line.

coursed ashlar. (*USA*) *See* REGULAR-COURSED RUBBLE.

coursing joint. *See* BED JOINT.

covalent bond. A type of bond between atoms achieved by the sharing of electrons. *Compare* IONIC BOND.

cove (coving). A MOULDING that joins a wall and ceiling with a concave curve. *See also* CORNICE.

cove lighting. A form of INDIRECT LIGHT from sources set above the COVE or CORNICE. The light is directed upwards at the ceiling from where it is reflected downward.

covenant. A legal agreement, such as a particular clause in a lease.

cover fillet (cover strip). A narrow strip of material covering a join between boards, such as wall or ceiling board.

cover flashing. *See* COUNTER FLASHING.

covering capacity (covering power, hiding power). The area that can be satisfactorily covered by a given quantity of PAINT or other coating. This ability is measured by the property of OPACITY.

cover strip. *See* COVER FILLET.

coving. *See* COVE.

cowl. A cover, often rotating, that is fixed on the top of a CHIMNEY to improve the flow of gases.

CP. *abbr.* CODE OF PRACTICE.

CP. *abbr.* Chromium plated.

CPA. *abbr.* CRITICAL PATH ANALYSIS.

CPM. *abbr.* Critical path method. *See* CRITICAL PATH ANALYSIS.

cPVC. *abbr.* Chlorinated polyvinyl chloride. A form of heat-resistant plastic, used for hot water pipes. *Compare* PVC.

cradle. (1) A movable platform suspended by rope or cables from the top of a building to allow access for cleaning, painting and general maintenance. (2) A temporary framework that carries a load during construction.

craft. *See* TRADE.

cramp. *See* CLAMP.

crampon. *See* NIPPERS.

crane. A tall machine from which heavy loads can be lifted with cables and moved around on site.

crawling. A defect where gloss paint or varnish shrink before drying and reveal the undercoat. CISSING is a mild form of crawling.

crawlway. A DUCT that is large enough, with pipes and cables installed, for a person to crawl through.

crazing. A general pattern of intersecting cracks in the surface of a coating such as concrete, plaster or paint.

creasing. Several courses of plain TILES laid flat in a brick wall. The line of tiles may project slightly from the wall to make a decorative feature.

creep. A gradual movement or deformation of a material or structure that is loaded. *See also* PERMANENT SET.

creosote. An oil obtained from coal tar and used as a PRESERVATIVE for rough woodwork, such as fencing.

CRI. *abbr.* Colour rendering index. *See* COLOUR RENDERING.

crib wall. A RETAINING WALL built of interlocking timber or concrete sections that are laid on top of one another and filled with earth or rock.

cricket. (*USA*) A small FLASHING designed to shed water into a gutter behind a CHIMNEY.

crinkle-crankle wall. *See* SERPENTINE WALL.

cripple. Any part of a frame that has been shortened to allow for an opening. Cripple studs are the short vertical STUDS that give extra strength over a door in a timber-framed wall. *See* WALLS diagram.

critical activity. In CRITICAL PATH analysis, an activity that is part of the 'critical path'. Any delay to a critical activity affects the completion of the entire project. *Compare* FLOAT TIME.

critical frequency. A particular frequency of sound at which the SOUND INSULATION of a partition, such as a wall, is less than theory predicts because of coincidence vibrations within the structure.

critical path analysis (critical path method). A NETWORK ANALYSIS method that graphically highlights those activities that will delay a whole project if they are delayed. *See also* ARROW DIAGRAM, CRITICAL ACTIVITY, FLOAT TIME, GANTT CHART, PRECEDENCE DIAGRAM.

critical temperature. The temperature below which gas can exist in the VAPOUR state.

critical velocity. For liquid flowing in a pipe, the velocity above which the nature of the flow changes from LAMINAR FLOW and starts to become TURBULENT FLOW. *See also* REYNOLD'S NUMBER.

crook. A bend along the edge of a length of timber.

cross cut. A saw-cut made at right angles to the grain of wood. *Compare* RIP.

cross-cut saw. A type of SAW with its teeth designed to cut across the grain of wood. *Compare* RIP SAW.

cross fall. A gradient that runs across the width of a structure rather than along the length.

cross garnet hinge. A type of STRAP HINGE in which the strap is hinged to a upright plate. The hinge may be used on a LEDGED AND BRACED DOOR.

cross grain. Grain in wood that does not run along the length of the piece. *Compare* STRAIGHT GRAIN.

cross hair. A fine line used for sighting within the telescope of a surveying instrument. *See also* RETICULE.

cross hatching. *See* HATCHING.

cross-head screw. A type of SCREW with crossed slots in the head that need to be turned with a special screwdriver. *See also* POZIDRIVE SCREW, PHILLIPS SCREW, RECESSED-HEAD SCREW. *Compare* SLOTTED-HEAD SCREW.

cross joint (head joint). A vertical mortar joint in brickwork, blockwork or stonework. *See also* PERPEND. *Compare* BED JOINT.

cross-lap joint. *See* HALVED JOINT.

cross linked polymer. A THERMOSETTING material.

cross section. The shape or drawing of an object when cut through at right angles to its length. *Compare* LONGITUDINAL SECTION.

cross wall. A form of construction for long buildings that uses connecting walls at intervals to distribute the load from the roof and floors. A TERRACE of houses uses cross walls between the dwellings.

cross welt (single-lock welt). A seam that joins two sheets of flexible roof material and usually runs across the roof, parallel to the ridge.

crow bar (*USA*: ripping bar). A long type of PINCH BAR used as a lever for moving heavy objects.

crown. The highest part of a curve, such as on an ARCH or a road.

crown silvered lamp. *See* CS LAMP.

crushing strength (crushing stress). The maximum load per unit area (COMPRESSION STRESS) that a sample of material can withstand under COMPRESSION. The values are used for comparing the strengths of bricks, concrete, stones and mortars. *See also* CUBE TEST, MORTAR-CUBE TEST.

CSI. *abbr.* (*UK*) Construction Surveyors' Institute.

CS lamp. Crown silvered lamp. A form of FILAMENT lamp in which the front of the bulb is silvered. The light is reflected back to a reflector fitting that produces a narrow beam of light.

CTF. (*UK*) Clay and Tile Federation.

Cu. Chemical symbol for COPPER.

cube test. A standard test for comparing the CRUSHING STRENGTHS of different CONCRETE samples. Cubes of concrete are made and compressed by standardized methods. *See also* COMPRESSION TEST.

cubing. The calculation of the volumes in a building, or of certain items in a BILL OF QUANTITIES.

culvert. A covered channel or large pipe that carries water beneath a building or road.

cup. (1) A warp in flat timber, such as floor boards, in which the long edges curl upwards to make a slight hollow. (2) A shaped ring that is designed to house the head of a SCREW and protect the surrounding material.

curb (kerb). The upright edge of a roadway or a flat roof.

curb cock. (*USA*) A STOP COCK under a sidewalk that can cut off the water supply pipe to a building.

curb rafter. A rafter in the flatter part of a MANSARD ROOF.

curing. (1) The control of temperature and humidity during the SETTING and HARDENING of concrete in order to achieve correct final strength. Curing often involves keeping concrete moist by water sprays or coverings. (2) The general chemical process that occurs when a polymer, such as paint or adhesive, hardens.

current (I). The flow of electrical energy in a circuit. The current is defined as the rate of flow of CHARGE past a point and measured in AMPERES. *See also* ALTERNATING CURRENT, AMMETER, DIRECT CURRENT, OHM'S LAW.

curtaining (sagging). A defect in a paint film usually seen on vertical surfaces when the application of too much paint causes the film to sag.

curtain wall (*USA*: enclosure wall). A system of CLADDING for a building that transfers no load from the rest of the structure and is often suspended from the frame of the building. The walling may be partly glazed and some high rise buildings are totally curtained in glass. *See also* SPANDREL.

curtilage. The total land belonging to a particular dwelling house.

cusec. A unit of liquid FLOWRATE such as in rivers. 1 cusec = 1 cubic metre per second in metric units, or 1 cubic foot per second in Imperial units.

cushion. *See* PADSTONE.

cutback bitumen. Ordinary solid STRAIGHT-RUN BITUMEN that is reduced in viscosity by the addition of a volatile oil such as kerosine or CREOSOTE. Cutback bitumens can be used at lower temperatures than straight-run bitumens and the hardening is slower. They are mainly used for road building, road repair, waterproofing, and paint preparation.

cut brick. A brick that is cut to shape with an AXE or BOLSTER. *Compare* GAUGED BRICK.

cut-in box. (*USA*) *See* PLASTERBOX.

cut nail. A heavy NAIL that has a rectangular cross-section and is made from steel plate instead of from wire.

cut-off. In general, a construction such as a wall or trench built below ground level and designed to keep water out.

cutout. An device in an electrical CIRCUIT, such as a FUSE or CIRCUIT-BREAKER that automatically breaks the circuit if excess current flows.

cutout box. (*USA*) *See* CONSUMER CONTROL UNIT.

cut string (bridge board, open string). An outer STRING or side of a stair that has steps cut in the upper edge. The TREADS of the stair rest on the string and overhang the side of the string.

cutting in. The formation of a clean edge or join between two painted areas.

cutting list. A list of the materials of fixed sizes, such as timber or steel bars, needed for a particular project.

cyan. A sky-blue colour. *See* SUBTRACTIVE COLOUR.

cybernetics. The study of control systems for machinery and the interaction of such systems with people.

cycle. In general, a repeating pattern in which conditions change and then return to the starting point. The length of time taken by one such change is the PERIOD. *See also* FREQUENCY.

cylinder. (1) (storage cylinder) A closed tank for storing hot water under pressure before it flows to the taps. The tank usually has cylindrical sides and domed ends. (2) (*USA*) Steel tubes driven through bad ground and filled with concrete to make pile foundations.

cylinder lock. An entrance door LOCK in which the mechanism is contained within a cylinder. It is usually operated by a key from the outside and by a knob from the inside. *See also* LATCH.

D

dado. A decorative treatment on the lower part of an internal wall that makes it distinct from the upper part. A dado may be formed by various methods which include wooden panelling or different paintwork. The top of the dado is usually finished with a dado capping or rail. *See also* SURBASE. *Compare* CHAIR RAIL, PICTURE RAIL.

dado joint. *See* HOUSED JOINT.

dais. A platform raised higher than floor level in a hall or large room. It is often used for speaking from.

Dalton's law. One of the GAS LAWS. The principle that in a mixture of gases, such as air, each gas exerts an individual 'partial pressure' that is independent of the other gases. The total pressure of the mixture is equal to the sum of the partial pressures. *Compare* BOYLE'S LAW, CHARLES'S LAW.

dam. A barrier constructed across a river to hold back water and form a RESERVOIR. *Compare* CAISSON. *See also* COFFERDAM.

damages (liquidated damages). Sums that the CONTRACTOR may be liable to pay if a project is not completed on time. The possible damages are sometimes detailed in the CONTRACT.

damp course. *See* DAMP-PROOF COURSE.

damper. A moveable plate in a flue or ventilator that can be adjusted to control the flow of air or other gases. *See also* REGISTER.

dampness. A moistness in the materials of a building, like those of external walls, which can have various causes such as CONDENSATION, rain penetration and RISING DAMP.

dampness meter. *See* MOISTURE METER.

damp-proof course (DPC). A barrier built into a wall to prevent the creep of moisture, such as RISING DAMP, through the materials. A DPC is formed from an impervious and durable material such as strong plastic sheet, bitumen felt, slate or engineering bricks. The courses are usually set in a wall at least 150 mm above ground level; they are also set above windows and doors, and below a coping. *Compare* TANKING.

damp-proof membrane. A continuous layer of impervious material, such as plastic sheeting, built into a roof or floor to prevent the entry of moisture. *Compare* DAMP-PROOF COURSE, TANKING.

dancing step (balanced step). A tapered step at a turn in a flight of stairs. It has a tread that is slightly narrower at one end. *Compare* WINDER. *See* STAIRS diagram.

Darby. A large flat FLOAT tool used for levelling surfaces such as plaster.

Darcy's formula. (1) One of the formulas that relates the flow of liquid in a full pipe to the size and friction of the pipe. The formula is normally used to predict the pressure lost by friction in a certain length of pipe. The loss of pressure head is proportional to the square of the velocity. *Compare* CHEZY FORMULA. (2) A formula that gives the velocity at which water percolates through saturated soil. The velocity is equal to the coefficient of friction multiplied by the HYDRAULIC GRADIENT.

datum. A line, point or level whose details are accurately known and used as a primary reference for other measurements. The datum for a building is often the nearest ORDNANCE BENCH MARK.

daylight. In lighting design, the natural light that reaches a room from the sky. For the purposes of lighting design, the overcast sky is usually taken as the source of daylight. *See also* DAYLIGHT FACTOR, STANDARD SKY.

daylight factor. A percentage comparison between the actual ILLUMINANCE from daylight at a point inside a room and the theoretical illuminance possible at the same point from an unobstructed hemisphere of sky. *See also* STANDARD SKY.

daylight factor components. A breakdown of the types of daylight that form a DAYLIGHT FACTOR. The daylight factor is equal to the SKY COMPONENT plus the externally reflected component plus the internally reflected component.

daylight factor protractor (BRE protractor). A special protractor, developed by the United Kingdom Building Research Establishment. It is used as an aid to the prediction of DAYLIGHT FACTORS from the design drawings of a building.

daywork. A method of payment in building work that is based on the hours worked by each person, the materials used and an agreed percentage. It is a small-scale form of COST PLUS CONTRACT.

dB. *abbr.* DECIBEL.

DC. *abbr.* DIRECT CURRENT.

DCF. *abbr.* DISCOUNTED CASH FLOW.

dead bolt. A type of door BOLT, with a square or rectangular cross-section. It is pushed home by turning a key in a lock.

dead knot. A KNOT that is not intergrown with the surrounding wood and is therefore easier to knock out. *Compare* LIVE KNOT.

dead leg. In a hot water system, a length of pipe that forms a spur from the circulating system. The water in the dead leg tends to remain stationary and therefore wastes heat.

dead light (fixed light, fixed sash). A window LIGHT that does not open. *Compare* OPENING LIGHT.

dead load (dead weight). The fixed weight of a structure itself and any permanent LOADS on it. *Compare* LIVE LOAD.

deadlock. A LOCK that only operates with a key on both sides. *Compare* LATCH.

dead shore. One of two vertical timbers that, together with a sole plate, supports a horizontal NEEDLE as a temporary support in a wall opening.

dead weight. *See* DEAD LOAD.

deal. A piece of sawn SOFTWOOD with a rectangular cross-section 50–100 mm thick and up to 300 mm wide. *Compare* PLANK.

death-watch beetle (*Xestobium rufovillosum*). An insect whose larvae tunnel and damage hardwoods such as oak, usually in decayed areas of old buildings. The adult beetle makes a ticking noise or 'death knock'. *See also* BEETLE.

decay. (1) A decomposition of wood usually caused by growth of FUNGI or attack by BEETLES. (2) A general wearing of a material, such as stone, caused by time and weather.

decibel (dB). In general, a comparative scale of measurement defined as one tenth of the logarithm, to base 10, of the ratio between two values. The scale is commonly used for comparing power in electrical circuits and for comparing sound strengths. *See also* A-SCALE, SOUND LEVEL, SOUND REDUCTION INDEX.

deciduous trees. Trees that shed and renew all their leaves once every year. These include all HARDWOOD and some SOFTWOOD trees.

decision tree. A branching type of FLOW CHART used in decision analysis. It shows all stages of a problem and all possible outcomes.

deck (decking). (1) A flat platform such as a bridge floor or a flat roof. (2) Sheets of material used to form floors or flat roofs.

declination. The variation in angle between the compass readings for magnetic north and true north. This declination varies from place to place and changes with time. *See also* MAGNETIC NORTH POLE.

decoration. The treatments and fittings in a building that make the environment appear finished and attractive. Paintwork and wallpaper are major item of decoration.

deducts. In MEASUREMENT, areas for doors and windows that are deducted from an area of brickwork, painting or other item.

deep bead (sill bead). An upright board fixed to the SILL and overlapping the bottom of a SASH WINDOW. When the sash is slightly raised it allows a little ventilation but prevents draughts.

deep-seal trap. An ANTI-SIPHON TRAP.

deep strip foundation. A form of continuous FOUNDATION in which a deep trench is filled with concrete and used as the footing for the walls. A deep strip foundation is often economic in time and costs when deep foundations are required in shrinkable clay soils. *See also* STRIP FOUNDATION.

deep well. A WELL that obtains water from below the level of the first impermeable layer. The classification of a well as deep or shallow depends on where it takes its water from and not on the depth of the bore.

defect. An irregularity in wood, such as decay, insect damage or a knot, that makes it less useful as a structural material.

defects liability period (maintenance period). A period following the completion of a contract during which the CONTRACTORS are required to repair any defects at their own expense. *See also* RETENTION MONEY.

deflection. (1) The movement of part of a structure that is under load. (2) In surveying, the angle between a line and the preceding line of a TRANSVERSE.

deformation. The change in shape of an object or structure when it is subjected to a FORCE.

defurring. The removal of deposits caused by FURRING on the inside hot water pipes or cylinders.

degree day value. A figure that describes the relative coldness of a locality. The number of days on which the air temperature falls below a standard figure is multiplied by the number of degrees by which it falls. The value is useful for calculating the annual fuel consumption of a building.

degree of compaction. A measure of the tightness or density of packing within a soil sample. It is calculated as the VOIDS RATIO in the loosest state minus the voids ratio of the sample divided by the voids ratio in the loosest state minus the voids ratio in the densest state.

degree of saturation. In a sample of wet soil, the percentage of the volume of VOIDS that are filled with water.

dehumidification. A reduction in the amount of water vapour in a sample of air. This effect may be measured by a lower RELATIVE HUMIDITY, lower DEW POINT, or lower VAPOUR PRESSURE.

dehumidifier. A mechanism, such as in an AIR CONDITIONING plant, that takes water vapour from the air and reduces the HUMIDITY. It may work by cooling the air or by a chemical drying agent.

de-ionized water (de-mineralized water). Water of high purity from which dissolved chemicals have been removed, usually by the process of ION EXCHANGE.

delamination. The deterioration of a LAMINATE caused by the separation of the layers.

deliquescence. The ability of some substances to absorb moisture from the air and to liquefy the moisture. Chloride salts that occur in brickwork or plaster may deliquesce and produce damp patches. The opposite process to EFFLORESCENCE.

delta connection. A method of connecting electrical devices to a THREE-PHASE SUPPLY without a common neutral cable. The three coils of the device make a triangular delta shape. *Compare* STAR CONNECTION.

de-mineralized water. *See* DE-IONIZED WATER.

demountable. Part of a construction, such as a partition, that is designed so that it can be taken down and re-erected at another location if necessary.

dense concrete. A heavy form of CONCRETE with few voids. *Compare* LIGHTWEIGHT CONCRETE.

density. The mass per unit volume of a material. The SI unit of density is the kilogram per cubic metre. *Compare* BULK DENSITY, RELATIVE DENSITY.

dentil. One of a series of rectangular ornaments on a CORNICE moulding.

derrick. A device like a CRANE which is used to lift and move heavy loads. A common arrangement is a lifting arm pivoted at the foot of a central tower.

desiccant. A substance that readily absorbs moisture and may be used as a dehumidifier. *See also* DEHUMIDIFICATION.

design-and-build contract. *See* PACKAGE DEAL.

designed mix. A method of specifying CONCRETE in which the quantities of ingredients for a mix are specially calculated to give the desired strength and other qualities needed for a particular project. *Compare* NOMINAL MIX, STANDARD MIX.

design strength. A material strength used in structural design with concrete or steel. The strength makes allowance for normal variations that occur in the manufacture of materials. The design strength is less than the CHARACTERISTIC STRENGTH, LIMIT STATE DESIGN.

detached house. A house that stands alone and shares no walls with other houses. *Compare* SEMI-DETACHED HOUSE, TERRACED HOUSE.

detail drawing. A drawing that contains enough detail for construction on site. *Compare* WORKING DRAWING.

detritus tank. A SEWAGE TREATMENT tank that collects the grit that is removed by settlement.

devil float (nail float). A hand FLOAT tool with projecting nails or spikes that scratch the surface of plaster and make a KEY for the next coat.

dew point. The temperature at which a fixed sample of moist air becomes saturated and CONDENSATION of the water vapour begins. The liquid water may appear as droplets that look like dew. *See also* HUMIDITY, SATURATION.

dew point gradient. A continuous plot of the change in DEW POINT temperatures that occur through the cross-section of a building shell such as a wall. For specified air conditions, CONDENSATION may occur in regions where the structural temperature falls below the dew point temperature.

dhw. *abbr.* Domestic hot water.

diagonal bond (raking bond). A form of brick BOND in which the courses are laid diagonally across the wall. *See also* HERRINGBONE BOND.

diamond saw. A CIRCULAR SAW with diamond-tipped teeth that give the hardness needed for cutting stone.

diaphragm plate. A flat-shaped element used to stiffen a structure, such as a plate between the girders of a bridge.

diaphragm pump. A device that pumps water by the action of a sheet of flexible rubber or other diaphragm moving back and forth.

diaphragm tank. In a sealed hot water system, a closed expansion tank divided by a rubber wall that expands with the water when the water is heated. The other side of the tank wall contains nitrogen gas which compresses easily and helps prevent the water pressure rising too much.

diaphragm valve. A type of VALVE in which the flow of water is controlled by the action of a flexible diaphragm.

diaphragm wall. (1) A CAVITY WALL with

an extra-wide gap, such as 500 mm, that is stiffened internally by a series brick connecting links between the two leafs of the wall. It is useful for tall windowless walls such as those in a warehouse. (2) An underground retaining wall that is built in a deep trench filled with mud, such as BENTONITE slurry, for initial support. The mud is replaced by concrete or is converted to concrete.

diatomite (diatomaceous earth). A white soil deposit formed from the ancient skeletons of tiny underwater organisms. It is used as an EXTENDER in paints and as an aggregate in lightweight building blocks and fireproof materials.

die. (1) A hard metal block with an internal screw that cuts a thread on to the ends of pipes. (2) A hole in a metal plate or block through which softer materials are forced into shape by EXTRUSION. (3) The square-shaped ends of a stair BALUSTER that fit into the stair base or rail.

differential pressure. The difference in air pressure between different areas of a building.

differential settlement. Uneven SETTLEMENT of a structure into the ground as different parts sink at different rates. Framed buildings can tolerate greater differences than brick or stone buildings.

diffraction. In general, a deviation in the direction of a wave when it meets the edge of an obstacle. Sound waves tend to be diffracted around objects. Diffraction of light waves by small apertures may be seen as light and dark fringes.

diffuser. A covering or LUMINAIRE that scatters the light from a LAMP. *See also* DIFFUSION.

diffuse reflection. A form of REFLECTION, such as by a matt surface, in which light is scattered in various directions. *Compare* SPECULAR.

diffusion. (1) The natural intermixing of the molecules of different liquids or gases to produce a uniform distribution. The process may be helped by a diffuser plate, as in air conditioning. (2) A scattering of light, such as by diffuse reflection or transmission through frosted glass.

diffusion treatment. A method of timber treatment in which the PRESERVATIVE is applied to the surface and allowed to penetrate naturally.

diffusivity. *See* VAPOUR DIFFUSIVITY.

digestion tank. The first stage of a SEPTIC TANK where ANAEROBIC bacteria start the breakdown of the raw SEWAGE.

digitizer. A device used in computer-aided design (*see* CAD) to transfer the features of a drawing on paper into the 'digital' form required by the computer.

dihedral cut. A cut made in some roof timbers to form a slot with sides that slope in different directions.

dikes. (1) *See* DYKE. (2) (*USA*) A type of cutting PLIERS.

dilapidations. The costs of damage or deterioration to a property during a tenancy, some of which are paid by the tenant.

dilapidations schedule. A list of the repairs needed to repair DILAPIDATIONS.

dimension. A distance measured between two points to give a length, width, height or thickness of an object. A particular value shown on a drawing for such a measurement.

dimensional coordination. The use of an agreed range of standard sizes for building components and related construction. *See also* MODULAR SYSTEM.

dimensional stability. The ability of a material or component to keep the same size or shape under different conditions.

dimension shingles. SHINGLES that are cut to a uniform size instead of to a random width.

dimensions paper. Paper that is ruled with vertical columns for the TAKING OFF process when compiling a BILL OF QUANTITIES.

dimension stone. *See* ASHLAR.

diminishing courses (graduated courses). Courses of SLATES where the GAUGE between slates decreases from the eaves to the ridge. *See also* RANDOM SLATES.

diminishing pipe. *See* TAPER PIPE.

diminishing stile (gunstock stile). A door STILE that is narrower at the top of the door, usually to allow a pane of glass to be fitted.

dimmer. A light switch that gives a continuous variation to the electrical supply and hence to the intensity of a light fitting.

DIN. *abbr.* Deutsche Industrie Normen. *See* STANDARDS ORGANIZATIONS.

direct cold water supply. A system of taps or other outlets that are fed by the incoming water MAINS rather than by a storage tank.

direct current (DC). A form of electrical current with a constant direction of flow. The outlet for the current, such as the terminals on a battery, remain permanently either positive or negative. *Compare* ALTERNATING CURRENT.

direct cylinder. A hot water CYLINDER where the stored water is directly heated in the boiler rather than by contact with a HEAT EXCHANGER as in the INDIRECT CYLINDER.

direct glazing. Glazing that is installed straight into an opening rather than into a frame in the opening.

direct heating. The use of a fuel to generate heat within the room or building where it is required. *Compare* CENTRAL HEATING, DISTRICT HEATING.

direct hot water system. A system in which water is directly heated in the boiler rather than by contact with a HEAT EXCHANGER.

direct labour. Craft operatives and labourers for a building project who are employed by the client or his agent rather than by the CONTRACTOR.

direct light. Light that reaches a surface direct from the source rather than by reflection. This effect may be measured by the VECTOR/SCALAR RATIO. *Compare* INDIRECT LIGHT.

direct radiation. Heat from the sun that enters a building by radiation through the windows rather than by conduction through the roof and walls.

direct ratio. A measure of the usable light from the light fittings installed in a particular room. It is the proportion of the total downward flux from the luminaires that falls directly on the working plane.

direct sound. Sound that is received by the most direct path rather than by reflections or by FLANKING TRANSMISSION.

dirty money. Extra payment made to construction workers for difficult or unpleasant conditions.

disability glare. A form of GLARE that lessens the ability to see detail, although it does not necessarily cause visual discomfort. *Compare* DISCOMFORT GLARE.

disappearing stairs. (*USA*) *See* LOFT LADDER.

discharge. (1) *See* FLOWRATE. (2) The removal of energy from a device such as a battery or SOLAR COLLECTOR.

discharge coefficient. The ratio of the actual, measured DISCHARGE to the theoretical value for a particular pipe or opening.

discharge lamp. A form of LAMP that initially produces light from an electric discharge current flowing through a gas or metal vapour kept at low pressure. *See also* CONTROL GEAR, FLUORESCENT LAMP, MERCURY LAMP, SODIUM LAMP. *Compare* INCANDESCENT LAMP.

discharge pipe. A pipe that leads waste products away from a SANITARY appliance in a bathroom.

discomfort glare. A form of GLARE that causes visual discomfort without necessarily lessening the ability to see detail. An unshielded lamp can give discomfort glare. *Compare* DISABILITY GLARE.

disconnecting trap. An INTERCEPTOR.

discontinuous construction. A form of construction with changes in materials that are installed to reduce the transmission of IMPACT SOUND or vibration through the structure of a building. Breaks in the structure are provided by air spaces, layers of insulating material, resilient pads and FLOATING FLOORS.

discounted cash flow. One of several accounting methods that take into account the changes of value in money with time. *See also* PRESENT VALUE.

disc sander. A portable SANDING MACHINE that smooths wood with glasspaper fixed on a rotating disc. The sanding disc can usually be attached to another tool such as an electric drill.

disinfection. The reduction to safe levels of micro-organisms that are harmful to humans. CHLORINATION and OZONE TREATMENT are methods of disinfection used in the final stage of the treatment of public water supplies.

dispersion. A system of very fine particles or droplets that are evenly distributed, but not dissolved, in a liquid. *See also* SUSPENSION.

displacement ventilation. A system of VENTILATION that supplies new air at low level and low velocity. This air displaces exhausted air which is then extracted at the highest point in the area.

distemper. An outdated water-based paint made from a PIGMENT of crushed chalk or 'whiting' and a BINDER of glue or casein.

distribution board, distribution box. *See* JUNCTION BOX.

distribution pipe. A pipe that carries water from a storage CISTERN.

distribution reservoir. *See* SERVICE RESERVOIR.

district heating. A system for heating a group of buildings from a central source, such as by the waste heat from an industrial process. *See also* COMBINED HEAT AND POWER.

district surveyor. *See* BUILDING INSPECTOR.

diurnal temperature. The daily cycle of temperature changes.

division wall. A wall that separates a room or building into several areas for the purpose of COMPARTMENTATION. The term may also mean a PARTY WALL.

DLOR. *abbr.* Downward light output ratio. *See* LIGHT OUTPUT RATIO.

doat. *See* DOTE.

dog. (1) In general, a spiked metal device for gripping or joining objects such as heavy timbers. (2) One of a pair of iron supports for burning wood in a fireplace.

dog ear fold. A corner in a metal sheet or tray formed by a fold in the metal rather than by a cut.

dog-legged stair. A stair that rises between floors with two parallel flights which turn back on one another without a stair well.

dog shore. A horizontal SHORE that runs between two buildings or other vertical surfaces.

dog-tooth course. A jagged effect in a brick wall formed by a course of bricks where each is laid diagonally with a projecting corner.

dolerite. Various forms of dark IGNEOUS rock.

dolly. (1) A flat platform with wheels used to move beams and other heavy items on a building site. (2) A block of wood placed over the top of a PILE to reduce damage to the head during pile driving.

dolomite. Calcium magnesium carbonate, $CaMg(CO_3)_2$, that occurs as a natural MINERAL. The whitish solid is used as a source of lime and for furnace materials.

door. *See* DOOR diagram.

door casing. The finishing materials around the opening of a door.

door closer. A spring or pneumatic device fixed between a door and the frame. It automatically closes the door and prevents slamming.

door frame. The structural frame that surrounds a door opening. It is made up of the DOOR JAMBS and DOOR HEAD.

door furniture. Any hardware associated with a door, such as the BOLT, DOOR CLOSER, HINGES and LOCK.

door head. The horizontal top member of a DOOR FRAME.

door jamb (door post). One of the vertical posts of a DOOR FRAME.

door lining. A surround of thin wood that may be fixed to the inside of an internal DOOR FRAME.

door post. *See* DOOR JAMB.

doorset. A collection of standard components that make up a door. It usually includes the door, frame, stops and furniture.

door sill. A horizontal timber at the foot of an outside door frame. *See also* SILL.

door stop. (1) (door trim). The casing or REBATE built in a door frame that prevents the door swinging through the frame and also makes a seal. *See also* SOLID STOP. (2) A device fixed on the floor to prevent the door from opening too far. (3) A device that catches and holds a door open.

dormer (dormer window). A vertical window built out from a sloping roof. It has its own side walls (cheeks) and roof. *See also* INTERNAL DORMER, LINK DORMER. *See* ROOF diagram.

dormer cheek. A vertical side of a DORMER WINDOW construction.

dormer window. *See* DORMER.

dot. (1) A SOLDER or LEAD covering for the heads of fastenings that secure lead sheet cladding. (2) Dabs of plaster used to cover fixings.

dote (doat, doze). Early signs of decay in timber such as dots of surface discoloration. The timber may then be termed 'doty'.

double-action hinge. A form of door HINGE, often mounted in the floor, that allows the door to swing open in either direction.

double ball catch. A cupboard fastening that uses two spring-loaded balls to grip a staple that projects from the door.

double connector. A short length of pipe with a thread at each end that is used to insert a connection into an existing pipe.

double door. A pair of doors that are hung so that they meet.

double-dovetail key (dovetail feather, hammer-headed key). A hardwood connector, shaped like two DOVETAIL pins, set into a butt joint between two timbers.

double eaves course. A double thickness of TILES or SLATES laid at the bottom edge of a roof. *Compare* EAVES COURSE. *See also* TILTING FILLET.

double Flemish bond. Brickwork in which the courses are laid so that a FLEMISH BOND is visible on both side of the wall. *Compare* SINGLE FLEMISH BOND.

double glazing. A form of MULTIPLE GLAZING made from two layers of glass with an air space between them. It is installed in a window or glazed door to reduce the transmission of heat or sound. For best THERMAL INSULATION the cavity is narrow and may be sealed with dry air or another gas. For SOUND INSULATION the air gap should be at least 150 mm with absorbent lining around the edges. *See also* SEALED UNIT, SECONDARY GLAZING.

double header. (*USA*) Two or more timbers fastened together to give extra strength in a structure, such as around a stair opening.

double-hung window. A SASH WINDOW.

double jack rafter. A form of RAFTER that runs from a VALLEY to a HIP.

double-pitch roof. (1) A roof that slopes in two different directions from a ridge. (2) A MANSARD ROOF.

double rebated. *adj*. Describing a frame with two REBATES cut in it.

double-return stair. A grand form of STAIR in which a wide flight leads to a landing. Two side flights, one on each side, continue the stair in the reverse direction to the next floor.

double Roman tile. A form of SINGLE-LAP TILE that has a roll up in the centre as well as at the edges. *See also* ROMAN TILE.

double roof. A form of pitched roof construction in which the COMMON RAFTERS are carried on PURLINS.

double time. An OVERTIME payment to workers of twice the hourly rate. It is usually for work outside normal working hours, such as on a weekend.

Douglas fir. A type of FIR whose SOFTWOOD is widely-used for general building.

dovetail (dovetail joint). A joint between pieces of wood made with protruding fan-shaped pins that key into matching slots. *Compare* COMBED JOINT. *See* TIMBER JOINTS diagram.

dovetail cramp. A connector, usually of metal with a double dovetail shape, that is used for joining stonework.

dovetail feather. *See* DOUBLE-DOVETAIL KEY.

dovetail joint. *See* DOVETAIL.

dowel. (1) A short length of wooden rod that is glued into drilled holes to join pieces of wood. (2) A short metal rod set into concrete to locate another component such as a post.

dowel pin. (*USA*) A barbed metal pin used to fasten MORTISE AND TENON JOINTS in windows and doors.

dowel plate. A metal plate perforated with holes in a range of sizes through which DOWELS can be trimmed and tested for size.

downdraught (*USA*: downdraft). A downward flow of air, usually in a CHIMNEY or FLUE that is badly designed.

downpipe (fall pipe, rainwater pipe, spouting, *USA*: conductor, leader). A pipe, usually vertical, that takes rainwater from a roof GUTTER down to ground level. *See also* RAINWATER HEAD, SHOE.

doze. *See* DOTE.

dozer. A small BULLDOZER.

DPC. *abbr.* DAMP-PROOF COURSE.

draft (drafted margin). (1) A smooth strip or margin around the face of an otherwise rough stone. (2) (*USA*) *See* DRAUGHT.

draftsman, draftswoman. A trained person who prepares DRAWINGS for use in construction.

draft stop. (*USA*) *See* FIRE CHECK.

drag. (1) (wirecomb) A metal plate or COMB with toothed edges that levels a surface, such as plaster, and produces a KEY for the next layer. (2) (pulling) In painting, a resistance to brushstrokes given by the paint.

dragline. A mechanical excavator in which a bucket hung from the end of a jib scoops material and is then pulled back by a cable.

dragon beam. A horizontal roof timber on which the HIP rafter rests. It runs from the corner wall junction to an ANGLE TIE.

dragon tie. *See* ANGLE TIE.

drain. A pipe or channel in the ground that carries storm water, waste water or SEWAGE.

drain cock. A small tap placed at the lowest point of a pipe system, such as a heating system. It allows the system to be drained for service.

drain pipe. A pipe designed to carry waste water or SEWAGE from a building. The pipes are made from durable, waterproof materials such as glazed CLAYWARE, concrete, metal or plastic.

drain plug. A device, such as a BAG PLUG, used temporarily to seal part of a DRAIN when testing for leaks.

drain rod. One of a connecting set of flexible rods that are pushed through an ACCESS EYE in a drain to clean and clear blockages. *See also* RODDING.

drain test. The testing of a drain for leakages. *See also* AIR TEST, HYDRAULIC TEST.

drain tile. *See* FRENCH DRAIN.

draught. (1) A current of air through a building or a chimney, which is usually caused by natural pressure differences. (2) The difference in alignment of the holes drilled for a DRAWBORE PIN.

draught strip. A continuous strip of sealing material, such as rubber or flexible metal, installed around the edge of a door or window frame to reduce unwanted heat loss and DRAUGHTS.

draw bolt. A fastening BOLT, such as fitted on a door, that is operated by hand rather than by key.

drawbore pin. A tapered steel pin that draws a TENON and MORTISE together. The pin is driven through holes in each piece that are slightly misaligned by a DRAUGHT. The drawbore pin may be replaced by a TRENAIL.

drawer. A storage container in the form of a deep tray that slides out of sight into a larger container.

drawing. A scaled representation of a building or parts of a building. *See also* CAD, DETAIL DRAWING, DRAFTSMAN, DRAFTSWOMAN, ELEVATION, PLAN, PROJECTION, SECTION, WORKING DRAWING.

drawing board. A smooth rectangular board used to hold paper in place while a DRAWING is worked on. The board may be fixed on a stand and have PARALLEL-MOTION EQUIPMENT attached. *Compare* PLOTTER.

draw-in system. An electrical installation in which the CABLES can be easily pulled through CONDUITS and ducts.

drawn glass. *See* SHEET GLASS.

draw-off pipe. A pipe from a hot water system though which water is taken for use, such as at the taps.

drencher system. A SPRINKLER SYSTEM.

dress. *vb.* (1) To use a PLANE or SANDPAPER to smooth a wooden surface. (2) To cut or trim stones into shape.

dressed. *adj.* In general, describing a material that has been given a worked finish. (1) Describing timber that has one or more planed surface. (2) Describing stone that has been squared all around and smoothed on the face.

dressed size (finished size). The size of timber after sawing and planing. The dressed size is about 6 mm less than the NOMINAL SIZE (3 mm for each WROUGHT FACE).

dressing. (1) A decorative finish, such as a detail in stonework or brickwork, that is different from the plain part of a building. (2) A coating of material, such as bituminous 'dressing compound' on roofing felt, that protects a surface. (3) The process of beating a sheet of metal to shape.

drier. An additive in paint or varnish that encourages OXIDATION of the DRYING OIL and speeds drying.

drift bolt. A thick steel pin for fixing heavy timbers. It is driven into holes bored into the timber.

drill. A hand or power tool that uses a rotating BIT to make a hole in a material.

drip. (1) (drip channel, drip edge, throat, throating) A groove or ridge beneath an overhang, such as a window SILL, that encourages water to drop clear rather than flow towards the building. (2) A step formed by the overlap of two sheets of FLEXIBLE ROOFING material.

drip cap. *See* DRIPSTONE.

drip channel, drip edge. *See* DRIP.

drip loop. A downward loop formed in overhead electric supply cables just before they enter a building. The loop prevents rainwater flowing along the cables.

dripstone (drip cap). A moulding of brick or stone that projects over a window or door and deflects rain.

driven pile. A PILE that is forced into the ground by blows on the top. *Compare* BORED PILE.

drive screw (screw nail). A threaded steel nail used for fixing roof sheets. It can be driven in with a hammer but is difficult to withdraw.

driving-rain index. A climatic measurement that combines the annual rainfall and average wind speed of a site. The index is used to predict the likelihood of rain penetration through brick or stone walls.

drop apron. A vertical strip of metal that runs around the EAVES of FLEXIBLE-METAL ROOFING.

drop ceiling. *See* SUSPENDED CEILING.

drop connection. A BACK DROP.

drop hammer. A metal block that is raised and dropped freely on to the head of a PILE in order to drive it into the ground. *See also* DOLLY.

dropped escutcheon. *See* KEYDROP.

drop system. A CIRCUIT in a heating system where the supply pipe rises to the highest level and feeds downward.

drop window. A form of SASH WINDOW that can slide down into a cavity beneath the SILL and leave the window area completely open.

drown pipe. A feed pipe for a CISTERN in which the inlet is placed below water level in order to reduce noise.

dry-bulb thermometer. An ordinary thermometer that measures air temperature. The dry-bulb reading on a HYGROMETER is compared with that of the WET-BULB THERMOMETER to determine HUMIDITY.

dry construction. A general method of building without those 'wet' trades that involve the large amounts of water contained in mortar and plaster. Dry partitions and prefabricated components help reduce the time needed for drying out.

dry density. The DENSITY of a material such as soil or aggregate after it has been dried. *Compare* BULK DENSITY.

dry hydrate. A form of HYDRATED LIME that is sold as a powder to be soaked before use.

drying oil. An animal or vegetable oil that hardens by OXIDATION when it is exposed to air in a thin layer. LINSEED OIL and TUNG OIL are the basis of traditional 'oil' PAINTS and VARNISHES. *Compare* DRIER.

drying shrinkage. The general process of shrinkage in materials such as concrete or timber that occurs as MOISTURE CONTENT decreases.

dry joint. A joint between materials that uses a GASKET rather than a MORTAR or MASTIC.

dry lining. The use of WALLBOARD and taped joints for the inside surface of a room without the use of wet plastering. *See also* DRY CONSTRUCTION.

dry mix (dry mortar, dry pack). (1) The ingredients of a CONCRETE or MORTAR MIX that are measured and mixed together ready for water to be added. (2) Concrete or mortar that contains enough water to set and harden but remains stiff and granular. It may be rammed into places such as the gap between UNDERPINNING piles and the building above.

dry partition. A type of internal wall that can be erected and finished without plastering.

dry resultant temperature. A COMFORT TEMPERATURE that combines the effect of air temperature, RADIANT TEMPERATURE and air movement. When air movement is low it may be calculated as the sum of half the radiant temperature and half the air temperature. *See also* EFFECTIVE TEMPERATURE, ENVIRONMENTAL TEMPERATURE.

dry riser. A vertical pipe built into a building for use by the fire-fighting services. The pipe is normally dry but water can be pumped into inlets at street level and drawn from outlet valves inside the building.

dry rot. A form of timber decay caused by a FUNGUS (such as *Merulius lacrymans* or *Serpula lacrymans*) which breaks down the wood and leaves it brittle and cracked in squarish patterns. Despite the name, 'dry' rot requires damp conditions but the threads (hyphae) of the fungus can penetrate brickwork and spread moisture. *Compare* WET ROT.

dry stone walling. A wall of stone or other masonry that is built without MORTAR.

drywall. *See* GYPSUM PLASTERBOARD.

dubbing out. The use of COARSE STUFF to fill hollows in wall surfaces before a finishing coat of PLASTER.

duckboard. A slatted board used to make a temporary path over wet or muddy ground.

duck-foot bend. A REST BEND.

duct. (1) A narrow passageway built into a structure to carry pipes, cables and other services. *See also* CRAWLWAY, TRUNKING. (2) A tube or channel, usually of metal or plastic, that carries air in a ventilation or air-conditioning system.

ductility. The ability of materials, especially metals, to be drawn out into a wire. Ductility is one aspect of a general property of TOUGHNESS. *See also* MALLEABILITY.

dumb waiter. A small LIFT or ELEVATOR, sometimes hand-operated, that carries food or supplies between floors.

dummy. A mass of lead or iron that is fixed on the end of a rod and used to straighten lead pipes.

dumper truck. An open motor vehicle that uses a large front hopper to carry excavated material on construction sites.

dumpling. In a large EXCAVATION, a mass of ground left in the middle of the site until the end of the dig. Trenches are dug around the sides of the dumpling which may be used as support for SHORING.

dumpy level. (1) A simple form of surveyor's LEVEL in which the telescope is rigidly fixed to a vertical axis and rotates on the levelling head. (2) In general, any small surveyor's LEVEL.

duodecimal system. A system of numbering with a base of twelve, rather than the base of ten used in the decimal system.

duplex apartment. (*USA*) A building divided into two living units, usually one above the other. *Compare* MAISONETTE.

dusting. The appearance of a powder on the surface of a material such as concrete.

Dutch arch (French arch). A brick ARCH that is flat both at top and bottom. It is constructed with ordinary bricks and uses few, if any, wedge-shaped bricks.

Dutch barn. An storage building without walls. It is usually constructed from a lightweight steel frame.

Dutch bond (English cross bond, Saint Andrew's cross bond). In masonry, a form of ENGLISH BOND in which the CROSS JOINTS of the headers are staggered.

Dutch door. (*USA*) *See* STABLE DOOR.

dutchman. An odd piece of wood, stone or other material used to fill a gap.

dwall. An internal partition that is not full height. *Compare* KNEE WALL, STUB WALL.

dwang. (1) NOGGINGS in a timber frame. (2) STRUTTING between floor joists.

dwarf wall. A low wall such as a PARAPET or SLEEPER WALL.

dwelling. A residential building used for family living. Other classes of buildings include commercial buildings used for shops and offices, warehouses used for storage, and industrial buildings used for manufacturing.

DWV. *abbr.* (*USA*) Drain-waste-vent.

dye. A colouring agent that, unlike a PIGMENT, is soluble and can penetrate a material. *See also* TONER.

dyke. (1) A long mound of earth built alongside a river to confine flood water. (2) A DRY WALL built of stone.

dynamic strength. The resistance of a structure to loads that are suddenly applied.

dynamo (generator). A device that uses the principle of electromagnetic induction to convert mechanical energy into electrical energy.

E

E. The symbol for the ELASTIC MODULUS.

ear. A flat projection on a pipe that is used to attach the pipe to a wall. *See also* LUG.

earth. (1) Soil or any excavated material. (2) Part of an electrical device or CIRCUIT that has an intentional connection to the ground via an EARTH ELECTRODE. It is a protective feature designed to operate a CIRCUIT BREAKER if the device accidentally comes into contact with the live supply.

earth electrode. The part of an EARTH circuit that is connected to the ground. It may be a metal plate or pipe buried in damp ground.

earthenware. Articles, such as clay pipes, that are fired to a lower temperature than other CERAMICS. *See also* STONEWARE.

earthing. The process of including an EARTH in an electrical circuit.

earth leakage. The electric current that flows in an EARTH circuit and operates a CIRCUIT BREAKER.

earth pressure. The force exerted by retained earth such as that behind a RETAINING WALL.

earthwork. A sizable excavation of the ground.

easement. (1) A legal right or permission to use another person's land, usually for rights of way, rights of light and the passage of service pipes and cables. (2) A curved MOULDING that makes the transition between two different surfaces.

easing. The process of trimming the edges of WINDOWS and DOORS so that they open easily and do not stick in the frames.

eastern closet. *See* SQUATTING CLOSET.

easy-clean hinge. *See* CLEANING HINGE.

eaves. The bottom edge of a roof where it usually overhangs a wall, and the area beneath this overhang. *See also* BOXED EAVES, EAVES COURSE, FASCIA BOARD, OPEN EAVES, SOFFIT BOARD, SPROCKET, TILTING FILLET. *See* ROOF diagram.

eaves board. *See* TILTING FILLET.

eaves course. The first course of plain tiles, slates or shingles at the bottom edge of a roof. *See also* UNDER-EAVES TILE. *Compare* DOUBLE EAVES COURSE.

eaves fascia. *See* FASCIA BOARD.

eaves flashing. A form of FLASHING or DROP APRON at the edge of some types of roof, such as a flat asphalt roof, that directs water into the EAVES GUTTER.

eaves gutter. The rainwater GUTTER that runs along the edge of a roof.

eaves plate. A WALL PLATE that runs between gaps at the top of walls or pillars and supports the RAFTERS.

eaves tile. A short type of roof tile used in the EAVES COURSE. *Compare* PLANE TILE.

ebonite. A hard black material made from rubber with a high proportion of sulphur. It is used as an ELECTRICAL INSULATOR.

ebony. A black stain or other finish on wood.

eccentric load. In general, a load that does not pass through the CENTRE OF MASS of a structure. It is often a vertical load that is off-centre at the top of a column.

echo. A repetition of sound caused by the return of sound waves reflected from a hard surface. The extra time taken to travel along the reflected path may cause the returning sound to be heard distinctly later than the DIRECT SOUND. *Compare* REVERBERATION.

economizer. An arrangement in a heating system that saves energy by using the hot flue gases to pre-heat the feedwater before it enters the boiler.

eddy flow. A form of TURBULENT FLOW in liquids or gases where the particles make small whirlpools.

edge-bedded. *See* FACE-BEDDED.

edge grain (vertical grain). The grain shown in QUARTER-SAWN WOOD when the GROWTH RINGS are at 45° to the face.

edge nailing. SECRET NAILING, usually of floor boards.

edging strip. (1) A wooden strip attached to the edge of a door. *See also* LIPPING. (2) A rubber channel that encloses the edges of a pane of glass used for SECONDARY GLAZING.

EDM. *abbr.* ELECTROMAGNETIC DISTANCE MEASUREMENT, electronic distance measurement, electronic distance meter.

effective ambient temperature. *See* ENVIRONMENTAL TEMPERATURE.

effective depth. In a REINFORCED CONCRETE beam or slab, the depth of concrete measured from the face of the compressed side to the centre of the stretched steel.

effective span. The distance between the load points where a beam or slab is supported. The effective span is usually greater than the usable CLEAR SPAN.

effective temperature. A COMFORT TEMPERATURE that takes into account the effect of air temperature, humidity and air movement. *See also* DRY RESULTANT TEMPERATURE, ENVIRONMENTAL TEMPERATURE.

efficacy. *See* LUMINOUS EFFICACY.

efficiency. A measure of the effectiveness of a system that converts energy from one form to another. Efficiency is the ratio of the output energy to the input energy and can never be greater than unity, or 100 per cent.

efflorescence. The formation of white powdery deposits on a brick or concrete surface while it dries. Brick clay, cement and lime contain soluble salts which are carried to the surface by water used in construction and are left in the form of crystals when the water evaporates. The effect is unsightly rather than harmful and stops when all salts have been removed. *Compare* DELIQUESCENCE.

effluent. Liquid or gas waste that flows from an industrial process or a sewage treatment plant.

eggshell (silk). A surface finish, usually on paint, that has more shine than MATT but not as much as a GLOSS finish.

egress. A place of exit from a building.

EL. *abbr.* Existing level.

elasticity. The ability of a deformed material to resume its original shape and size when the forces acting on it are removed. *Compare* PLASTIC. *See also* ELASTIC MODULUS.

elastic limit. A value of STRESS or deforming force applied to a material above which ELASTICITY is lost. The material therefore does not recover its original size and shape when the stress is removed.

elastic modulus (coefficient of elasticity). A measure of the ELASTICITY or stiffness of a particular material. It is given by the ratio of STRESS divided by STRAIN. If the strain is measured as a change in length then the elastic modulus is known as YOUNG'S MODULUS. Measuring the strain as twist or shear gives the rigidity modulus; measuring the strain as change in volume gives the bulk modulus. *See also* POISSON'S RATIO.

elastomer. Any form of RUBBER, or similar material, that can regain its original size after undergoing a large deformation such as that caused by stretching to twice its length.

elbow (knee). A sudden right-angled turn in a pipe or road. It is sharper than a 'bend'.

ELCB. *abbr.* Earth leakage circuit breaker. *See* EARTH LEAKAGE.

electrical insulation. Materials that are poor CONDUCTORS of electricity. They have few FREE ELECTRONS and high ELECTRICAL RESISTANCE. Insulators used in electrical installations include plastics, rubber, glass, porcelain and dry air.

electrical resistance. The ability of a conductor to oppose the flow of electric CURRENT. It varies with the type of material and usually increases with temperature. The unit of measurement is the OHM. *See also* IMPEDANCE.

electric arc welding. *See* ARC WELDING.

electric current. *See* CURRENT.

electric discharge lamp. *See* DISCHARGE LAMP.

electric immersion heater. *See* IMMERSION HEATER.

electricity distribution. The part of an electricity supply system that supplies individual customers from a system of transformers and cables in the streets. *See also* SUBSTATION.

electricity supply. A system of equipment and organization that includes the generation of bulk electrical power, its transmission over long distances and distribution to users.

electric panel heater. A form of PANEL HEATER that is heated by the electrical resistance of buried cables.

electric storage heater. *See* STORAGE HEATER.

electrochemical series. An arrangement of the metals in order of their electrical potential or ability to attract electrons. Reactive metals such as potassium and calcium are at one end of the series; unreactive metals such as gold and silver are at the other end. The series also determines the ease with which a metal corrodes. *See also* GALVANIC SERIES, NOBLE METALS.

electro-copper glazing. A method of fixing panes of glass between CAMES of copper. Extra copper is deposited on the cames by a process of ELECTROLYSIS.

electrode. A conductor by which CURRENT enters or leaves an electrolyte during ELECTROLYSIS. A terminal of an electric cell or battery. *See* ANODE, CATHODE.

electrode boiler. A large electric boiler that heats by passing an alternating current directly through the water.

electrolier. A light fitting or LUMINAIRE that hangs from a ceiling.

electrolysis. The chemical decomposition of a dissolved or molten material (an ELECTROLYTE) by an electric CURRENT. The current is connected to the material by ELECTRODES and passes through the material by the movement of charged atoms (IONS).

electrolyte. A liquid that conducts electric current in the process of ELECTROLYSIS. Electrolytes may be acids, bases or salts that dissolve in water.

electrolytic corrosion. A common form of CORROSION in which the decomposition of metal is caused by the process of ELECTROLYSIS. The current required for the electrolytic action may be generated by combinations of metals that form a SIMPLE CELL. *See also* GALVANIC SERIES.

electromagnetic distance measurement (electronic distance measurement, EDM). In surveying, the measurement of length by instruments that transmit light or radio waves. Electronic circuits measure the time taken for signals to be received or reflected and also convert the results to lengths. *See also* LASER. *Compare* OPTICAL DISTANCE MEASUREMENT.

electromotive force (emf). A source of electrical energy, such as a generator or battery, that causes an electrical CURRENT to flow if it is connected into a circuit. The unit of emf is the VOLT. *Compare* POTENTIAL DIFFERENCE.

electron. One of the fundamental particles from which all atoms are made. Electrons have a negative electric charge and a very small mass compared to the proton and neutron. They occupy positions surrounding the nucleus and the behaviour of electrons determines the type of BONDS between atoms. *See also* ELECTRICAL INSULATION, FREE ELECTRONS.

electronic air cleaner. *See* ELECTROSTATIC FILTER.

electronic distance measurement. *See* ELECTROMAGNETIC DISTANCE MEASUREMENT.

electronic moisture meter. *See* MOISTURE METER.

electro-osmosis. An electrical method of controlling the creep of water through ground soil and in walls, such as RISING DAMP. When a direct current is passed through such damp materials, the water tends to move towards the CATHODE. The method is not recommended by some authorities.

electroplating. The formation of a thin coating of one metal on another by ELECTROLYSIS. The object to be plated is used as the CATHODE in an electrolyte containing a dissolved salt of the metal that is to be deposited.

electrostatic filter (electronic air cleaner). A method of filtering air by the use of electrically-charged plates that attract and hold small particles of dust.

electrovalent bond. An IONIC BOND.

element. (1) A part of a building or structure with a distinct function. Examples include a foundation, wall or roof. (2) One of the fundamental forms of matter, such as hydrogen or carbon, from which all other materials are made. (3) An electrical heating device such as the IMMERSION HEATER used in a hot water cylinder.

elevation. (1) A DRAWING that gives a flat view of an object such as from the front of a building. It shows features projected on to a vertical plane without perspective. *Compare* PLAN. (2) The ALTITUDE of a point.

elevator. *See* LIFT.

elliptical arch. An arch that follows the curve of an ellipse rather than a circle.

elliptical stair. A STAIR with a well in the shape of an ellipse rather than a circle.

elm. A HARDWOOD tree of the genus *Ulmus*. The dull-brown wood has been used in making furniture.

elongation. The increase in length of a material or structural member while it is under tensile STRESS.

embankment. A long mound of earth or rock built to carry a road or railway, or to hold back water.

embossed. *adj.* Describing a decorative surface that is raised above the surrounding surfaces.

emergency lighting. A secondary system of lighting for a public building that provides the minimum lighting to allow safe escape from the building in an emergency such as fire. The emergency lighting can continue to operate if the main electricity supply fails.

emery cloth. *See* GLASSPAPER.

emf. *abbr.* ELECTROMOTIVE FORCE.

eminently hydraulic lime. A type of HYDRAULIC LIME that has a similar composition and effect as PORTLAND CEMENT.

emissivity. The fraction of heat radiated by a surface compared to that given off by a technically perfect emitter, known as a black body. The value varies with the temperature (wavelength) of the heat source and building surfaces are usually rated for low temperature radiation. *Compare* ABSORPTIVITY, REFLECTIVITY. *See also* SURFACE RESISTANCE.

emulsifier system. A SPRINKLER SYSTEM for fire fighting that uses a high pressure water spray to fight oil fires. The powerful spray emulsifies the escaping oil and suppresses burning by coating the droplets with water.

emulsion paint. A form of paint where the BINDER is carried as an emulsion or SUSPENSION of tiny droplets dispersed in water. Modern emulsion paints use synthetic polymers rather than oils. The paint dries by evaporation of the water rather than by the oxidation process of OIL PAINTS. Most emulsion paints are slightly porous and are not recommended for kitchen and bathroom surfaces.

enamel. (1) A durable form of GLOSS PAINT whose properties are a result of having a high proportion of the BINDER or varnish. (2) (vitreous enamel) A hard, glass-like surface produced on a metal object, such as a bath, by coating the surface with powdered glass and firing in a furnace.

encase. *vb.* To cover with another material or lining. For example, steelwork is encased with concrete for increased FIRE RESISTANCE.

encastre (encastered beam). A beam that has ends fixed or BUILT-IN. *See also* FIXED END.

enclosed fuse. (*USA*) *See* CARTRIDGE FUSE.

enclosed stair. *See* CLOSED STAIR.

enclosure wall. (*USA*) *See* CURTAIN WALL.

end-bearing pile. A PILE that transmits most of its load through to hard ground at the base of the pile. *Compare* FRICTION PILE.

end grain. The surface of timber exposed when it is CROSS-CUT, such as when a tree is felled.

end-lap joint. A type of joint between two timbers at right angles formed by halving each timber for a length equal to the width of the other.

energy. The ability to do work. Energy exists in various forms such as heat, mechanical, chemical and electrical energy. It can be converted from one form to another but the total energy of a system remains constant. The SI unit of energy is the JOULE. *See also* CALORIFIC VALUE, PRIMARY ENERGY.

energy audit. An assessment of the use of energy in a building and the costs of that energy. The audit usually includes recommendations for the reduction of energy use and costs.

engineering brick. A hard, dense and semi-vitreous variety of BRICK that meets high limits for crushing strength and water absorption. Engineering bricks are suitable for brickwork that has to be underground or carry heavy loads.

English bond. A brickwork BOND in which alternate courses contain only HEADERS and then only STRETCHERS. *See* WALL diagram. *See also* DUTCH BOND, ENGLISH GARDEN WALL BOND. *Compare* FLEMISH BOND, STRETCHER BOND.

English cross bond. *See* DUTCH BOND.

English garden wall bond (American bond, colonial bond, Scotch bond). A form of ENGLISH BOND brickwork in which headers are laid in fewer courses, such as every sixth course, instead of in alternate courses. *See* WALL diagram.

English roof tile. (*USA*) A type of SINGLE-LAP roofing tile that overlaps and interlocks at the each edge so that the surfaces remain level.

enthalpy. The total heat energy content of a sample, with reference to 0°C. For example, steam at 100°C has a higher enthalpy than water at 100°C. Enthalpy includes the SENSIBLE HEATS and the LATENT HEATS.

entity. In computer-aided design (*see* CAD), a part of a drawing that is created or manipulated by one process. Lines and circles are simple entities.

environmental temperature. (1) A temperature for conditions inside or outside a building which take account of several factors. (2) A form of COMFORT TEMPERATURE inside a room that combines the effect of air temperature with the RADIANT TEMPERATURE from the surrounding surfaces. It can be calculated as the sum of two-thirds the radiant temperature and one-third the air temperature. *Compare* DRY RESULTANT TEM-

PERATURE, EFFECTIVE TEMPERATURE, SOL-AIR TEMPERATURE.

epoxy resin. A class of SYNTHETIC RESIN used as a basis for strong ADHESIVES and paints that are often supplied in TWO-PACK form.

epsom salts ($MgSO_4.7H_2O$). A naturally-occurring mineral that causes HARD WATER. *See also* PERMANENT HARDNESS.

equilibrium. The state of a body or structure that does not move. In stable equilibrium a body tends to return to its original position after a movement. A body in unstable equilibrium tends to move further from its original position.

equilibrium moisture content. The value of MOISTURE CONTENT in timber, or other materials, that remains constant for given conditions of temperature and humidity.

equivalent continuous sound level (L_{eq}). A method of assessing exposure to noise such as that found in industrial environments. The noise may vary in level with time. L_{eq} is that continuous SOUND LEVEL, in dB(A), that gives the same total sound energy as the varying level being measured. The risk of hearing damage corresponds to this total energy input or 'noise dose'.

equivalent temperature. A type of ENVIRONMENTAL TEMPERATURE.

erection. The process of assembling parts of a building, especially the frame.

ergonomics. The study of interactions between people and their working environment. Such a study leads to efficient designs and positions for machines and furniture.

erosion. The gradual wearing away of a surface layer by environmental effects such as weather or by mechanical abrasion.

escalator. A system of moving stairs for carrying people between different levels in public places such as shops. The steps are mounted on a motor-driven belt that moves in an endless loop.

escape route. A means of exit from a building, such as a special stairway, that is required by law as an escape path in case of fire.

escutcheon (scutcheon). A metal plate fixed on the door around a keyhole. *See also* KEY DROP.

escutcheon pin. A small brass NAIL used to fix an ESCUTCHEON or other small items of hardware.

espagnolette bolt. A long bolt system for fastening tall windows or CASEMENT DOORS. The action of a central handle drives home bolts at both top and bottom.

estimating. The process of assessing the total cost of building work by taking an established RATE for particular activities or areas and multiplying by the QUANTITIES involved. *See also* BILL OF QUANTITIES.

estimator. A person qualified in ESTIMATING. *See also* QUANTITY SURVEYING.

etching. The process of cutting into a surface such as glass, metal or concrete, by the action of acid. Decorative patterns may be produced by coating the surface with wax that resists the acid except where it is scratched.

European projection. *See* ORTHOGRAPHIC PROJECTION.

eutectic salt. A substance that absorbs significant amounts of heat when melting and releases that heat when freezing. This mechanism can be used for the storage of heat when the salts melt at a temperature in the range 15–30°C.

evaporation. (1) The conversion of a liquid into its vapour form. Evaporation usually occurs on the surface of the liquid at temperatures below the boiling point. (2) The mechanism by which some surface coatings, such as EMULSION PAINT or LACQUER, dry and harden. *Compare* OXIDATION.

evaporator. A container in which a substance changes from a liquid to a vapour, a process that requires latent heat which can be taken from the surroundings of the evaporator. This effect is put to use by REFRI-

GERATORS and HEAT PUMPS. *Compare* CONDENSER.

event. In CRITICAL PATH ANALYSIS, a junction or node that represents a stage or activity when a project is represented on a network.

even-textured. *adj.* Describing a type of timber GRAIN that shows little difference between spring and summer growth.

excavation. The digging out and removal of soil or rock from a site.

excavator. A machine that does excavation work. *See also* BACKACTOR, BACKHOE, JCB.

exfoliated vermiculite. An aerated mineral product made by heating VERMICULITE which then expands. It is used as an AGGREGATE and a THERMAL INSULATOR.

exfoliation. The flaking of a surface, such as that of stone that is exposed to the weather.

expanded clay (sintered clay). Particles of clay that have been burnt in a kiln to produce air-filled pellets that can be used as LIGHTWEIGHT AGGREGATE.

expanded metal. Sheet steel or other metal that is cut with slots and expanded to form a mesh. It is used as metal LATHING to support plasterwork and for lightweight concrete reinforcement.

expanded plastic (foamed plastic). A cellular form of a PLASTIC produced by the introduction of air to make a lightweight THERMAL INSULATION material. *See also* CLOSED CELL.

expanded polystyrene. A type of EXPANDED PLASTIC that is used as a thermal insulating material. It is commonly available in the form of beads, blocks and sheets. *See also* POLYSTYRENE.

expanded polyurethane. A type of EXPANDED PLASTIC that can be foamed on site for the insulation of cavity walls. *See also* POLYURETHANE.

expansion bay. A space built into the side of a pipe DUCT to allow for EXPANSION JOINTS in the pipe.

expansion bolt (anchor). A device for making a fixing to concrete, brick or stone. After the bolt has been installed into a pre-drilled hole, parts of the mechanism are forced to expand inside the hole to prevent withdrawal.

expansion joint. *See* MOVEMENT JOINT.

expansion sleeve (expansion pipe). A sleeve or tube built into a wall or floor through which another pipe can pass. The inner pipe can then expand or contract without damage to the structure.

expansion strip. A length of resilient material installed in a joint between two parts of a structure. The compression and expansion of the strip absorbs movement in the structure.

expansion tank. A tank on the PRIMARY CIRCUIT of a heating system that allows the water to expand when it is heated. *See* HEATING diagram.

expert witness. A qualified person who is engaged to give a professional assessment of a situation to help resolve a dispute.

exposed aggregate. A finish applied to the surface of concrete work in which the coarse aggregate is exposed. The sand and cement of the surface are removed by hammering, ETCHING with acid or by blasting with grit.

exposure. The extent to which a building site is affected by the effects of weather. Methods of assessment include the DRIVING-RAIN INDEX.

extended price. A price in a BILL OF QUANTITIES produced by multiplying a RATE for a project or item by the QUANTITY involved.

extender. In general, an inert substance that is added to paints, plastics and glues to help modify the properties and to lower the cost. Paint extenders are commonly made from crushed DIATOMITE, kaolin, silica and mica. They have low HIDING POWER but help to adjust the working properties.

extending ladder. A telescopic LADDER that is made longer by sliding out extra sections of ladder.

extension bolt (monkey-tail bolt). A door or gate BOLT with a long bent handle.

exterior plywood (exterior-grade plywood). A form of PLYWOOD that is laminated together with moisture-proof glue. The plywood can be used outdoors, providing that is protected by paint. *Compare* INTERIOR PLYWOOD.

externally reflected component. *See* DAYLIGHT FACTOR COMPONENTS.

external wall. A wall that has at least one face exposed to the weather or to the earth.

extra (extra work). Work that was not included in the original CONTRACT. It is usually ordered by the architect by a written VARIATION ORDER.

extractor fan. An electric fan that drives out air in an EXTRACT SYSTEM.

extract system. A ventilation system in which the contaminated air is sucked out by extractor fans. Because the pressure in the building is lowered, fresh air enters from outside through whatever gaps are made available. *Compare* INPUT SYSTEM.

extrados. The upper surface of stones or bricks that form an ARCH. *Compare* INTRADOS.

extrapolate. To estimate a value on a graph or table at a point outside the range of existing values. *Compare* INTERPOLATE.

extra work. *See* EXTRA.

extruded section. Structural sections made from light alloy by EXTRUSION.

extrusion. The process of forming long tubes or other sections by forcing material such as metal or plastic through a DIE.

eye. (1) An ACCESS EYE. (2) An opening in a metal object such as a ring bolt. (3) The hole in the head of a hammer into which the handle fits.

eyebrow dormer. A form of low DORMER WINDOW that is covered by a curve of the roof.

Eyring formula. An acoustic formula for predicting REVERBERATION TIME in a room with high absorption, such as a broadcasting studio. In such cases the formula gives better results than SABINE'S FORMULA.

F

f. *abbr.* Fixed.

fabric. (1) The shell of a building separating the inside from the outside. (2) A form of reinforcement made from wire mesh and used for concrete slabs.

fabrication. The preparation in the workshop of steel members of a building frame to make it ready for assembly on site.

fabric heat loss. The heat lost from a building by transmission through parts of the external FABRIC, such as the walls and roof. *See also* U-VALUE.

facade. The front face of a building.

face. (1) The exposed or the working surface of a wall, of a building, or of a material such as stone, timber, or plasterboard. (2) The cutting edge on the tooth of a saw.

face-bedded. Stonework that is laid so the NATURAL BED or grain is vertical. Arch stones have to be face-bedded but faced-bedded stones may flake.

face brick. (*USA*) *See* FACING BRICK.

faced wall. A wall in which a face material such as that of FACING BRICKS is bonded together with COMMON BRICKS. *Compare* CAVITY WALL.

face hammer. A hammer with a notched head used for dressing the surface of stone.

face joint. The edge of a brickwork CROSS JOINT that is visible on the face of a wall.

face marks. Pencil marks made on wood by a joiner to indicate the planed sides from which further measurements are to be made.

face plate. The surface of a MARKING GAUGE that slides against the wood.

face putty. The triangular strip of GLAZING PUTTY on the outside of window glass. *Compare* BEDDING PUTTY, BEADING.

facia. *See* FASCIA BOARD.

facing brick (*USA*: face brick). A type of BRICK specially selected for its appearance and often used to cover COMMON BRICKS.

factor of safety (safety factor). (1) In structural design, an allowance for variations in the strength of materials or for variations in the loading. *See also* CHARACTERISTIC STRENGTH, LIMIT STATE DESIGN. (2) A comparison of the maximum strength of a component or structure and the working load imposed on it. It can be calculated as the ultimate STRESS divided by the maximum permissible stress. *Compare* LOAD FACTOR.

fadding. The process of applying a SHELLAC polish with a pad called a 'fad'.

fading. The gradual loss of colour in a painted surface caused by AGEING or exposure to weather.

Fahrenheit temperature. A non-metric scale of TEMPERATURE with units in 'degrees', based on the melting point of ice as 32°F and the boiling point of water as 212°F. *Compare* ABSOLUTE TEMPERATURE, CELSIUS TEMPERATURE.

faience. A glazed form of TERRA COTTA that has been used for a facing on walls.

fair cutting. The cutting of facing brickwork with a bolster or trowel rather than with a saw. In a BILL OF QUANTITIES it is measured in linear metres, assuming a 102 mm thickness for a half brick. *Compare* ROUGH CUTTING.

fair-faced. *adj.* Describing a wall of brickwork or blockwork that is neatly laid and smoothly jointed. A fair-faced internal surface is usually left unplastered.

fall. (1) A slope given to a surface such as a roof or yard in order to carry water to a drain. (2) The gradient of drains, rivers, roads.

fall bar. The bar of a simple door latch such as a THUMB LATCH.

fall pipe. *See* DOWNPIPE.

false body. The condition of a paint that temporarily becomes less stiff when stirred or brushed. The effect appears similar to THIXOTROPIC paint although the mechanism is different.

false ceiling. *See* SUSPENDED CEILING.

false tenon. A TENON of hardwood inserted to make a stronger joint.

false work. (1) FORMWORK and other temporary support for concrete. (2) Temporary scaffolding and covers used to support and protect a building during construction.

fan. (1) A temporary scaffolding platform projecting over the street from a building during repair or demolition. The platform slopes slightly upwards and is meant to deflect and catch any falling objects. (2) A revolving system of blades used to move air in ventilation ducts.

fanlight. A small window LIGHT set above a door, usually as part of the door frame. It was originally semi-circular in shape, like a fan. *See* WINDOW diagram.

fan truss. A roof TRUSS that contains both inclined and vertical struts. *Compare* FINK TRUSS. *See* ROOF TRUSS diagram.

fascia board (eaves fascia). (1) A vertical board fixed to the ends of the rafters or wall plate. The EAVES GUTTER is fixed to the board. *See* ROOF diagram. (2) A wide board fixed to the face of a building for carrying a name, such as a shop title.

fastener (fixing). In general, a device for holding different building components together. They are in various forms such a timber CONNECTOR or GANG NAIL. *See also* RESTRAINT FIXING.

fast track. A general method of project management in which the construction of a building starts before all details are finalized.

fat board. A board used by bricklayer for carrying MORTAR.

fat edge. A ridge of wet paint that collects on the lower edge of a newly-painted area because the paint flows too freely or is applied too thickly.

fatigue. A reduction in strength of a material or structural member caused by repeated variation in STRESS.

fat lime. *See* HIGH CALCIUM LIME.

fat mix. *See* RICH MIX.

fat mortar. A mix of MORTAR that sticks to the trowel. *Compare* LEAN MORTAR.

fattening. An increase in the VISCOSITY of paint (though not enough to make it unusable) which occurs during storage.

faucet. (*USA*) A valve that controls the flow of water at an outlet, such as in the kitchen. *See also* JUMPER, TAP, WASHER.

FCEC. *abbr.* (*UK*) Federation of Civil Engineering Contractors.

Fe. Chemical symbol for IRON.

feather. (1) The process of tapering one surface into another. (2) (spline) A tongue or projection on the edge of a board that fits into a matching slot on the edge of the adjacent board, such as in floor boards.

feather edge. An edge that tapers from one side to another. Feather-edged board is used for WEATHER-BOARDING or CLOSE-BOARDED FENCING.

feathering compound. *See* LEVELLING COMPOUND.

feather joint (feathered joint, ploughed and tongued joint). A joint made between adjacent boards by inserting a cross tongue into the slotted edges of the boards.

feebly hydraulic lime. LIME that is made from limestone that contains a small proportion of clay.

feed cistern. A CISTERN or tank that supplies cold water to the hot water system and to taps not intended for the supply of drinking water. *See* HEATING diagram.

feeder. An electrical cable or water pipe that connects a branch appliance to the main run of cables or pipes.

fees. The payments made by a client to members of a building design team, such as architects.

felt. *See* BITUMEN MATERIALS, BUILT-UP ROOFING.

felt nail. *See* CLOUT NAIL.

female thread. A THREAD on the inside of a pipe or tube. *Compare* MALE THREAD.

fence. (1) A system of upright boards (pales) or panels that enclose a site. (2) A built-in guide to control the direction of timber as it passes through a SAW BENCH. (3) A screen that protects people from the moving parts of a machine.

fender. (1) A device such as a mat, bar or post that is designed to absorb impacts from vehicles and ships. (2) A large BAULK of timber laid on the ground to protect the upright poles of scaffolding from the wheels of traffic.

fender wall. A form of DWARF WALL that supports the hearth slab in a suspended ground floor.

fenestration. The arrangement of windows and other openings in the walls of a building.

ferritic steel. A form of STAINLESS STEEL that contains a higher proportion of chromium than usual.

ferro-concrete. An older term for REINFORCED CONCRETE.

ferrous metal. A metal or ALLOY containing a significant amount of IRON.

ferrule. (1) A metal band around the handle of a tool, such as a chisel, to prevent splitting. (2) In plumbing, a short length of tube such as a SLEEVE PIECE.

festoon lighting. A chain of light fittings hung in a curve between two points.

festoon staining. A pattern of staining seen on external walls and caused by different flows of water down the surfaces. *See also* PATTERN STAINING.

fettle. The process of removing roughness from a metal casting and generally finishing off work.

FFL. *abbr.* Finished floor level.

FHA. (*USA*) Federal Housing Administration.

fibreboard. A variety of boards or sheets that are made by bonding wood and other fibres together with resins or glues. Medium density and hardboards are compressed during manufacture while insulating or softboards are made less dense. Boards may be treated to improve fire and moisture resistance. *See also* CHIPBOARD, MDF, MFC.

fibreglass. *See* GLASS FIBRE.

fibre-reinforced concrete. Concrete, such as GLASS REINFORCED CONCRETE, made with a fibrous aggregate that gives useful characteristics such as low density or high strength.

fibre saturation point. A condition of wood when the cell cavities are empty of water but the cell walls still hold their maximum water. For most timbers this point is between 25 and 30 per cent MOISTURE CONTENT. Drying below this point causes the timber to shrink.

fibrous concrete. *See* FIBRE-REINFORCED CONCRETE.

field drain (French drain, land drain). Porous or perforated pipes that are laid in the ground without sealed joints. They are designed to admit ground water and drain the surrounding land.

fieldstone. Natural stone in sizes used for RUBBLE WALLS.

figuring. (*USA*) The process of taking off QUANTITIES from drawings to produce an ESTIMATE.

filament lamp. A form of INCANDESCENT LAMP such as the common light bulb. An electric current heats a filament of tungsten wire to about 2800 K where it glows and gives off light. To prevent rapid burning of the metal the filament is surrounded by unreactive gases contained within a glass envelope. *See also* GLS LAMP, PAR LAMP.

file. A tool with a rough metal surface used to scratch or to smooth the materials. *See also* RASP.

fill. Material such as earth that is used in the construction of an EMBANKMENT and similar constructions. *Compare* SPOIL.

filler. (1) A paste used to fill indentations in rough surfaces, such as wood, prior to painting. (2) A material added to plastics and paints as an EXTENDER.

fillet. (1) A narrow strip of material, usually wood, used to fill or to cover joins. (2) *See* CEMENT FILLET.

filling piece. A small piece of timber set in a large piece so as to smooth the surface.

fillister. (1) The REBATE on a window frame or glazing bar that receives the glass. (2) A type of PLANE with a narrow blade used for cutting groove or REBATES.

film. A thin coating of material such as PAINT.

film adhesive. Adhesive supplied in the form of a thin sheet and used when continuous adhesion is needed, such as for VENEERS.

film building (film forming). The ability of paint to form a FILM that is adhesive, strong, continuous and flexible.

filter bed. An arrangement for the FILTRATION of water or sewage in which the liquid is passed through a mechanical or bacteriological medium. *See* SAND FILTER.

filtration. The general process of removing solids and bacteria from liquids by passing them through a form of strainer such as a FILTER BED. *See also* COARSE FILTER, MICROSTRAINER, SAND FILTER.

final certificate. The document that authorizes final payment to the CONTRACTOR for the construction of a building. *See also* INTERIM CERTIFICATE, PENULTIMATE CERTIFICATE.

fine aggregate. A type of AGGREGATE, such as sand or grit, that is used in a mix of concrete or asphalt and fills the spaces between particles of COARSE AGGREGATE. Fine aggregate for concrete is defined as that which passes through a sieve screen with holes 5 mm square. Fine aggregate for ASPHALT is that which passes through a sieve screen with holes 3 mm square.

fine cold asphalt. *See* COLD ASPHALT.

fineness. The relative size of particles making up a particular cement, sand or pigment. It is found by passing a sample through standard screen sizes and it may be expressed as a single-figure fineness modulus or as a GRADING CURVE.

fine solder. A high quality SOLDER used for making BLOWN JOINTS. It contains a higher proportion of tin and has a lower melting point than plain plumbers' solder.

fine stuff. The type of plaster used for the FINISHING COAT.

finger joint. An in-line joint formed between the ends of two timbers by a series of interlocking fingers in each piece.

finger plate. A smooth, cleanable plate fixed on the surface of the door near the latch to protect the door from marks.

finial. An architectural ornament, such as

ball or point, fixed to the top of a roof, gable or pinnacle.

finish. (1) The appearance and quality of a component, decoration or completed building. (2) (*USA*) JOINERY that is fixed.

finished size. *See* DRESSED SIZE.

finishing coat. The top coat of a multi-layered system for applying plaster or rendering. *See* THREE-COAT WORK.

finishings. The materials and processes used for the joinery, plaster, decoration and other features that produce the FINISH of a building.

finishing trade. One of the activities, such as plastering and decorating, that produce the final appearance of a building.

fink truss (French truss, W-truss). A roof TRUSS in which all the STRUTS are at an angle rather than vertical. *Compare* FAN TRUSS. *See* ROOF TRUSS diagram.

fir. Evergreen coniferous trees, especially those of the genus *Abies* such as SILVER FIR.

fire alarm. A device that warns people of the danger of fire. It may be operated manually or automatically activated by the presence of smoke or heat. *See also* SMOKE DETECTOR.

fireback. The rear wall of a fireplace.

fire break. *See* FIRE STOP.

fire brick. A brick made of FIRECLAY that can withstand high temperatures.

fire cell. *See* COMPARTMENTATION.

fire check door. *See* FIRE DOOR.

fireclay. Forms of clay that contain a high proportion of silica and alumina. *See* FIRE BRICK.

fire compartment. *See* COMPARTMENTATION.

fire door (fire check door). A door constructed to have FIRE RESISTANCE for a minimum period of time such as 30 minutes.

fire extinguishing equipment. Portable or fixed equipment designed to fight fires in buildings. Equipment includes FIRE HYDRANTS, SPRINKLER SYSTEMS and hand-held sprays.

fire hydrant. A permanent outlet on the water main, inside or outside the building, designed for use by the Fire Service. *Compare* DRY RISER.

fire load. A measurement of the total fuel for a fire provided by the contents of a building and its construction materials. One system of rating is by the joules generated per cubic metre of building.

fireplace. A place in a room where a fire can be made to heat the room. It is usually a cavity in a wall that is connected to a CHIMNEY. *See also* HEARTH.

fire prevention. The general business of preventing outbreaks of fire, reducing risk of fire spread and of avoiding danger from fire to people or property. *Compare* FIRE PROTECTION.

fireproof. A term that should be replaced by the more precise idea of FIRE RESISTANCE because few materials or constructions are unaffected at all by fire.

fire protection. The features of design, systems or equipment in a structure that reduce the danger to people and property by detecting, extinguishing or containing fires. *Compare* FIRE PREVENTION.

fire resistance. The ability of a structure, or part of a structure, to resist fire for a specified period of time. The resistance is measured by (a) stability (resistance to collapse), (b) INTEGRITY (resistance to flame penetration) and (c) INSULATION (resistance to temperature rise).

fire-resisting. *adj.* The property of an element of a building, such as a door, that passes the tests of FIRE RESISTANCE for a suitable period of time such as 30 minutes.

fire stop (cavity barrier). A barrier designed to prevent fire or smoke passing through a roof or cavity wall.

fire testing. The assessment of materials for a variety of behaviours in a fire. The relevant BRITISH STANDARD tests include combustibility, fire propagation, and surface spread of flame.

fire vent (smoke outlet, smoke vent). An area of roof or wall designed to release smoke and hot gases in the event of fire. High temperatures cause the vent to open or melt THERMOPLASTIC roof sheets. Fire venting helps prevent the spread of fire in large open buildings where COMPARTMENTATION is not possible.

firewall. (*USA*) A wall built with a high FIRE RESISTANCE as part of a system of COMPARTMENTATION.

firing. The process of heating BRICKS and other clay products to a high temperature in a kiln.

firmer chisel. A CHISEL used for ordinary carpentry and joinery.

firring. (1) Strips of timber used to pack out uneven construction, such as floor joists, and produce an even surface. (2) Tapered strips of timber used to give a slope to a roof.

first angle projection. *See* ORTHOGRAPHIC PROJECTION.

first fixings. (1) Structural timber work in a building such as JOISTS and RAFTERS. (2) Fittings, such as plugs, that are installed to carry later joinery work. *Compare* SECOND FIXINGS.

first floor. (1) (*UK*) The floor just above the ground level floor. *Compare* GROUND FLOOR. (2) (*USA*) The floor at ground level.

fish glue. An old form of ADHESIVE made from fish waste. *See also* ANIMAL GLUE.

fish plate. A form of timber CONNECTOR.

fish tail. A shape put on the ends of posts to give better stability when they are embedded in concrete. The fish tail shape is made by splitting the end and twisting the two halves in opposite directions.

fitch. A small, long-handed brush used for painting areas that are otherwise difficult to decorate. *See also* LINING FITCH.

fitment. An item of furniture that is fixed in a building.

fitting. (1) An item of pipework such as a coupling, bend, elbow or union. *See also* CAPILLARY JOINT, COMPRESSION JOINT. (2) The process of trimming a component and installing it in a particular position.

fixed end. A fixing at the end of a beam or column that restrains the member from turning. *See also* FIXING MOMENT. *See also* ENCASTRE. *Compare* HINGE.

fixed light. *See* DEAD LIGHT.

fixed-price contract. A CONTRACT in which construction is completed for a fixed sum without provision for FLUCTUATIONS. *See also* LUMP-SUM CONTRACT.

fixed sash. *See* DEAD LIGHT.

fixing. *See* FASTENER.

fixing brick (fixing block, nog). A brick or block made from sawdust, lightweight concrete or some other material that can be nailed. It is used to fasten joinery.

fixing fillet (fixing slip, pallet, wood slip). A slip of wood inserted into the mortar joint of brickwork and used to fasten joinery.

fixing moment. The BENDING MOMENT that is needed at the FIXED END of a beam in order to prevent it rotating.

fixing slip. *See* FIXING FILLET.

fixture. An item fixed to a building and regarded as the landlord's property. Examples included sanitary fittings and other plumbing.

flagstone (flag, flagging). A slab of stone or concrete used as material for paving or as a cover for a pit.

flaking. The loss of thin particles from a layer of paint, stone or other facing material.

flame cleaning. The use of a flame to remove MILL SCALE and other surface defects from structural steelwork that has weathered.

flame cutting. The process of cutting thick sections of steel and other metals by a hot flame such as oxy-acetylene.

flame retardant. A treatment for materials that increases the resistance to flames spreading across the surface of a material.

flame spread. *See* SURFACE SPREAD OF FLAME.

flammable. *adj.* Capable of burning easily. This term is recommended instead of 'inflammable', which may cause confusion.

flange. In general, a projecting rim or strip used to fasten or strengthen a component. (1) A disc attached to the end of a pipe to attach it to another pipe or component. (2) The wide strips at the top and bottom of a GIRDER or ROLLED-STEEL JOIST. *Compare* WEB.

flanking transmission. The passage of sound into a room by a variety of indirect pathways that bypass the route of the DIRECT SOUND.

flank wall. A wall at one side of a building.

flash. The general technique of using FLASHING to make a watertight joint between external coverings.

flash drying. The rapid drying of paint by applying radiant heat for a short time.

flashing. A strip of flexible material used to waterproof a joint between two external surfaces of a building, such as the roof and a wall. Lead, zinc, copper and waterproof felts are common materials for flashings.

flashover. A stage in the development of a fire when there is a sudden spread of flame which results in burning throughout a space.

flash point. The lowest temperature at which the vapour of a combustible liquid will catch fire.

flash set. An over-rapid setting of cement, usually in concrete.

flat. One of a number of self-contained dwellings within a single building. *Compare* MAISONETTE. *See also* CONDOMINIUM.

flat arch (jack arch, straight arch). A type of ARCH built with a horizontal undersurface (INTRADOS). It may be used above windows and doors. *See also* SOLDIER ARCH.

flat coat. A coat of paint with a non-glossy finish, such as an UNDERCOAT.

flat cost. The cost of labour and material only without other costs such as overheads and profits.

flat grain. The grain of timber that has been FLAT SAWN. Because of the parallel cuts most of the annual rings will have an angle less than 45° to the face of the timber.

flat plate collector. A device that uses a black, heat-absorbing plate to collect radiant energy from the sun.

flat roof (platform roof). A roof surface that appears to be almost flat. It has a slope between 1 in 40 and 1 in 80 to allow for drainage. *Compare* PITCHED ROOF.

flat-sawn timber (plain-sawn timber). Timber from logs that has been 'converted' by the use of parallel saw cuts. *See also* SLASH GRAIN. *Compare* QUARTER-SAWN TIMBER.

flat slab. A structural slab of reinforced concrete that spans a space in two directions. *Compare* BEAM.

flatting down. *See* RUBBING DOWN.

flaunching. A triangular strip of mortar or a FILLET used to seal the top of a chimney stack around the chimney pot. *See* ROOF diagram.

Flemish bond. A brickwork BOND in which every course contains HEADERS and STRETCHERS laid alternately. *See* BRICKWORK diagram. *See also* DOUBLE FLEMISH

BOND, SINGLE FLEMISH BOND. *Compare* ENGLISH BOND, MONK BOND.

Flemish diagonal bond. A brickwork BOND in which a complete course of STRETCHERS alternates with every course of mixed HEADERS and stretchers.

Flemish garden wall bond (Sussex garden wall bond). A brickwork BOND in which each course is laid with a sequence of three STRETCHERS and one HEADER, each header being placed above or below the middle of the stretchers.

fletton. A distinctive variety of COMMON BRICK with bands of pink on their faces. Made chiefly from clay in Bedfordshire and Northamptonshire, UK.

flex (flexible cord, *USA*: lamp cord). An electric CABLE used to make the final connection with a portable appliance such as a domestic heater or lamp. For flexibility, the conductors are composed of a large number of copper strands.

flexible metal. A rolled form of sheet metal such as lead, copper, zinc and aluminium that is used for roofing or FLASHING. *See also* SEAM.

flexure. A bending effect. *See also* CONTRAFLEXURE.

flier (flyer). A STEP in a straight flight of STAIRS. *Compare* WINDER.

flight. A series of STAIRS that runs between floors or landings.

flight hole. The hole in the surface of wood where an adult insect, such as the FURNITURE BEETLE, has emerged from tunnels in the wood.

flint. Dark grey lumps of natural SILICA found in chalk.

flint glass. A form of glass containing lead compounds.

flint wall. A wall built of FLINT stones set in mortar.

flitch. (1) A large section of timber intended for CONVERSION. (2) A timber from which VENEERS are cut, or a stack of cut veneers.

flitch beam (sandwich beam). A compound BEAM built-up from two timber beams bolted together with a steel plate (FLITCH PLATE) sandwiched between them.

flitch plate. The steel plate used inside a FLITCH BEAM.

float. (1) (floater, floating trowel) A hand tool with a flat wooden or metal surface used for smoothing PLASTER or SCREED. *Compare* DARBY. (2) *See* FLOAT TIME.

floater. *See* FLOAT.

float glass. Modern high-quality window GLASS that is manufactured by floating the hot glass on a molten metal. *Compare* SHEET GLASS. *See also* ARMOUR-PLATE GLASS, PLATE GLASS.

floating coat. An undercoat of plaster.

floating floor. A form of DISCONTINUOUS CONSTRUCTION used to reduce the transmission of IMPACT SOUND through a floor. The upper surface of the floor, whether timber or screed, is separated from the load-bearing part by a resilient layer such as mineral wool. To be effective the floating floor must not be bridged by any rigid contact such as a nail.

floating screed. A type of FLOATING FLOOR with a wearing surface of mortar rather than boards.

floating trowel. *See* FLOAT.

float switch. An electric switch operated by a water float that stops and starts a pump when the water level rises or falls.

float time (float). In CRITICAL PATH analysis, the time that a non-CRITICAL ACTIVITY has to spare without affecting the completion time of the project. *See also* INDEPENDENT FLOAT, INTERFERING FLOAT, TOTAL FLOAT.

float valve. *See* BALL VALVE.

flocculation. A process, usually gentle stirring, that encourages fine particles in water to stick together in a 'floc' that can be removed. *See also* COAGULATION.

floodlighting. A system of artificial lighting that uniformly illuminates a large area, usually outdoors.

floorboards. Boards that span the spaces between floor joists to form the structure of the floor. Boards may be TONGUE AND GROOVED to give improved strength, fire resistance and airtightness. Floor boards may also be made from CHIPBOARD.

floor brad. *See* BRAD.

floor clip (bulldog clip, sleeper clip). A steel clip that is pushed into the surface of a concrete floor or screed before it sets. The wooden BATTENS or SLEEPERS of the floor structure are then attached to the clips.

floor cramp. A tool used to force floorboards together and to hold them before nailing.

floor joist. One of the COMMON JOISTS that forms the structure of a floor.

floor plan. A drawing of a particular floor of a building that shows the layout of the rooms and other details.

floor stop. A form of DOOR STOP that is fixed in the floor to prevent a door swinging open too far.

floor strutting (herringbone strutting). A system of cross struts inserted between floor joists at the middle of their spans to give them stiffness and to resist buckling. The struts may be solid pieces or several lengths criss-crossed to make a herringbone pattern.

flow chart. A form of graphical aid that shows the links between the stages of a project or parts of a problem. *Compare* CRITICAL PATH ANALYSIS. *See also* DECISION TREE.

flow lines. (1) The paths traced by particular particles of a moving fluid, such as in a pipe. (2) Lines on a diagram that represent the direction of flow of water through the soil near a dam.

flow meter. A device that measures the FLOW RATE in a pipe. *See also* VENTURI EFFECT.

flow pipe. A pipe taking hot water from a boiler to a storage cylinder or heating panel. *Compare* RETURN PIPE. *See* HEATING diagram.

flowrate (discharge). The quantity of liquid or gas that flows per unit time in a pipe or channel.

flr. *abbr.* Floor.

fluctuation. An allowance in a CONTRACT for changes in the cost of labour and materials during the course of construction.

flue. The duct that carries smoke and gases away from a fireplace or boiler. A brick-lined flue is included within the construction of a traditional brick chimney and flues can also be built from ready-made flue blocks or pipes. *See also* BALANCED FLUE.

flue gas. The gases, such as carbon dioxide, produced by the process of combustion in a boiler fire. The composition of the gases can be recorded as a measure of the efficiency of the boiler.

flueless appliance. A type of fuel-burning device, such as a gas cooker, where the products of combustion stay in the room rather than leave by a FLUE. The water vapour produced by such appliances can be a cause of CONDENSATION in buildings.

flue lining (flue liner). Ready-built sections of fireclay, concrete or metal that are used to line the inside of a flue. The linings need to be waterproof since the combustion of gas and oil produces water vapour, which may condense in the cool upper part of the chimney and cause damage.

fluid. A substance that takes the shape of its container. Although liquids and gases are different states of matter they are both fluids and sometimes behave in a similar manner. *See* HYDRAULICS.

flume. An open channel made of concrete, steel or wood that carries water.

fluorescent lamp. A modified form of DIS-

CHARGE LAMP in which most of the light comes from a layer of fluorescent material. The phosphors in this layer absorb invisible ULTRA-VIOLET RADIATION emitted by the discharge and re-emit it as visible light. Tubular fluorescent lamps, based on a mercury vapour discharge, are a common method of lighting shops and offices. Different mixtures of phosphors are used to give different colour qualities to the light. *See also* COLOUR TEMPERATURE.

fluorescent paint. Paint containing pigments that absorb energy from the ULTRA-VIOLET RADIATION present in natural light. The energy is re-emitted as visible light which gives the painted surface a bright quality. *Compare* PHOSPHORESCENT PAINT.

fluorinated hydrocarbon (fluorocarbon). A compound formed by replacing some or all of the hydrogen atoms in a HYDROCARBON by fluorine. *See also* FREON.

fluorocarbon. *See* FLUORINATED HYDROCARBON.

flush. (1) The general effect of having different surfaces, such as walls and doors, lined up in the same plane. (2) *vb.* To clean out unwanted material. WCs, pipes and channels are flushed by a rush of clean water

flush door. A door with a continuous smooth face of plywood or hardwood that covers a solid or cellular interior. *See also* HOLLOW-CORE DOOR.

flush hinge. A type of HINGE that does not have to be recessed into a door frame like a BUTT HINGE. *See* DOOR diagram.

flush joint. The POINTING of a mortar joint so that it is flush with the face of the bricks. *See* BRICKWORK diagram.

flushing cistern. A small tank above a WC that holds water to flush the WC pan. *See also* AUTOMATIC FLUSHING CISTERN, LOW LEVEL FLUSHING CISTERN.

flushing valve. In a FLUSH CISTERN, a valve that automatically releases a FLUSH when the water pressure in the cistern builds up.

flushometer valve. *See* FLUSHING VALVE.

flush pipe. The pipe that connects a FLUSHING CISTERN to a WC or urinal.

flushvalve. *See* FLUSHING VALVE.

flux. (1) A substance applied to metal surfaces that are being joined by soldering, braising or welding. The flux helps dissolve the films of oxide that prevent good adhesion. (2) *See* LUMINOUS FLUX.

fly ash. *See* PULVERIZED FUEL ASH.

flyer. *See* FLIER.

flying bond. (1) *See* ENGLISH GARDEN WALL BOND. (2) *See* MONK BOND.

flying scaffold. A form of SCAFFOLD hung by ropes or cables from an outrigger that projects from the building.

flying shore (horizontal shore). A horizontal beam or other structure that is fixed above ground between two walls. *See also* NEEDLE. *Compare* RAKING SHORE.

foamed concrete. *See* AERATED CONCRETE.

foamed plastic. *See* EXPANDED PLASTIC.

foamed slag. A lightweight AGGREGATE made from blast-furnace slag. The molten slag is usually foamed with water to produce a cellular structure.

foil. A thin flexible form of material such as aluminium, copper and zinc that is usually less than 0.15 mm in thickness. *See also* ALUMINIUM FOIL.

foil-backed board. Plasterboard with a layer of ALUMINIUM FOIL bonded on to its rear surface. The aluminium acts as an insulator against radiant heat and as a VAPOUR BARRIER.

folded plate construction. A method of constructing large spans, usually roofs, with a series of reinforced concrete slabs set at angles to one another. The multiple pitched roofs give additional stiffness.

folding casement. Usually a pair of CASEMENT doors or windows that are hinged so that they can be opened in a confined space.

folding shutter. *See* BOXING SHUTTER.

fondu. *See* HIGH-ALUMINA CEMENT.

foot block. *See* ARCHITRAVE BLOCK.

foot bolt. A TOWER BOLT that is set vertically at the foot of a door.

foot-candle. (*USA*) A non-metric unit of ILLUMINANCE. 1 foot-candle = 1 LUMEN per square foot.

foot cut. (*USA*) *See* BIRDSMOUTH JOINT.

footing. The base construction of a wall where it is widened to spread the load over a larger area. The term may also include a concrete FOUNDATION.

foot plate. (1) A horizontal roof timber laid on top of the WALL PLATE joining the foot of a rafter and the ashlaring. (2) *See* SOLE PLATE.

footprints (combination pliers). An adjustable WRENCH with serrated jaws that are used in pipe fitting.

force. An influence that produces a change in the motion or shape of a body on which it acts. Force can be calculated as the product of mass and acceleration. The SI unit of force is the newton.

force cup. A flexible rubber cup attached to a wooden handle. The cup is placed over household waste pipes and pushed up and down to produce a change in pressure that often clears blockages.

forced circulation. A heating system in which water is moved through the pipes by a pump rather than by GRAVITY CIRCULATION. Most hot water central heating systems use forced circulation.

forced drying. The rapid drying of paint by exposure to moderate heat, not above about 65°C.

fore plane. A general-purpose PLANE used for working long pieces of wood.

fore sight. The last sighting made in a surveying series before an instrument is moved. This point becomes the first reading (the BACK SIGHT) from the new position.

forge. (1) *vb.* To shape hot metal, such as steel or wrought iron, by pressing or hammering. (2) A place where forgings are made.

formaldehyde. A gas with an irritating smell. It is used in the production of certain SYNTHETIC RESINS from which glues and insulating materials are made.

formica. A trade name for a form of LAMINATED PLASTIC.

form of contract. The particular type of standardized contract used between a client and a contractor.

form of tender. Documents, such as a BILL OF QUANTITIES, that are issued by the architect and used by competing contractors to submit their proposed prices.

formwork (casing, shuttering). A temporary construction of boards or sheets used to contain and shape concrete when it is placed and while it hardens. *See also* MOULD.

Forstner bit. *See* CENTRE BIT.

fossil fuel. FUELS such as coal and oil that are found in the earth as the remains of ancient plants or small organisms that have been preserved as fossils.

fossil resin (fossil gum). A natural RESIN, such as copal, found in buried in the ground. Fossil resins have been used as a base for varnish.

foul water. WASTE WATER, but not rainwater, discharged from foul and waste fittings into the drains of a building.

foundation. (1) The ground on which a building or other structure rests. (2) The part of a construction that transfers the load of the building to the ground beneath. *See also* PILES, RAFT FOUNDATION, STRIP FOUNDATION.

foundation stone. An ornamental stone set in a building near the ground to record details of the construction. Names and dates are usually carved on the surface and objects such as coins may be placed inside.

four-coat system. A method of painting new wood. The system includes a PRIMER, UNDERCOAT and a final coat. The additional coat may be another undercoat or gloss coat depending on the type of paint and the situation.

foxiness. A darkened appearance of wood that is beginning to decay.

frame. A number of timber, steel or concrete MEMBERS that are joined together to form a stable construction on to which other components can be fastened. The frame may be the surround for a door or the complete skeleton of a high-rise building.

frame construction. A construction made with timber structural members. *See also* BALLOON FRAME.

framed door. A door built on a rigid frame that consists of horizontal RAILS fixed between two vertical STILES at the side.

framed, ledged and braced door. A door with a visible construction of boards attached to a frame of rails or LEDGES fixed between vertical STILES and stiffened by diagonal braces. *Compare* LEDGED AND BRACED DOOR.

framed partition. A partition, such as an internal wall, made by from a frame of STUDS and other members.

frame saw (gang saw). A power SAW with vertical blades fixed in a frame. It is used to cut wood or stone.

framing square. *See* STEEL SQUARE.

frass. The excrement and debris left by BEETLES and other insects.

free electrons. Loosely-bound ELECTRONS that are able to move from one atom to another. The presence of significant numbers of free electrons increases electrical and thermal conductivity.

free field. In acoustics, an area in which the sound waves spread out without meeting any objects.

freestone. Building stone, such as limestone and sandstone, that is fine-grained and sufficiently uniform to be worked in any direction.

free stuff. *See* CLEAR TIMBER.

French arch. *See* DUTCH ARCH.

French door. *See* CASEMENT DOOR.

French drain. (1) *See* FIELD DRAIN. (2) Sections of round tile or perforated pipe that are laid alongside the foundations of a building to collect and remove excess ground water.

frenchman. A cutting tool used to trim mortar when JOINTING or POINTING brickwork. It is used with the help of a straight edge such as a JOINTING RULE.

French polish. A coating that is built-up in layers to give a high polish to furniture. It is made from SHELLAC dissolved in methylated spirits. *See also* SPIRITING OFF.

French roof. *See* MANSARD ROOF.

French truss. *See* FINK TRUSS.

French window. *See* CASEMENT DOOR.

freon. A common form of REFRIGERANT such as that used in domestic machines. It is a mixture of organic compounds containing carbon, chlorine and fluorine.

frequency. In general, the number of events occurring in a given time such as one second. The SI unit of frequency is the hertz (Hz). 1 Hz = 1 cycle per second. (1) In sound, the number of complete vibrations per second. The pitch of a note depends on frequency. (2) In ALTERNATING CURRENT, the number of complete positive and negative peaks of current. The public electricity supply has a frequency of 50 Hz in the United Kingdom and 60 Hz in North America.

frequency distribution. An arrangement of statistical data into groups of magnitude and the number of times (frequency) that such magnitude occurs. The results of a large number of results can be displayed on a graph as a curve on which shows the frequency rising to a maximum value and falling back to zero. The tightness of the spread can be measured by the STANDARD DEVIATION. *See also* GAUSSIAN DISTRIBUTION.

fret. *See* FRETWORK.

fret saw. A fine, narrow-bladed SAW used to cut curves and sharp corners in decorative woodwork often termed 'fretwork'. The saw may be a hand tool or power operated.

fretwork (fret). Decorative woodwork that is cut by a FRETSAW.

friable. *adj.* Describing soil that easily crumbles into fine particles.

friction-grip bolt. *See* HIGH-STRENGTH FRICTION-GRIP BOLT.

friction losses. The loss of energy or pressure HEAD caused by friction between a liquid and the surfaces of a pipe.

friction pile. A PILE in which the load is entirely resisted by friction with the earth surrounding the pile. *Compare* END-BEARING PILE.

frieze. The upper part of a wall, just below the cornice.

frog. An indentation, often V-shaped, that is moulded into one of the flat faces of a brick or block. A wall is stronger if the bricks are laid with frogs facing upwards so that they become filled with mortar.

frontage line. *See* BUILDING LINE.

frost. The fall in outside air temperature to below freezing point and the resulting white deposit of ice particles on objects.

frost action. The effect of FROST on materials such as concrete, stone and brick if they contain moisture. When cold weather causes water to freeze the water must increase in volume; this expansion is often destructive.

frostat. A THERMOSTAT that reacts when the temperature of the outside air falls to near freezing point.

frost heave. The expansion of ground, usually upwards, caused by the freezing of water within a soil such as CLAY.

frost line. The maximum depth to which soil becomes frozen.

FRP. *abbr.* Fibre-reinforced plastic.

fs. *abbr.* Full size (used on drawings).

FSP. *abbr.* FIBRE SATURATION POINT.

fuel. A source of ENERGY that can be converted to a usable form, usually by the process of burning. *See also* FOSSIL FUEL, PRIMARY ENERGY.

fugitive colour. A colour that tends to fade when exposed to light or weather.

fulcrum. *See* LEVER.

full coat. In paintwork, the thickest coat that can be applied.

fuller's earth. Various types of clay that readily absorb oil and grease. Its applications in decorating include the removal of grease from surfaces before painting. *See also* BENTONITE.

full-wave valve. A COCK or valve in a pipe that opens in a way that does not obstruct the flow of water. It is used in systems where the pressure is low.

fundamental frequency. In sound, the predominant frequency of a note. The fundamental is usually accompanied by OVERTONES.

fungi. *See* FUNGUS.

fungicidal paint. *See* FUNGICIDE.

fungicide. A chemical substance that resists the growth of FUNGUS on a surface. Fungicides are used in washes for treating surfaces infected with mould and as additives in fungicidal paints which discourage mould growth.

fungus (*pl.* fungi). A type of plant growth that feeds on organic material such as timber. Fungus spores (seeds) are carried by wind and can lie dormant until growth is encouraged by warm and humid conditions. *See also* DRY ROT, HYPHAE, MOULD, MYCELIUM, WET ROT.

furniture beetle (*Anobium puntatum*). An insect that is destructive to timber and furniture. It spends most of its life cycle as a larva that tunnels and feeds within the wood. It leave FLIGHT HOLES of about 1.5 mm in diameter. *See also* BEETLE.

furring. (1) An effect of HARD WATER which deposits salts as a crust inside hot-water pipes, tanks and boilers. (2) *See* FIRRING.

fuse. A safety device that quickly breaks an electric circuit if excessive current flows. Simple fuses use wire that melts at the critical current. *See also* CARTRIDGE FUSE, CIRCUIT BREAKER.

fuse box (fuse board). A housing, such as a CONSUMER UNIT, that contains fuses.

fused plug. An electrical plug that contains a FUSE. When fitted to the flexible cable for an appliance it provides extra protection against excessive current flow.

fusible link. A piece of metal designed to melt in a fire. It is used on devices such as a FIRE DOOR that is held open by the link but swings shut if the link melts.

fusible plug. A special metal plug fitted in a boiler. The plug is designed to melt and release steam if the water level falls below a safety level.

fusion welding. A form of WELDING in which the join between metals is made by molten material rather than by pressure.

G

gabion. A cylindrical container filled with stones, used to construct the sides of a small COFFER DAM.

gable (gable end). The upper part of a outside wall where the slopes of the roof give the gable a triangular shape. *See* ROOF diagram.

gable board. *See* BARGE BOARD.

gable coping. A COPING that caps a GABLE WALL if it projects above the roof surface.

gable end. *See* GABLE.

gable post. A short post set at the top of a GABLE into which the BARGE BOARDS are fixed.

gable roof. A roof with a GABLE at each end.

gable shoulder. The overhang at the foot of a gable coping formed by the GABLE SPRINGER.

gable springer (skew corbel). An overhanging stone set at the foot of a GABLE COPING to form the transition between the slope of the roof end and the vertical wall.

gablet. A small gable-like shape set over a DORMER window or other opening.

gable wall. An outside wall that is topped by a GABLE and forms the gable end.

gage. (*USA*) *See* GAUGE.

gallet (garnet). A chip of rock or stone.

galleting. The setting of GALLETS or small pieces of tile into mortar for decoration or for filling spaces between roof tiles.

gallows bracket. An ANGLE BRACKET formed with a triangular frame.

galvanic corrosion. *See* ELECTROLYTIC CORROSION.

galvanic series. A practical form of the ELECTROCHEMICAL SERIES in which commonly-used metals and alloys are arranged in order of the ease with which they corrode. Magnesium and zinc are high in the order of the corrosion series while copper and tin are low; the exact order can vary with the type of environment.

galvanized nails. NAILS given a protective coating of zinc by the process of GALVANIZING.

galvanized wire. Wire given a protective coating of zinc by the process of GALVANIZING.

galvanizing. The process of coating a metal, usually steel, with a layer of zinc in order to prevent corrosion of the steel. Because zinc is higher in the GALVANIC SERIES it corrodes before the steel (iron) corrodes. *See also* SACRIFICIAL PROTECTION.

gambrel roof. (1) A roof with a GABLET near the ridge that forms a hipped roof below. (2) (*USA*) *See* MANSARD ROOF. *See* ROOF diagram.

gamma radiography. A form of NON-DESTRUCTIVE TESTING for concrete in which RADIOACTIVE gamma rays are passed through the concrete and detected by methods similar to those used for X-rays.

ganger. The person in charge of a group of workers such as those who excavate or concrete. *See also* NAVVY.

gangnail. *See* CONNECTOR.

gang saw. *See* FRAME SAW.

gangway. A temporary path of SCAFFOLD BOARDS laid over rough ground on building sites.

gantry. A bridge-like framework from which loads can be suspended.

gantry crane. A type of CRANE set on a tall moving platform that can straddle vehicles to lift objects from them.

gantt chart. A horizontal BAR CHART used in NETWORK ANALYSIS.

gap-filling glue. A form of ADHESIVE designed to stick together surfaces that are rough or cannot be brought into close contact. *Compare* CLOSE-CONTACT GLUE.

Garchey system. A special drain system that uses water to collect and carry domestic waste to a central disposal point.

garden wall bond. In general, an economical form of brick bonding that has a high proportion of stretchers. *See* ENGLISH GARDEN-WALL BOND, FLEMISH GARDEN-WALL BOND.

gargoyle. A grotesque face or other ornament that projects near the roof of a building and often acts a waterspout.

garland drain. A shallow trench designed to intercept surface and subsoil water before it reaches an excavation.

garnet. *See* GALLET.

gas circulator. A gas-fired water heater used for heating water stored in a storage cylinder.

gas concrete. *See* AERATED CONCRETE.

gas discharge lamp, gaseous discharge light. *See* DISCHARGE LAMP.

gas instantaneous water heater. *See* INSTANTANEOUS WATER HEATER.

gasket. A strip or sheet of flexible material used to seal joints between components in a building. Examples include rubber rings used to join water pipes and neoprene strips between wall panels.

gas laws. A group of related physical laws describing the behaviour of gases under variations of pressure, temperature and volume. *See* BOYLE'S LAW, CHARLES'S LAW, DALTON'S LAW.

gas pliers. An adjustable form of WRENCH. They are like FOOTPRINTS but have curved jaws.

gate hook. A pin that is set in a gate post to point upwards and receive the gate hinge. *See also* GUDGEON.

gate pier. A substantial form of gate post built of concrete, brick or stone.

gate valve. A form of STOP COCK that closes a pipe by inserting a plate across the direction of flow. The open gate valve gives minimal resistance to the flow.

gathering. The section of a chimney where the passage reduces in size from the fireplace to the flue. *See also* THROAT.

gauge. (1) The proportions of materials contained in a mix of mortar or plaster. (2) The process of mixing several materials in set proportions. (3) The distance between centres of battens on which TILES or SLATES are laid. This distance is the same as that of the MARGIN or exposed length between the bottom edge of one course and the bottom edge of the next course. *See also* LAP. (4) *vb.* To accelerate the setting of plaster by the addition of ADDITIVES. (5) A strip of wood or metal placed at the boundary of asphalting to show the correct thickness.

gauge board. A square board used for carrying plaster and tools. *See also* SPOT BOARD.

gauge box. *See* BATCH BOX.

gauged arch. A type of ARCH that is built of GAUGED BRICKS laid with very fine joints. *Compare* AXED ARCH.

gauged brick (rubber). A soft brick that can be sawn or rubbed to an exact shape. *Compare* AXED BRICK.

gauged mortar. A form of sand and cement MORTAR to which lime is added to give a more workable mix and a more flexible joint.

gauging. (1) The cutting of stones and bricks to make them uniform in size. (2) The reduction of sawn timber to a uniform thickness and width.

gauging plaster. *See* HEMIHYDRATE PLASTER.

gaul. A hollow formed in the FINISHING COAT of plaster.

Gaussian distribution (normal distribution). In statistics, a theoretical pattern of distribution that conforms with the actual scatter of many observed results. When the pattern is plotted as a FREQUENCY DISTRIBUTION it gives a characteristic bell-shaped curve.

gazebo. A summer house or other small garden building that is sited so as to give a pleasant view.

G-cramp. A clamping device in the shape of the letter G. It is used in joinery to hold items together while they are glued.

generator. *See* DYNAMO.

geodesy (geodetic surveying). A SURVEY of land on a large scale so that the curvature of the earth must be allowed for. *Compare* PLANE SURVEYING.

geodetic construction. *See* STRESSED-SKIN CONSTRUCTION.

geodetic surveying. *See* GEODESY.

geology. The scientific study of the origin and the composition of the Earth's crust.

geometric stair. A form of STAIR with a continuous STRING around a curve. It has no newel posts or landings.

Georgian glass. *See* WIRE GLASS.

geotechnics. The study of the engineering properties of the ground.

German siding (novelty siding). A form of WEATHER-BOARD where the lower board fits behind a rebate in the upper board. It is similar to SHIPLAP boarding.

geyser. A form of INSTANTANEOUS WATER HEATER that is installed above a fitting such as a sink.

GFCI. *abbr.* (*USA*) Ground fault circuit interrupter. A CIRCUIT BREAKER that monitors the current in a circuit and breaks the circuit if a leak to earth is detected.

gib board (Gibraltar board). Trade name for a form of GYPSUM PLASTERBOARD.

giga-. A standard prefix that can be placed in front of any SI unit to multiply the unit by 1,000,000,000. The abbreviated form is 'G' as in GJ for gigajoule. *See also* KILO-, MEGA-, MICRO-, MILLI-, TERA-.

gimlet (screw starter). A hand tool with a threaded point used to bore small holes in wood in readiness for a screw. Unlike an AWL the handle is at right angles to the point. *See also* TWIST GIMLET.

gimp pin. A small pin with a large flat head. The pins are often used to fix furniture fabrics to timber frames.

girder. A large steel or concrete BEAM that supports a heavy load within a structure. *See also* CHORD, WEB.

girth (girt). The distance measured once around the circular border (circumference) of a round timber.

GJ. *abbr.* Gigajoule, where 1 GJ = 1,000,000,000 joules. *See also* GIGA-, JOULE.

GL. *abbr.* Ground level.

gland. In general, a flexible sleeve or washer that is compressed in a joint to make a seal. (1) (olive) The soft copper or brass ring used in a COMPRESSION JOINT. (2) A waterproof seal used where an electrical cable enters an appliance.

glare. A discomfort or a decrease in vision caused by an excessive range of brightness in the visual field. *See also* DISABILITY GLARE, DISCOMFORT GLARE, GLARE INDEX.

glare index. A numerical rating of GLARE that takes account of the strength of the source and the relative positions of the source and viewpoint. A recommended maximum glare index for office lighting is 19.

glass. A hard brittle material used to make translucent window panes. Glass contains a mixture of metal silicates obtained form raw materials such as sand and limestone. Glass is theoretically a strong material but its structural use is limited by brittleness unless it is made in threads (GLASS FIBRE) and combined with another material such as a binder. *See also* FLOAT GLASS, HEAT-RESISTANT GLASS, PLATE GLASS, SILICA, SODA ASH, TOUGHENED GLASS.

glass block (glass brick). A hollow block of glass used to build translucent partitions. *Compare* PAVEMENT LIGHT.

glass fibre. Flexible fibres of GLASS that can be used as reinforcement for materials, or to trap air for thermal insulation. *See* GLASS REINFORCED CONCRETE, GLASS REINFORCED PLASTIC.

glasspaper (emery cloth). A paper or cloth with an abrasive surface that is used to smooth surfaces such as wood. The abrasive surface is made from hard particles, not always glass, in different grades of fineness. *See also* SANDPAPER.

glass-reinforced concrete (GRC). A form of FIBRE-REINFORCED CONCRETE in which the addition of GLASS FIBRE makes a COMPOSITE material with improved strength and impact resistance. It is useful for concrete components with thin sections, such as cladding panels and pipes.

glass-reinforced plastic (GRP). A form of COMPOSITE material made by binding GLASS FIBRES with a synthetic resin. It is used to make building components such as rooflights and cladding panels.

glass stop. A form of GLAZING BEAD.

glass wool. *See* GLASS FIBRE.

Glauber's salt. Sodium sulphite decahydrate, $Na_2SO_4.10(H_2O)$, a EUTECTIC SALT.

glaze. (1) To fit glass into windows. (2) A coating applied to ceramic objects, such as bathroom fittings, before they are fired in the kiln. The glaze gives a glassy protective surface. *See also* SALT GLAZE, VITREOUS.

glaze coat. A thin transparent coat of coloured paint that has the effect of enhancing the colour beneath.

glazed ware. Drainage pipes and fittings that have been given a GLAZE, such as by adding common salt during firing in the kiln. *See also* CLAY WARE.

glazier's chisel. A form of PUTTY KNIFE shaped like a chisel.

glazier's putty. A flexible paste of WHITING and linseed oil that is used to make a waterproof seal between the outside of window glass and the frame. *See also* FILLET.

glazing. (1) The glass fitted into a window or LIGHT. (2) The general arrangement of glass in windows or the process of fitting the glass.

glazing bar (window bar). A timber or metal bar that supports and joins panes of glass within a window LIGHT. *See* WINDOW diagram.

glazing bead (glazing fillet). A small strip of wood, metal or plastic that holds the glass in a window frame.

glazing block (setting block). A small block of resilient material placed in the rebate of a window frame to position and support a sheet of glass.

glazing brad. *See* GLAZING SPRIG.

glazing compound. A flexible sealant used to hold glass in place. The term can include GLAZIER'S PUTTY.

glazing fillet. *See* GLAZING BEAD.

glazing gasket. An extruded strip of flexible material with a slot that is used to hold glass in a window. It can be used instead of a GLAZING BEAD.

glazing rebate. A REBATE that receives glass in window frames or glazing bars.

glazing sprig (glazing brad). A small headless nail used to hold glass in a frame while the GLAZIER'S PUTTY is applied and hardens. The sprigs are left buried beneath the bead of putty.

globe thermometer. A thermometer fixed inside a blackened metal globe. The device is sensitive to radiant heat and can be used to assess the RADIANT TEMPERATURE component of a COMFORT TEMPERATURE.

gloss. A surface finish, usually on paint, that reflects most light falling on it and appears shiny. *Compare* EGGSHELL, MATT.

GLS. *abbr.* General lighting service. *See* GLS LAMP.

GLS lamp. A FILAMENT LAMP in its common form as a light bulb. The code letters may be used on drawings.

glue. A liquid ADHESIVE used to bond pieces of material such as wood. Traditional glues are made from animal gelatine; modern glues are based on synthetic POLYMERS.

glue block. An ANGLE BLOCK in a stair.

glue line. The thin layer of GLUE between two joined surfaces.

GMS. *abbr.* Galvanized mild steel.

going. In a staircase, the horizontal depth of the TREAD that is stepped on. *Compare* RISE. *See* STAIRS diagram.

going rod. A rod used to set out the GOING of a flight of stairs.

gold size. An ADHESIVE paint used to fix gold leaf. The leaf is applied while the gold size is in a tacky state.

goose neck. (*USA*) *See* SWAN NECK.

gouge. (1) A CHISEL with a curved edge that is used for grooving and carving wood. (2) A tool used for carving stone.

gouge bit. A BIT with a rounded end.

gpm. *abbr.* Gallons per minute.

grab. An excavating device that is hung from a crane or other arm. A split and hinged bucket is dropped on to the material to be lifted.

grade. (1) (*USA*) *See* GRADIENT. (2) The process of clearing and levelling ground ready for construction. (3) The process of smoothing the surface of an unsealed road.

graded aggregate. AGGREGATE that contains specified proportions of different particle sizes in order to obtain a dense mix.

graded filter. Layers of gravel and sand arranged in order of fineness so that water flowing through one material does not carry particles into another material and clog the drainage through the filter. *See also* FILTRATION.

gradient (*USA*: grade). A measure of slope or inclination from the horizontal. It is usually expressed as one vertical unit divided by the number of horizontal units needed to give that vertical rise.

grading curve. A curve formed when the percentages of each group of particles in a sample, such as an AGGREGATE, are plotted against their size. The curve can be used to assess the FINENESS of a sample.

graduated courses. *See* DIMINISHING COURSES.

graffito. *See* SGRAFFITO.

grain. The visible pattern in wood caused by the arrangement of ANNUAL RINGS and fibres in the tree. *See also* INTERLOCKING GRAIN.

graining. The painting of surfaces in patterns that resemble the grain of wood or marble. The pattern is made in the wet coat of paint by the use of graining combs, rags and other implements. *See also* MARBLING, MOTTLING.

granite. An IGNEOUS ROCK composed of relatively large crystals of quartz, feldspar and other minerals. Granite has a high crushing strength.

granolithic concrete (granolithic screed). A hard-wearing concrete surface that contains AGGREGATE selected to give a durable and non-slip surface.

grate. A metal frame that holds the solid fuel in a fireplace or furnace.

graticule. *See* RETICULE.

gravel. A natural form of coarse AGGREGATE, often found as small rounded stones. *Compare* SHINGLE.

gravel board. A horizontal board fixed at the bottom of the vertical boards in CLOSE-BOARDED FENCING. The gravel board help to protect the end grain of the vertical boards from rot and is easier to replace if necessary.

gravel roof. (*USA*) A flat roof of BITUMEN material with a layer of GRAVEL on top to protect the bitumen from solar radiation.

gravity circulation. A heating system in which water moves through the pipes by natural CONVECTION currents when gravity causes colder water to fall and warmer water to rise. Some hot water systems use gravity circulation between the boiler and the storage cylinder. *Compare* FORCED CIRCULATION.

gravity dam. A massive form of DAM that is prevented from overturning by its own weight rather than by its shape.

gravity filter. A filter that passes water through the filtration material by pressure that is supplied by the height or 'head' of water rather than by a pump. The slow SAND FILTER is a form of gravity filter.

gravity main. A supply pipeline in which the water moves because of the gradient of the pipe.

GRC. *abbr.* GLASS REINFORCED CONCRETE.

grease trap. A form of drainage GULLEY designed to prevent grease from kitchens entering and blocking the drain. The grease solidifies in the trap and floats to the surface where it can be removed by raising a built-in tray.

green concrete (green mortar). Concrete (or mortar) that has set but is not fully hardened. It often has a dark, greenish colour that lasts several days.

greenhouse effect. An increase in temperature caused by radiant heat from the sun that enters through a glass window but does not escape. The shortwave radiation from the sun is absorbed by objects and re-radiated as longer-wave radiation that cannot pass through glass. The Earth's atmosphere produces a similar effect.

green mortar. *See* GREEN CONCRETE.

green timber. Timber that has not undergone SEASONING.

grid bearing. An angle on a map or drawing measured between a GRID LINE, usually north-south, and a required direction.

grid line. One of the lines drawn on a plan to form the GRID PLAN.

grid plan. A layout of straight lines drawn on a plan to give a square or rectangular layout that usually coincides with the edges of major features on the plan.

grillage foundation. A FOUNDATION designed to distribute concentrated loads such as beneath columns. One form consists of several layers of steel beams on a bed of concrete. The beams in each layer are laid at right angles to the next layer and covered in concrete.

grille. (1) A screen made from wood or metal strips that are crossed to form a grid that can be partially seen through. It may be used for decoration or for security such as protecting windows. (2) A grating through which air can pass, usually into a ventilation duct.

grinder. A rotating wheel with an abrasive surface that is used to sharpen the cutting edges of tools.

grindstone. A form of GRINDER wheel made from natural sandstone and turned at a slow speed to prevent overheating and spoiling of cutting edges.

grinning. The failure of a coat of paint to

cover an undercoat. The colour beneath is said to be 'grinning through'.

grit blasting. A method of cleaning discoloured concrete, stone or metal by abrasion with a spray of fine particles.

groin. (1) The curved edge formed by the intersection of two VAULTS. (2) *See* GROYNE.

grommet (grummet). A type of washer used to make a pipe connector watertight.

ground. (1) A general term to describe a surface before it is painted or plastered. (2) *See* EARTH. (3) *See* COMMON GROUND.

ground beam. A BEAM at ground level that supports the base of a wall. The beam usually rests on foundation piers at each end.

ground brush. A large round brush with projecting bristles. It is sometimes used to paint rapidly over large areas such as for a GROUND COAT.

ground coat. An opaque coat of paint that is covered by further treatment such as a GLAZE.

ground floor (*USA*: first floor). (*UK*) The floor nearest ground level.

ground heave. *See* HEAVE.

ground plate. The SOLE PLATE of a timber building frame.

ground plug. (*USA*) *See* THREE-PIN PLUG.

ground sill. *See* SOLE PLATE.

groundwater (interstitial water). Water that occurs naturally in soil or rocks at a level below the WATER TABLE. *See also* SPRING WATER.

groundwork. Construction work, such as excavation and drain laying, that takes place in the ground.

grout. (1) A material pressed into the joints between tiles or other components to make the surface complete. (2) A liquid SLURRY of cement or other material that is poured into the joints of brickwork for extra strength, or injected into the ground for improving the ground quality.

growth ring. *See* ANNULAR RING.

groyne. A low wall built from a river bank or seashore into the water to stabilize movement of the shore.

GRP. *abbr.* GLASS-REINFORCED PLASTIC, glass-reinforced polymer.

grub screw (set screw). A short headless screw with parallel sides. It is commonly used to fasten doorknobs on spindles to prevent them slipping.

grummet. *See* GROMMET.

GS. *abbr.* General structural. A marking found on some STRESS-GRADED TIMBER. *See also* STRENGTH CLASS.

guard board. *See* TOE BOARD.

guard rail. A hand rail fitted along the edge of a high level platform or walkway such as a scaffold.

gudgeon. (1) A pin for holding two blocks of stone together. (2) A hook or ring for holding a gate closed. *See also* GATE HOOK.

gullet. (1) A narrow trench dug to start a large earth or rock cutting. (2) The space between the cutting teeth in a saw or file.

gulley (gully). A ground-level entrance to a DRAIN into which pipes can empty and surface water can enter through a grating and TRAP. *See also* GREASE TRAP.

gully. *See* GULLEY.

gum. A sticky substance that oozes from certain trees and is soluble in water. It is similar to RESIN; the term 'gum' can include both gums and resins.

gunite (shotcrete, sprayed concrete). Concrete or mortar that is thrown on to a surface, such as formwork, by a high pressure jet of air.

gunmetal. (1) A cheap type of BRONZE that

contains a high percentage of COPPER. (2) Various forms of dark grey ALLOYS.

gunstock stile. *See* DIMINISHING STILE.

gusset. (1) A metal plate used to join or reinforce members of a structure. (2) A piece of sheet metal, often triangular, used to cover a junction in flexible metal roofing.

gutter. A rainwater channel built along the edge of a roof or road. It is designed to gently slope to a nearby drain. *See also* BOX GUTTER, EAVES GUTTER, SECRET GUTTER. *See* ROOF diagram.

gutter bearer (gutter board). One of a number of short timbers that support a BOX GUTTER at the edge of a roof.

gutter plate. (1) A wall plate below a metal gutter. (2) One side of a VALLEY GUTTER.

guy (guy rope). One of a number of securing ropes or cables that run outward from a tower or temporary structure, usually to the ground.

gymnosperms. Flowering plants whose seeds are unprotected by an ovary. Many species of SOFTWOOD trees are gymnosperms. *Compare* ANGIOSPERMS.

Gyproc. (*UK*) Trade name for a form of GYPSUM PLASTERBOARD.

gypsum. Calcium sulphate, $CaSO_4.2H_2O$. The material used to make GYPSUM PLASTERS. In the natural form it usually has two molecules of WATER OF CRYSTALLIZATION. *See also* HARD WATER.

gypsum lath (gypsum plank). Lengths of GYPSUM PLASTERBOARD about 600 mm wide used to line walls and ceilings.

gypsum plaster. Various types of PLASTER in common use that are made by heating GYPSUM to eliminate some or all of the WATER OF CRYSTALLIZATION. *See also* ANHYDROUS GYPSUM PLASTER, HEMIHYDRATE PLASTER, PLASTER OF PARIS.

gypsum plasterboard (gypsum wallboard, plasterboard, drywall). Various types of pre-formed building board made with a core of GYPSUM PLASTER sandwiched between two sheets of heavy paper. After installation the board may be skimmed with RETARDED HEMIHYDRATE PLASTER or decorated directly.

gypsum wallboard. *See* GYPSUM PLASTERBOARD.

H

ha. *abbr.* HECTARE.

HAC. *abbr.* HIGH-ALUMINA CEMENT.

hacking. (1) The process of making a surface rough to give better adhesion for plaster or rendering. (2) A course of RUBBLE WALLING that consists of alternate single stones on top of double stones. (3) Brickwork or stonework in which each course is slightly set in from the course above. *Compare* CORBEL.

hacking knife. A knife used to remove old PUTTY from a window frame before new glass is fixed.

hacksaw. A saw with a thin steel blade stretched across a frame. It is used for cutting metal and may be a handsaw or mechanical saw.

haematite (hematite, Fe_2O_3). One of the forms of iron oxide found in iron ore. *See also* MAGNETITE.

haft. The handle of a knife or other small tool.

ha-ha (haw-haw). A boundary formed by a ditch that allows an unobstructed view from one side.

hair. Animal hair that was traditionally used as reinforcement in lime plaster undercoats.

hair cracks. Fine cracks that run at random over the surface of paint, plaster or concrete but do not penetrate through the coat.

hair hygrometer. A form of HYGROMETER that uses changes in the length and tension of animal hair as a measure of HUMIDITY.

half bat. A half brick that is cut into two along its length.

half-brick wall. A wall with a thickness equal to the width of one brick. Such a wall is in STRETCHER BOND and is used in a CAVITY WALL. *Compare* WHOLE-BRICK WALL.

half hatchet. A form of HATCHET with a notch for pulling out nails.

half landing (half-space landing). A platform where a flight of STAIRS turns and continues in the opposite direction. The landing runs across the width of both stairs. *Compare* QUARTER LANDING.

half-sawn stone. Stone that has been cut with half the cost of the cutting charged to each side of the original stone.

half-span roof. *See* LEAN-TO ROOF.

half-timbered. In traditional buildings, a wall in which the spaces between a heavy timber frame are filled with brick, stone or other material.

halogen lamp. *See* TUNGSTEN-HALOGEN LAMP.

halved joint (cross-lap joint, halving). A joint between two timbers formed by cutting away half the thickness of each piece. *Compare* END-LAP JOINT. *See* TIMBER JOINTS diagram.

halving. *See* HALF JOINT.

hammer. A hand tool with a flat steel head fitted at right angles to a handle. The flat face of the head is often used to drive nails and the other end, or PEEN, may be rounded. *See also* CLAW HAMMER, CLUB HAMMER, JOI-

NER'S HAMMER, PEEN HAMMER. *See* WOODWORKING TOOLS diagram.

hammer beam. One of a pair of horizontal beams that project from a wall and carry vertical struts to support the roof.

hammer beam truss. A mediaeval timber roof TRUSS based on HAMMER BEAMS.

hammer finish. A painted finish that looks like metal after it has been hammered. *See also* HAMMERING.

hammer-headed chisel. A CHISEL with flat steel head used in masonry. It is struck with a hammer rather than a mallet.

hammer-headed key. *See* DOUBLE-DOVETAIL KEY.

hammering. The process of shaping or finishing the surface of metal by working with a hammer.

hand brace. A form of BRACE used by carpenters.

hand drill. A hand tool of relatively small diameter used for boring holes in wood and metal.

handed. *adj.* A door or window is right-handed or left-handed depending on the direction in which it swings open. In the United Kingdom a door is right-handed if it has hinges on the right when viewed from the side into which the door opens. This sense of door and window fittings varies.

hand float (skimming float). A flat wooden tool held in the hand and used for smoothing the finishing coat of plaster.

handrail. A horizontal bar at waist level that runs along the top of the BALUSTRADE of a balcony or stair. *See also* WREATH. *See* STAIRS diagram.

handrail bolt (joint bolt). A bolt that is threaded at both ends and used to connect two ends of a HANDRAIL.

handrail punch. A tool used to tighten the nut at the end of a HANDRAIL BOLT.

handsaw. A SAW held in the hand and often used by carpenters and joiners.

hang. (1) To fit a door or window into its frame by the use of hinges. (2) To stick wallpaper to a surface. (3) To suspend curtains.

hanger. (1) A vertical structural member, usually steel, that carries a suspended load such as a suspended ceiling. (2) A STIRRUP STRAP.

hanging gutter. An EAVES GUTTER that is fixed to the ends of rafters or to the fascia.

hanging sash. *See* SASH WINDOW.

hanging stile. The STILE on which the hinges of a door or casement window are fixed.

hardboard. A type of FIBRE BOARD with a hard smooth surface.

hard burnt. *adj.* (1) Describing a clay brick or tile that has been fired at high temperature in the kiln. The firing gives high compressive strength and low absorption. *See also* VITREOUS. (2) Describing a hard type of plaster, such as KEENE'S CEMENT.

hardcore. Lumps of hard stone, old brick and concrete used to form a stable base beneath pathways and roads.

hard dry. The state of a paint film when it is dried throughout its thickness and another coat can be applied.

hardener. An additive that speeds the setting and hardening of synthetic resins such as adhesives.

hardening. (1) The gain in strength that occurs in concrete after the INITIAL SETTING TIME. (2) A process that increases the HARDNESS of metals. Steel, for example, can be hardened by heating and rapid cooling.

hard finish. A finishing coat of HARD PLASTER.

hardness. (1) A mechanical property of materials that provides resistance to indentation, cutting and abrasion. *See also* BRINELL HARDNESS TEST, MOHR'S HARDNESS SCALE,

ROCKWELL HARDNESS TEST, SCLEROMETER. (2) *See* HARD WATER.

hardpan. A compacted soil that has become cement-like in nature and is difficult to excavate.

hard plaster. A form of plaster with a high resistance to impact, such as cement and sand rendering or KEENE'S CEMENT. Lime plasters are softer than gypsum plasters.

hard putty. *See* HARD STOPPING.

hard solder. A form of SOLDER that contains copper and melts at a higher temperature than SOFT SOLDER. *See also* BRAZING.

hard standing. A surface suitable for parking vehicles.

hard stopping (hard putty). A paste used to fill deep holes and cracks ready for painting. Traditional mixes contain plaster of Paris while modern materials use synthetic resins.

hardware (ironmongery). General term for small metal fittings such as those used for doors and windows.

hard water. Water that contains certain dissolved salts, mainly of calcium and magnesium, which make it difficult to obtain a lather with soap. Natural waters tend to be hard if they are collected over areas of LIMESTONE or GYPSUM rocks. If the soluble salts are deposited in pipes or boilers they cause FURRING. Hard water can be treated by various forms of WATER SOFTENING. *See also* PERMANENT HARDNESS, TEMPORARY HARDNESS. *Compare* SOFT WATER.

hardwood. Wood from trees that are usually broad-leaved and deciduous. Not all hardwood trees produce timber that is hard. *Compare* SOFTWOOD. Common hardwood trees include ASH, BEECH, ELM and OAK.

harl, harling. *See* ROUGH CAST.

harmonic. *See* OVERTONE.

hasp. A hinged plate with a slot used for fastening in a HASP AND STAPLE.

hasp and staple. A fastener for securing door and gates. It consists of a HASP plate that fits over a STAPLE and is locked by a pin or padlock.

hatchet. A small axe that can be used with one hand. It is used for splitting or rough-dressing timber.

hatching (cross hatching). In construction drawing, a pattern of parallel lines used to fill areas and indicate different materials.

haunch. (1) The thickened base part of a projecting TENON in a HAUNCHED TENON. (2) The side of an arch or the edge of a roadway.

haunched tenon. A type of TENON where the projecting tongue of wood is thicker at the base than at the tip.

haunching. (1) Concrete that is poured around a buried stoneware pipe to support it. (2) A MORTISE that receives a HAUNCHED TENON.

haw-haw. *See* HA-HA.

hawk (mortar board). A square board with a handle beneath, used to carry mortar or plaster.

HD bolt. *abbr.* HOLDING-DOWN BOLT.

head. (1) The upper horizontal part of a timber door frame, window frame or opening in a stud wall. (2) The hitting end of a hammer. (3) A measure of the energy content of a fluid due to its height, pressure and velocity. The term is often used to mean the height of the free water level in a tank compared to a reference point such as tap. *See also* STATIC HEAD.

head casing. The part of the external ARCHITRAVE that is above a door.

header. A brick laid lengthwise across a wall so that the shorter end or 'head' is seen on the face of the wall. *Compare* STRETCHER. *See also* BOND.

header joist. (*USA*) *See* TRIMMER JOIST.

head flashing. A FLASHING that acts as a small GUTTER around a projection from a roof.

heading. A small pilot tunnel that is later enlarged to a full tunnel.

heading bond. A brickwork BOND made only of HEADERS. It is particularly used for curved walls.

heading course. A course of HEADERS in brickwork.

head joint. A CROSS JOINT in brick or stonework.

head nailing. A method of fixing SLATES with a nail hole near the top of the slate. Each nail is covered by two slates but the method is less stable for exposed sites than CENTRE NAILING.

headroom. The clear height to the ceiling, measured from the floor or the NOSING of stairs.

headwater. The source waters of a stream or river.

hearing threshold. *See* THRESHOLD OF HEARING.

heart. The wood at the centre of a log.

hearth. The slab of concrete or stone that is built beneath a FIREPLACE and projects into a room to protect the floor.

heartshake. A timber defect in the form of a radial SHAKE that runs from the heart of a log.

heartwood. Timber taken from the interior of a log where the wood is denser and usually darker than the outside SAPWOOD.

heat. One of the forms of ENERGY. The SI unit of heat is the JOULE. *See also* BRITISH THERMAL UNIT, CONDUCTION, CONVECTION, RADIATION. *Compare* TEMPERATURE. *See also* entries under THERMAL.

heat-absorbing glass. A type of window GLASS that passes most visible light but restricts the passage of heat RADIATION from the sun. It does not usually reduce heat CONDUCTION.

heat balance. The extra heat energy that must be supplied, or extracted, so that the total HEAT LOSSES from the building equal the total HEAT GAINS.

heat bridge. *See* COLD BRIDGE.

heat capacity. *See* THERMAL CAPACITY.

heat exchanger. A device with a large area of fins or other metal shapes that transfer heat to a surrounding fluid, such as air or water. A heat exchanger such as a CALORIFIER can be used for heating or for cooling. *See also* THERMAL WHEEL, WETBACK. *See* HEATING diagram.

heat gains. The heat energy that a building gains from the sun, from activities in the building and from appliances such as lighting. *See also* CASUAL HEAT GAIN, HEAT BALANCE, SOLAR HEAT GAIN.

heating. *See* HEATING diagram.

heating element. The part of an electric heater that gives off heat when an electric current flows in a special type of cable.

heating panel (radiator). A room heaters in a hot-water CENTRAL HEATING system. Most of the heat is transferred from the hot water by the processes of CONDUCTION and CONVECTION. *See* HEATING diagram.

heat insulation. *See* THERMAL INSULATION.

heat losses. In buildings, the heat energy that is lost by transmission through the fabric to the outside and by INFILTRATION of cooler air. *See also* FABRIC HEAT LOSS, HEAT BALANCE, U-VALUE.

heat of hydration. Heat given off during the HYDRATION of cement.

heat pump. A device that extracts heat from a low temperature source such as the ground and upgrades it to a higher temperature. It uses a standard REFRIGERATION CYCLE in which the EVAPORATOR collects the heat and the CONDENSER coils give off the heat. *See also* COEFFICIENT OF PERFORMANCE.

heat recovery. The general process of capturing and using heat that would otherwise be wasted, such as that contained in waste water or stale air. HEAT EXCHANGERS and HEAT PUMPS may be used.

heat-resistant glass. A form of GLASS with a low coefficient of expansion that allows the glass to be heated with a low risk of stress and breakage. Boro-silicate glass is a common form of heat-resistant glass; Pyrex is one tradename.

heat transfer. The movement of heat energy between objects or areas at different temperatures. *See* CONDUCTION, CONVECTION, RADIATION.

heat treatment. A method of altering the properties of metal by cycles of heating and cooling, such as ANNEALING and HARDENING.

heat wheel. A form of HEAT EXCHANGER sometimes used for pre-heating air in an air conditioning system. A rotating metal wheel of high thermal capacity absorbs heat from warm exhaust air during one part of its cycle and transfers the heat to cooler air in a nearby duct.

heave (ground heave). The rise in level of a soil surface, or foundation, caused by expansion of a soil such as CLAY. *Compare* SUBSIDENCE.

heavyweight concrete. *See* HIGH-DENSITY CONCRETE.

hectare (ha). A metric unit for large areas of land. 1 hectare = 10,000 square metres or approximately 2.5 ACRES.

heel. (1) The portion of a beam or rafter that rests on a support. (2) The bottom part of the HANGING STILE of a door. (3) The base of a dam or retaining wall, on the side of the structure in contact with the soil or water.

heel strap. A steel strap that joins a rafter to the tie beam of a roof truss.

helical hinge. A HINGE for a swing door.

helidon. A lamp that is moved in a controlled way to imitate the motion of the sun around a model of a building. It can be used to predict the pattern of direct sunlight falling on a building. *Compare* ARTIFICIAL SKY.

Helmholtz resonator. A CAVITY ABSORBER.

helve. The handle of an axe, sledge hammer or other heavy tool.

hematite. *See* HAEMATITE.

hemihydrate plaster (gauging plaster, $CaSO_4.1/2H_20$). A form of GYPSUM PLASTER made by heating GYPSUM ($CaSO_4.2H_20$) to drive out most of the WATER OF CRYSTALLIZATION. As 'plaster of Paris' it is very quick-setting and is usually treated with an additive such as KERATIN. This retarder slows setting and gives RETARDED HEMIHYDRATE PLASTER.

hemlock. A SOFTWOOD timber used for general construction purposes.

herringbone bond. A type of DIAGONAL BOND formed by laying each alternate course of bricks, or similar elements, in opposing directions to form a herringbone pattern.

herringbone strutting. *See* FLOOR STRUTTING.

hertz (Hz). The SI unit of FREQUENCY. 1 hertz = 1 cycle of vibration per second.

hessian (burlap). A coarse fabric woven out of jute or hemp fibre. It is used as wall decoration and to make reinforced fibrous plaster, asphalt materials and sacks.

hew. *vb.* To shape timber roughly with an axe or hatchet.

hew stone. Stone that has been roughly dressed by a hammer.

hex head. The head of a BOLT or NUT with six sides instead of four.

Hg. Chemical symbol for MERCURY.

H-hinge (parliament hinge, shutter hinge). A HINGE with plates that project on each side and make a shape like the letter H. The hinge is able to turn through 180° and is used for SHUTTERS and some doors. *See* DOOR diagram.

hickey. (*USA*) A portable tool used to bend steel conduits and other tubing.

hickory. A strong, flexible North American wood used for the handles of tools and similar applications.

hidden lines. In drawing, dashed lines that represent those edges of an object that are not seen from a particular viewpoint. *See also* WIRE FRAME.

hidden nailing. *See* SECRET FIXING.

hiding power. *See* COVERING CAPACITY.

high-alumina cement (ciment fondu, fondu, HAC). A rapid-hardening type of CEMENT that is made at higher temperature and contains more aluminium compounds than PORTLAND CEMENT. It can carry a load after 24 hours but is not now recommended for structural use as it may suffer from CONVERSION.

high-build paint. A paint that can give a dense coating with high COVERING CAPACITY.

high-calcium lime (fat lime). A form of LIME with a high content of calcium oxide. After SLAKING, high-calcium limes are mainly used for making mortars, renderings and putty.

high carbon steel. A form of plain CARBON STEEL with a carbon content of 0.5–1.5 per cent. High carbon steel is the hardest of plain carbon steels and generally used to make dies and cutting tools. *Compare* LOW CARBON STEEL, MILD STEEL.

high-density concrete (heavyweight concrete). CONCRETE made from high-density AGGREGATES such as iron ore and scrap iron.

highlighting. Techniques of painting or lighting that emphasize the raised parts of surface.

high-pressure sodium lamp. *See* SODIUM LAMP.

high-pressure system. A central heating system with small pipes in which hot water is circulated at pressures above 10 atmospheres. *Compare* LOW-PRESSURE SYSTEM.

high-strength friction-grip bolt. A high-tensile bolt used instead of rivets for fixing steelwork. The bolt is tightened against high-tensile steel nuts and washers so that the connected members are held by friction between their surfaces rather than by the bolt. *See also* LOAD INDICATING BOLT.

high-tensile steel. Steel that is used to construct bridges and similar structures. It has a higher YIELD POINT than MILD STEEL.

hinge. (1) A metal connector fixed between a door or window and its frame that allows the door or window to swing open and shut. *See* BAND-AND-STRAP HINGE, BUTT HINGE, H-HINGE, RISING BUTT HINGE, T-HINGE. (2) A joint in a structure that can rotate slightly and eliminate BENDING MOMENTS at the joint. *Compare* FIXED END.

hinge-bound door. A door that is difficult to close because the HINGES are set too deeply in the frames.

hip. The external angle formed at the intersection of two sloping roof surfaces, or where the triangular end of a HIPPED ROOF joins a slope of the main roof. *See* ROOF diagram. *Compare* VALLEY.

hip capping. The strip of roofing felt or other material fixed over a roof HIP.

hip hook. *See* HIP IRON.

hip iron (hip hook). A curved metal bar fixed to the bottom of the HIP RAFTER. It holds the final HIP TILE in place and prevents the line of hip tiles from sliding downward.

hip knob. An ornament, such as a FINIAL, fixed to the top of a roof HIP.

hipped gable roof. A roof that has a HIP running part of the way down from the ridge to a GABLE.

hipped roof. A roof with sloping ends instead of vertical GABLES. The sloping 'hipped ends' are triangular.

hip rafter. A rafter that forms a HIP on to which the JACK RAFTERS are fixed.

hip tile. A angled tile that covers the edges of roof tiles at the join in roof surfaces of a HIP. The tiles are usually made of clay or concrete with a sharp-angled or rounded shape. *See also* HIP IRON.

histogram. A statistical diagram in which the FREQUENCY DISTRIBUTION of quantities is shown as columns of different areas.

HMSO. *abbr.* (*UK*) Her Majesty's Stationery Office.

hod. A tray shaped like a box cut in two across the diagonal. It is fixed to a long handle and used to carry bricks or mortar.

hod carrier. A bricklayer's assistant who carries bricks and mortar in a hod.

hog. *See* CAMBER.

hoggin. Graded gravel or a fine BALLAST that is used as a base for paths and roads.

hogging. The formation of a CAMBER on a surface or structure. *Compare* SAGGING.

hogsback tile. A RIDGE TILE with a section that is parabolic rather than semicircular.

hog wire. (*USA*) Wire used to provide REINFORCEMENT in concrete.

hoist. A device that raises and lowers a load by a cable passed over a pulley. The cable is wound and unwound on a motor-driven drum.

holderbat. A circular fixing that clamps around a pipe and holds it clear of a surface.

holding-down bolt. A bolt cast into concrete foundations and used as an ANCHOR for fixing steel and timber elements to the base.

holding-down clip. A clip used to secure FLEXIBLE-METAL roofing to the roof deck. The top of the clip is shaped like a CAPPING.

hole saw (annular bit, tubular saw). A drilling tube with serrated edges used to cut a large circular REBATE in wood or metal.

holidays (skips). Areas of decoration accidentally left unpainted.

holing. The punching of holes in SLATES.

hollow. A surface rounded inward into a concave shape.

hollow-backed flooring. Floorboards with their undersides hollowed out to give firmer fixing and extra ventilation.

hollow bed. A BED JOINT in brickwork or stonework that is not filled in the middle.

hollow block. *See* CELLULAR BLOCK.

hollow clay tile. *See* CELLULAR BLOCK.

hollow-core door. A type of FLUSH DOOR where the facing surface of plywood or hardboard is attached to a framework or skeleton. It is lighter and cheaper than a SOLID DOOR.

hollow partition. (1) A PARTITION built of CELLULAR BLOCKS. (2) A discontinuous partition constructed with two leaves of brick or plasterboard.

hollow roll (seam roll). A joint along the fall of a slope in FLEXIBLE-METAL roofing. Adjacent sheets are joined together and bent around in a roll.

hollow-tile floor (pot floor). A concrete floor that uses hollow tiles or blocks to reduce the volume and weight of concrete in the floor. A common construction uses rows of precast beams spanned by hollow blocks and covered by a layer of concrete.

hollow wall. (1) *See* CAVITY WALL. (2) (*USA*) A wall built as two leaves joined together by bricks rather than by wall ties.

hollow-web girder. *See* BOX BEAM.

hone (oilstone, whetstone). A stone of fine quartz or silicon carbide used to sharpen the final edge of a cutting tool.

honeycomb. A repetitive pattern of voids in a material or structure.

honeycomb fire damper. A method of automatically closing a ventilation duct in the case of fire to give COMPARTMENTATION. The honeycomb damper uses an INTUMESCENT PAINT that froths when heated and sets into a foam many times its original volume.

honeycomb wall. A form of HALF-BRICK WALL built with gaps between the ends of stretchers. The gaps allow air to pass through the wall. This construction is often used for SLEEPER WALLS beneath suspended timber floors.

honing gauge. A clamp that holds a CHISEL at a fixed angle while it is sharpened on a HONE.

hood. A projection, such as a DRIPSTONE, set above a window or other opening in order to deflect water.

hook and eye. A CABIN HOOK and SCREW EYE used to fasten a door or window.

hook bolt. A galvanized BOLT that is bent into a U-shaped bend at the unthreaded end. It is used to fix roof sheets to the angled edges of steel rafters or purlins.

Hooke's law. The principle that, for materials below their ELASTIC LIMIT the deformation produced by a load is proportional to the size of that load. A structural expression of Hooke's law is that STRAIN is proportional to the applied STRESS.

hook joint. An airtight type of joint used between the edges of adjacent CASEMENT doors and windows. Each edge is cut with a profile in the shape of a letter F and the edges lock together.

hook strip. A BATTEN fixed on a wall as a base for cup hooks.

hoop iron. Thin strips of iron, steel or other material used to bind packages of bricks or blocks and sometimes used to reinforce brickwork joints.

hoop stress. *See* RING TENSION.

hopper. The triangular window panes at the sides of a HOPPER LIGHT.

hopper head. *See* RAINWATER HEAD.

hopper light. An area of a window hinged at the bottom and opening inward. It usually has triangular HOPPERS on each side of the opening to help prevent draughts.

horizon. A plane at right angles to a PLUMB LINE.

horizontal circle. The circular scaled plate on a THEODOLITE that is used to read angles.

horizontal control. In land surveying, the connection of points with known TRIANGULATION values.

horizontal shore. *See* FLYING SHORE.

horn. An extension of a timber member at a joint in a frame, such as in a window. It protects the joint during transport and may be cut off on site or used to give extra fixing. *See* WINDOW diagram.

horse. A temporary wooden frame such as a TRESTLE.

horsepower (HP). A non-metric unit of POWER. 1 horsepower = 746 WATTS.

hose cock. A water tap, usually outside a building, with a fitting that will connect to a hose pipe.

hose reel. A length of hose tubing wound on a circular frame. When it is part of fire-fighting equipment it is permanently attached to a water supply.

hot-air seasoning. The use of a KILN to dry timber.

hot-air stripper. A hand tool, like a hair drier, that blows electrically-heated air onto painted surfaces, used for BURNING OFF old paint. Its lower temperature is safer than a BLOW LAMP.

hot-dip coating. The application of a protective coating to a component by dipping it in molten metal such as zinc. *See also* GALVANIZING.

hot pressing. A method of gluing materials such as plywood between the heated plates of a press. It is often used with THERMOSETTING glues.

hot rolling. A method of HOT WORKING such as that used for making steel beams. Hot steel bars are passed through pairs of heavy rollers to give them their final shape. *See also* ROLLED-STEEL JOISTS, UNIVERSAL BEAMS. *Compare* COLD ROLLING.

hot-water cylinder. A CYLINDER used for storing hot water. It often contains an IMMERSION HEATER or HEAT EXCHANGER. *See* HEATING diagram.

hot-wire anemometer. An ELECTRICAL form of ANEMOMETER that can measure relatively low-speed currents of air, such as those caused by convection in rooms. The movement of the air is detected by its cooling effect on an electrically-heated probe.

hot working. The shaping of metal objects at a temperature high enough to avoid the stresses that may occur with COLD WORKING. *See also* HOT ROLLING.

housed joint (dado joint). A joint between timbers formed by sinking one member into a shallow 'housing' in the other. It is used for the treads of a stair.

house drain. A drain that takes waste water or sewage from a house to the public SEWER, or other point of disposal.

housed string. *See* CLOSED STRING.

house longhornbeetle (*Hylotrupes bajulus*). An insect, about 15 mm long, that is destructive to timber, especially sapwood. It spends much of its life cycle as a larva that bores tunnels and feeds within the wood. It leaves FLIGHT HOLES of 6–10 mm in diameter. *See also* BEETLE.

housemaid's sink. *See* BUCKET SINK.

housing. The shallow hole or groove in a HOUSED JOINT between timbers.

Howe truss. A type of roof TRUSS with intermediate members that are upright and usually made of steel.

HP. *abbr.* HORSEPOWER.

HTHW. *abbr.* High temperature hot water.

H-type gasket. A type of GASKET used in glazing. It has a cross-section in the shape of a letter H.

hub. (*USA*) The bell-shaped end of a pipe that has been enlarged to receive the straight end of another pipe.

HUD. *abbr.* (*USA*) Department of Housing and Urban Development.

hue. The basic tint of a COLOUR. The appearance of a hue also depends on its intensity or SATURATION.

humidifier. The part of an AIR-CONDITIONING system that, if necessary, adds moisture to the air and increases the HUMIDITY.

humidistat. A device that reacts to changes in HUMIDITY and is used to control air-conditioning plant.

humidity. The presence of moisture in the air in the form of water VAPOUR. It may be specified by measurements such as ABSOLUTE HUMIDITY, DEW POINT, RELATIVE HUMIDITY, VAPOUR PRESSURE. *See also* SATURATED AIR.

humus. (1) A dark organic material found in topsoil. It is produced by rotting vegetation and is not suitable for foundations. (2) Solid material that flows out in the EFFLUENT from sewage filters.

hung ceiling. *See* SUSPENDED CEILING.

hungry. A surface that absorbs too much paint and causes the paint coat to be thin and non-uniform.

hurlinge. A form of BUTT HINGE where one flap is smaller than the other and closes into the larger flap. The stile and frame do not need rebates cut for the hinge.

HVAC. *abbr.* Heating, ventilation and air conditioning.

hydrant. A permanent outlet on the water main, such as a FIRE HYDRANT.

hydrated lime. *See* SLAKED LIME.

hydration. The chemical reaction between water and a material such as cement powder. During the process the cement paste sets and hardens while heat is given off.

hydraulic cement. A form of CEMENT that can harden under water, like PORTLAND CEMENT.

hydraulic friction. The resistance to the flow of fluid in a pipe or channel caused by roughness of surface and obstructions. It may be expressed by the LOSS OF HEAD.

hydraulic gradient. For water moving in a pipe or channel, the difference in the height or pressure HEAD between the two ends divided by the horizontal length of the flow.

hydraulicity. The ability of certain cements, limes and mortars to harden under water.

hydraulic jump. A sudden change of depth when fluid flowing in a channel undergoes changes in velocity. A WEIR produces an artificial jump that may be used to measure the flowrate of a river.

hydraulic lift. A passenger or goods LIFT that is raised by the action of a ram pushing from beneath, rather than lifted by a winding cable. The ram is operated by a pressurized liquid.

hydraulic lime. A type of LIME containing aluminium silicate that hardens by combining with water. *Compare* HIGH-CALCIUM LIME.

hydraulic mean depth. For fluid flowing in a channel, a measure of fluid contact with the channel surface. It is expressed as the cross-section of the flow divided by the wetted perimeter of the channel.

hydraulics. The study of flow in FLUIDS, such as water and air. Hydraulics is usually concerned with practical applications while 'hydrodynamics' is the theoretical study of fluid behaviour. *Compare* HYDROSTATICS.

hydraulic test (water test). A DRAIN TEST in which the drain is filled with a set head of water and the level observed for an hour to detect any changes.

hydrocarbon. A chemical compound that contains only carbon and hydrogen. Petroleum oils and other products are hydrocarbons. *See also* FLUORINATED HYDROCARBON, METHANE.

hydrodynamics. *See* HYDRAULICS.

hydro-electric power station. A POWER STATION that uses the energy of moving water to turn the TURBINES. The water is usually stored behind a large DAM.

hydrogen ion. A positively-charged hydrogen atom. The presence and concentration of hydrogen ions are responsible for the properties of ACIDS.

hydrograph. A graph that shows variations in the DISCHARGE rate of a body of water.

hydrological cycle. A sequence of processes by which a certain proportion of the world's natural water evaporates and returns as rainfall.

hydrometer. An instrument that measures RELATIVE DENSITY.

hydrostatics. The study of pressures and forces in FLUIDS that are at rest. *Compare* HYDRAULICS.

hygrograph. A device that records changes in HUMIDITY.

hygrometer (psychrometer). An instrument for measuring water vapour in the air (HUMIDITY). *See also* HAIR HYGROMETER, REGAULT HYGROMETER, WET BULB, WHIRLING SLING.

hygrometry. The measurement of HUMIDITY in the air by the use of HYGROMETERS.

hygroscopic. The readiness of some materials to absorb moisture, sometimes directly from the air.

hygroscopic water. A film of moisture that surrounds the particles of a soil.

hypalon. (*USA*) A rubber-based roof coating.

hyperbolic paraboloid roof. A roof of

SHELL CONSTRUCTION that covers a square but has the shape of a two curves that cross. Although the curving is continuous the roof can be constructed from straight structural members.

hyphae. The fine branching strands that make up the body of a FUNGUS.

Hz. *abbr.* HERTZ.

I

I. *abbr.* (1) Electrical CURRENT. (2) *See* MOMENT OF INERTIA.

IAAS. *abbr.* (*UK*) Incorporated Association of Architects and Surveyors

IC. *abbr.* Inspection chamber (used in drainage).

ICBO. *abbr.* (*USA*) International Conference of Building Officials.

ICE. *abbr.* (*UK*) Institute of Civil Engineers.

ID. *abbr.* Inside diameter.

idle time. Time during which a worker attends and is paid but has to wait for better weather, supplies of materials or for another trade to finish.

IED. *abbr.* INTEGRATED ENVIRONMENTAL DESIGN.

IEE. *abbr.* (*UK*) Institution of Electrical Engineers.

igneous rocks. Rocks, such as granite, that come from molten 'magma' forced out from the mantle of the earth. The molten material sometimes forms an igneous 'intrusion' into spaces between existing rocks. *Compare* METAMORPHIC ROCKS, SEDIMENTARY ROCKS.

ignitability. The ability of a material to start burning when a small flame is applied.

ignition. The process that causes COMBUSTION to start.

IL. *abbr.* INVERT level (used in drainage).

illuminance. The amount of light falling on a surface. It is measured as the density of LUMINOUS FLUX on an area. The SI unit of illuminance is the LUX. A non-metric unit is the FOOT CANDLE.

illuminance meter. *See* LIGHT METER.

illuminant. The light falling on an object or the light source itself.

illumination. The effect of light falling on a surface which then becomes illuminated. The amount of illumination is specified by the ILLUMINANCE.

imbex. *See* ITALIAN TILES.

imbricated. A surface covered with overlapping pieces like tiles or slates.

IMechE. *abbr.* (*UK*) Institution of Mechanical Engineers.

IMinE. *abbr.* (*UK*) Institution of Mining Engineers.

immersion heater. An electrical heating element set inside a HOT WATER CYLINDER. It may be the only means of heating water in the cylinder or it may supplement the heat from a boiler.

impact load. An imposed or LIVE LOAD with an increased effect because it is applied suddenly, such as the load produced by a dropped object. *Compare* STATIC LOAD.

impact sound. In the study of sound insulation, sound that is generated on the partition being assessed. Footsteps on a floor are a type of impact sound. *See also* AIRBORNE SOUND.

impact strength. The ability of a material to

withstand sudden loads. The opposite of BRITTLENESS.

impedance. Opposition to the flow of ALTERNATING CURRENT. It can include ELECTRICAL RESISTANCE.

impermeability. The resistance of a material to the passage of liquid or gas through its structure. *See also* PERMEABILITY.

impermeability factor (run-off coefficient). A measure of the rainwater that runs off a surface. It is the ratio of the amount of rain collected from a surface compared to that which falls on it.

imposed load. *See* LIVE LOAD.

impost. The top of a pillar or pier that carries an arch.

impregnation. (1) In general, the process of saturating a material. (2) A timber treatment that uses pressure or vacuum to force preservative into the pores of the wood.

improved wood. Wood that has been treated or laminated in order to improve its strength or other properties.

improvement line. The proposed edge of a new or realigned road which may affect the BUILDING LINE.

inbark. BARK that has become embedded within wood with growth of the tree.

incandescence. The state of a material that gives off light when heated to high temperature.

incandescent lamp. A lamp, such as the common FILAMENT LAMP, that produces light by heating a material to a temperature at which it glows. *Compare* DISCHARGE LAMP.

incentive system. A wages system where an extra 'bonus' payment is made for work that is faster or of higher quality.

incinerator. A furnace for burning rubbish.

incise. The cutting or carving of wood or stone.

inclined shore. *See* RAKING SHORE.

inclinometer. *See* CLINOMETER.

incombustible. *See* NON-COMBUSTIBLE.

increaser. A coupling that increases in diameter and is used for joining pipes of different sizes.

incrustation. (1) The hard deposits left on hot water pipes, tanks and boilers that are FURRED. (2) A layer of airborne waste that builds up on walls in industrial areas.

indent. A gap left in a course of brickwork or stonework so that a BOND can be made with future work. *See also* TOOTHING.

indented joint. A joint that lengthens timbers by notching the wood to match a fish plate connector bolted to the two timbers.

indenting. *See* TOOTHING.

indenture. A contract or legal agreement. The traditional indenture between an APPRENTICE and 'master' has matching shapes in the documents.

independent float. A type of FLOAT TIME in CRITICAL PATH ANALYSIS. It is the spare time available for a particular activity that does not affect the float time of other activities.

indeterminate frame. A frame that contains some redundant structural members. The stresses in the frame are therefore difficult to calculate.

indicator bolt. A type of door BOLT with a slider that shows whether the room is vacant or engaged. It is used for WC cubicles.

indirect cylinder. A hot water CYLINDER where the stored water is heated by contact with a HEAT EXCHANGER rather than directly heated in the boiler. The arrangement prevents FURRING by hard water. *Compare* DIRECT CYLINDER. *See also* PRIMARY CIRCUIT, SECONDARY CIRCUIT.

indirect heating. *See* CENTRAL HEATING.

indirect light. Light that reaches a surface largely or completely by reflection, rather than as DIRECT LIGHT. For example, COVE LIGHTING is reflected from a ceiling. Indirect lighting may be measured by the VECTOR/SCALAR RATIO.

industrialized building (system building). A method of planning and construction using standard-sized components that are prefabricated where possible. The aim is to reduce costs and to improve reliability and speed of construction.

inert filler. A FILLER in paints or plastics that does not react chemically with other ingredients.

inert gases. *See* NOBLE GASES.

inertia. (1) In general, the natural tendency of a body to oppose any change to its state of rest or state of uniform motion. (2) The resistance of a BEAM to bending. *See also* MOMENT OF INERTIA.

inert pigment. A PIGMENT in paints or plastics that does not react chemically with other ingredients.

infill (infilling). (1) Material installed inside a framework or partition to give insulation, fire resistance, stiffness and weather protection. Brickwork is a traditional infill material while lightweight materials may now be used. (2) The addition of a new building between existing properties.

infiltration. (1) The entry of air into a building for VENTILATION, usually in an uncontrolled way. (2) The process of rainwater soaking into the ground. *Compare* RUN OFF.

inflammable. *adj. See* FLAMMABLE. The term 'inflammable' is not recommended as it may cause confusion in an area where confusion can be dangerous.

inflexion point. *See* CONTRAFLEXURE.

infra-red drying. *See* STOVING.

infra-red radiation. A form of electromagnetic radiation with wavelengths slightly greater than those of red light. It can be felt as heat RADIATION from the sun or from RADIANT HEATING devices. Although infra red radiation is not visible, it can be detected by photographic film and other devices that are then used to see heat patterns. *Compare* ULTRA-VIOLET RADIATION.

ingo. The REVEAL of a window or door.

ingot. A piece of metal cast into a simple shape while molten.

ingrown bark. *See* INBARK.

inhibiting pigment. A PIGMENT in a paint that is applied directly to metals to help prevent corrosion. Zinc and lead compounds have been used, although the use of lead is now restricted for health reasons. *See also* RED OXIDE.

inhibitor. A material added to compounds such as paints in order to prevent or delay a chemical reaction such as oxidation. Inhibitors are added to paints, for example, to help prevent skinning.

initial setting. The process that takes place when freshly-mixed concrete or mortar becomes stiff. The 'initial setting time' is the time taken for a standard paste to become stiff. *See* VICAT APPARATUS.

injection moulding. A simple method used to make PLASTIC articles by melting the raw material and passing it into a mould.

inlaid parquet. PARQUET flooring that is glued to a wood backing before being fixed to the floor.

inlay. A decorative pattern made on a surface by the insertion of a contrasting material into the background material in a way that keeps the surface smooth. It is usually applied to wood and LINOLEUM. *See also* MARQUETRY, VENEER.

inorganic. Materials that are based on MINERALS rather than on carbon which is the base of organic compounds.

input system. In general, a ventilation system where the air supply is pumped into a room rather than sucked from it. *Compare* EXTRACT SYSTEM.

insert. (1) A plug or fastening cast into concrete as a fixing for further construction. (2)

A VENEER of wood that fills a knot hole or other defect in plywood.

inside glazing. Glass that is inserted into the window frames and secured from the inside rather than from the outside of a wall. *Compare* OUTSIDE GLAZING.

inside trim. The ARCHITRAVE around the inside of door and window openings.

in situ. *adj.* Describing work done on the building site rather than prefabricated at a factory.

inspection chamber. A shallow type of MANHOLE installed at junctions and bends in a drain to allow clearance of blockages. The chamber may be built of brick, or be precast in concrete or plastic.

inspection eye. *See* ACCESS EYE.

inspection junction. A short branch from a large drain to the surface. It can be used to inspect the flow.

instantaneous water heater. A device that heats water only when the hot water tap is turned on. The water can be heated by gas burner or electric element and no storage tank is needed. *See also* MULTI-POINT HEATER.

insulating materials. In general, materials that provide a significant amount of INSULATION. (1) Heat insulating materials are usually porous and lightweight. *See also* THERMAL INSULATION. (2) Insulating materials for AIRBORNE SOUND are usually heavyweight rather than lightweight. *See also* SOUND REDUCTION INDEX. (3) Electrical insulating materials, such as plastic, have few FREE ELECTRONS. *See also* ELECTRICAL INSULATION.

insulation. A material or method that restricts the transmission of heat, sound, vibration or electricity. *See also* ELECTRICAL INSULATION, SOUND INSULATION, THERMAL INSULATION.

insulator. A material that provides INSULATION. *Compare* CONDUCTOR.

intaglio. A design engraved into a hard surface, such as on an intaglio tile.

integral waterproofing. The WATERPROOFING of concrete by the design of the mix or by the use of ADMIXTURES in the concrete rather than by a separate membrane.

integrated environmental design (IED). A method of designing buildings in which each specialization is involved in allied disciplines and in the overall design. IED helps give better designs and quicker results at no additional cost.

integrity. In FIRE RESISTANCE, the completeness of a structure. The absence of cracks and gaps that might allow flames to penetrate the structure.

intensity. (1) Intensity of colour, *see* SATURATION. (2) Intensity of light sources, *see* LUMINOUS INTENSITY. (3) Intensity of sound, *see* SOUND LEVEL.

intercept. In surveying, the length of a staff that is visible between the two STADIA HAIRS of a telescope.

interceptor (disconnecting trap). A form of TRAP installed between a drain and the sewer. The trap was designed to keep the sewer air separate from the building drain, but they are not used in modern systems.

interface. The junction between the surfaces of two different materials or components.

interference. The effect when wave motions are added or subtracted. Interference in sound can lead to 'beat' notes. The colour patterns seen on the surface of thin films of oil or digital recording disks are caused by interference of light waves.

interfering float. A type of FLOAT TIME in CRITICAL PATH analysis. It is the difference between the total float time and the total spare time (FREE FLOAT).

intergrown knot. *See* LIVE KNOT.

interim certificate. A document that authorizes partial payment to the CONTRACTOR during the construction of a building. *Compare* FINAL CERTIFICATE.

interior plywood (interior-grade plywood). A type of PLYWOOD made with glue that is not resistant to moisture, unlike EXTERIOR PLYWOOD.

interlaced fencing (woven fencing). A fence formed from wide slats of thin timber that are flexible enough to be woven alongside one another.

interlocking grain. Wood GRAIN in which the fibres slowly change their slope in each successive ring.

interlocking joint. A joint in ASHLAR stonework where the projection from one stone fits a groove in the adjacent stone.

interlocking tile. *See* SINGLE-LAP TILE.

intermediate rafter. *See* COMMON RAFTER.

internal dormer. A vertical window that is recessed into a sloping roof rather than built out from the roof like a standard DORMER. The cheeks of the dormer are within the roof line and a flat space in front of the window must be waterproofed.

internally reflected component. *See* DAYLIGHT FACTOR COMPONENTS.

International Organization for Standardization (ISO). *See* STANDARDS ORGANIZATIONS.

international system of units. *See* SI UNITS.

interpolate. To estimate a value on a graph or table at a point between existing values. *Compare* EXTRAPOLATE.

intersection angle. A DEFLECTION angle in surveying.

interstitial condensation. A form of CONDENSATION in buildings where water vapour condenses to liquid within a building element such as a wall or roof. The resulting dampness can damage structural materials such as steelwork and make insulation less effective. *Compare* SURFACE CONDENSATION. *See also* VAPOUR BARRIER.

interstitial water. *See* GROUNDWATER.

intertie. *See* NOGGING.

intrados. The curved surface on the inside of an arch or vault. *Compare* EXTRADOS.

intrusion. *See* IGNEOUS ROCKS.

intumescence. The ability of certain materials to expand permanently when heated. Strips of intumescent materials are used, for example, around the edges of FIRE DOORS.

intumescent paint. A paint that, when heated, swells and gives a layer of insulation that may increase FIRE RESISTANCE.

inv. *abbr.* INVERT.

inverse square law. For a point source, the intensity of its effect decreases in inverse proportion to the square of the distance from the source. For example, increasing the distance three times reduces the effect by nine times. This law has effect in areas such as intensity of sound and illumination of surfaces.

invert. The lowest point in the bottom surface of a drain, sewer, tunnel or channel.

inverted roof (upside-down roof). A form of flat roof in which the rainproof layer and drainage system are actually underneath a layer of thermal insulation. The insulation must be of a suitable form such as an expanded plastic with a closed cell structure. This type of construction is often the result of upgrading the insulation of an existing roof. An inverted roof is also a WARM ROOF.

ion. An atom or group of atoms that have become electrically charged. The movement of ions plays a part in CONDUCTION and ELECTROLYSIS. *See also* GAS DISCHARGE LAMP, ION EXCHANGE, SMOKE DETECTOR.

ion exchange. A method of purifying water to give DE-IONIZED WATER. The dissolved chemicals in water form IONS which can then be removed.

ionic bond (electrovalent bond). A type of BOND between atoms where the complete transfer of electrons forms IONS whose charges attract one another. *Compare* COVALENT BOND.

ionization smoke detector. *See* SMOKE DETECTOR.

IR. *abbr.* Infra red. *See* INFRA-RED RADIATION.

iroko. A HARDWOOD timber often used as a substitute for TEAK.

iron. (1) A pure metal element (chemical symbol Fe) with a white silvery colour and high density. Its mechanical properties are greatly modified by small amounts of carbon and other metals. *See also* CARBON STEEL, CAST IRON, WROUGHT IRON. (2) A heavy tool or machine for smoothing ASPHALT. (3) *See* PLANE IRON.

iron core. A steel bar that supports a wooden handrail on the top of BALUSTERS.

ironmongery. *See* HARDWARE.

iron oxides. Chemical compounds of IRON and oxygen which are found naturally in ores. *See* HAEMATITE and MAGNETITE. Several types of reddish or dark paint PIGMENTS are based on iron oxides. *See also* RUST.

irregular coursed rubble. Walling built of RUBBLE arranged in COURSES of varying depth.

ISE. *abbr.* (*UK*) Institution of Structural Engineers.

ISL. *abbr.* Internal silvered lamp. A form of FILAMENT lamp that directs the light into a beam.

ISO. *abbr.* International Organization for Standardization. *See* STANDARDS ORGANIZATIONS.

isobar. A line on a map or chart that joins points at the same BAROMETRIC PRESSURE. It is commonly used for studying weather.

isocyanate. A toxic compound from which POLYURETHANES are formed. It is used as the basis of TWO-PACK paints and of adhesives like instant 'superglues'.

isolating strip. An EXPANSION STRIP.

isolator. Part of an electrical circuit, such as a copper bar, that can be removed to make the circuit inactive.

isometric projection. A form of drawing PROJECTION giving a three-dimensional view of a building in which the dimensions are shown in true proportions. All vertical lines remain vertical and horizontal lines are commonly drawn at 30° to the horizontal. *Compare* AXONOMETRIC PROJECTION, ORTHOGRAPHIC PROJECTION, PERSPECTIVE DRAWING.

isotherm. A line on a map or chart that joins points of the same temperature.

isotope. Atoms that are of the same element but have a different atomic mass. Isotopes are identical in chemical properties but some are unstable and show RADIOACTIVITY.

isotropic. *adj.* Describing substances that have the same physical properties in all directions. *Compare* ORTHOTROPIC.

IstructE. *abbr.* (*UK*) Institution of Structural Engineers.

Italian tiles (Spanish tiles, *USA*: mission tiles, pan-and-roll tiles). A system of roof TILES that uses two different shapes of tile. The channel tile on the bottom is flat while the over-tile or imbex is curved.

item. A description of work and material, or labour, that is listed in a BILL OF QUANTITIES together with an appropriate quantity. A space is left for a contractor to insert a proposed price alongside each item.

J

J. *abbr.* JOULE.

jack. A mechanical device that raises heavy loads for a short distance by pushing from beneath. Smaller jacks may be operated by a screw and larger devices by hydraulic ram.

jack arch. (1) See FLAT ARCH. (2) A small brick or concrete ARCH built on a steel rail or rolled-steel joist. It was formerly used to carry heavy loads or floors.

jacked pile. A PILE that is forced into the ground by jacking downward from an existing building in order to UNDERPIN it.

jack plane. A general-purpose type of PLANE used for the initial working of wood.

jack rafter. A short roof RAFTER fixed between a HIP RAFTER and eaves, or between a VALLEY and ridge. *See also* VALLEY JACK.

jalousie window. A type of WINDOW formed from overlapping horizontal glass slats that are rotated to admit air. *Compare* VENETIAN BLIND.

jamb. (1) The vertical edge of an opening in a wall that shows the full thickness of the wall. *See also* REVEAL. (2) A vertical member in the side of a door or window frame. *See* WINDOW diagram.

jamb lining. A lining that covers the edge of a wall JAMB.

japan. *See* BLACK JAPAN.

Japanese lacquer (Japanese wax). Type of LACQUER made from the berries of the sumac tree from Japan and China.

japanning. *See* BLACK JAPAN.

JCB. *abbr.* Trade name for a BACKACTOR multi-purpose excavator.

JCT. *abbr.* (*UK*) JOINT CONTRACTS TRIBUNAL.

jemmy. A type of small PINCH BAR.

jet freezing. A method of temporarily freezing a section of pipe in a central heating or water system so that pipework can be modified without draining. The equipment may use liquid nitrogen or carbon dioxide as the refrigerant.

jetty (jutty). (1) A deck built on piles that projects into the sea. (2) An upper floor, or other part of a building, that overhangs the base of the building.

jib (boom). A horizontal arm-like structure, such as on a crane. The end of the jib is used to support equipment for lifting or cutting.

jib crane. A CRANE that uses a JIB for lifting.

jib door. A door designed to fit flush into the wall.

jig. A device like a clamp that holds work securely, or guides tools to give consistent results during repetitive work.

jig saw. A power SAW that cuts with an up-and-down motion. It is used to cut sharper curves than a BAND SAW.

jobber. A semi-skilled person who can carry out most types of house repair including joinery, painting, plumbing, bricklaying and plastering.

jog. *See* OFFSET.

joggle. (1) In block or stonework, a projection on a block that fits a projection in an adjoining block. (2) The mortar that fills the recess between adjacent blocks. (3) A small TENON or HORN for a wooden joint.

joggle post. (*USA*) *See* KING POST.

joiner. A person who makes and installs JOINERY.

joiner's gauge. *See* MARKING GAUGE.

joinery. (1) The finished woodwork of a building, such as window and door frames, skirting boards, architraves. *Compare* CARPENTRY. (2) (*USA*) Finishing carpentry. *Compare* FINISH.

joint. In general, a connection made between two components. (1) The gap between adjacent bricks, blocks or stones that is filled with mortar. *See* BED JOINT, CROSS JOINT. (2) *See* MOVEMENT JOINT.

joint bolt. *See* HANDRAIL BOLT.

Joint Contracts Tribunal (JCT). (*UK*) A coordinating body of building organizations that include architects, contractors, local authorities and surveyors. It produces several standard forms of building CONTRACT that are widely used.

jointer. (1) A bricklayer's tool used to finish the surface of mortar JOINTS. (2) *See* JOINTER PLANE.

jointer plane (jointer). A long form of PLANE used to plane long boards.

jointer saw. A power saw for cutting stone.

joint fastener. *See* CORRUGATED FASTENER.

joint filler (joint sealant). A MASTIC type of compound used to seal the construction joints.

jointing. The process of trimming and shaping the surface of a mortar joint before it has fully hardened (GREEN MORTAR). *See also* WEATHER-STRUCK JOINT. *Compare* POINTING.

jointing compound. *See* JOINTING MATERIALS.

jointing materials (jointing compound, jointing tape). Waterproof materials worked into the threads of a pipe joint to complete the seal. Materials range from traditional pastes to PTFE tape.

jointing rule. A bricklayer's straight edge that is used with a JOINTER or FRENCHMAN.

jointing tape. *See* JOINTING MATERIALS.

jointless flooring (composition flooring). Various types of flooring laid without joints except those necessary for expansion or shrinkage. Materials used include ASPHALT, GRANOLITHIC CONCRETE, MAGNESITE, TERRAZZO and PITCH MASTIC.

joint rule. A steel RULE used by plasterers to form MITRES at junctions such as those of cornice mouldings. It has one end cut at an angle of 45°.

joint runner. A rope of fibrous material used to pack the outside of a pipe joint.

joint sealant. *See* JOINT FILLER.

joint tape. A strip of paper, or other material, used to cover joins between wallboards such as GYPSUM PLASTERBOARD.

joist. One of a series of parallel timber or steel BEAMS that support a floor, ceiling or flat roof. Timber joists may be termed common joists and steel joists may be termed rolled-steel joists (RSJs). *See* ROOF diagram. *See also* TRIMMED JOIST, TRIMMER, TRIMMING.

joist anchor. *See* WALL ANCHOR.

joist hanger (joist stirrup, wall hanger). A steel strap or plate that supports the end of a JOIST. It may be fixed into a wall or attached to a TRIMMING JOIST. *Compare* STIRRUP STRAP.

joist trimmer. A steel plate that is fixed on a JOIST and carries another joist at right angles. *See also* TRIMMER.

joule (J). The SI unit for any form of work or ENERGY including heat. 1 joule is defined as 1 newton of force multiplied by 1 metre of distance moved. *Compare* BRITISH THERMAL UNIT, KILOWATT HOUR.

journeyman, journeywoman. A person who has learned a TRADE as an APPRENTICE.

jumbo brick. (*USA*) A brick larger than the standard size.

jumper. (1) A temporary connection that bypasses part of an electrical circuit. (2) Inside a water TAP OR FAUCET, a brass part that carries the WASHER and lifts when the tap is turned on.

junction box (conduit box, distribution box). A box installed in an electrical installation. It is used to make joints between cables and to protect them. The box may also contain switches and fuses.

junction point. In surveying, a point on a curve where a circular part joins a straight or non-circular part.

jutty. *See* JETTY.

K

K. *abbr.* KELVIN.

k. *abbr.* (1) KILO-. (2) COEFFICIENT OF THERMAL CONDUCTIVITY. (3) RADIUS OF GYRATION.

kaolin (kaoline, china clay). A fine white CLAY used to make PORCELAIN.

KAR. *abbr.* KNOT AREA RATIO.

Kata thermometer. A form of ANEMOMETER that can measure relatively low-speed currents of air, such as those caused by convection in rooms. The movement of the air is detected by its cooling effect on a dry bulb thermometer.

kauri (kauri gum). A large conifer tree of New Zealand. Its fossilized gum was used in the manufacture of traditional varnish.

Keene's cement. *See* ANHYDROUS GYPSUM PLASTER.

keeper. (1) A socket or guide on a door jamb into which the bolt of a lock slides. (2) A metal loop on a THUMB LATCH that limits the movement of the latch.

kelvin (K). The unit of temperature in the ABSOLUTE TEMPERATURE SCALE. The symbol K is not followed by a degrees sign. For practical purposes 0°C = 273 K and 100°C = 373 K.

Kentish rag. *See* RAGSTONE.

keratin. A type of retarder that is added to HEMIHYDRATE PLASTER (Plaster of Paris) in order to slow setting. Keratin is obtained from protein in animal horns, hoofs and hair.

kerb. *See* CURB.

kerf. A cut made by a SAW. *See also* SET.

kerfed beam. A beam with some partial saw cuts (KERFING) that allow the beam to be bent.

kerfing. A method of making several saw cuts on one side of a piece of wood. The cuts do not pass right through the wood and they allow the piece to be bent. Kerfing is used for the risers of some stairs and in a KERFED BEAM.

kettle scale. *See* FURRING.

kevil. *See* CAVIL.

key. (1) A rough or serrated surface that gives a good mechanical BOND for a coat of material like adhesive, concrete, mortar, paint or plaster. (2) A small piece of hardwood inserted inside a joint to give extra strength. *See also* DOUBLE-DOVETAIL JOINT, FEATHER. (3) A tool used by bricklayers to make a KEYED JOINT.

key block. *See* KEYSTONE.

keydrop (dropped escutcheon). A small metal plate that covers the keyhole when the key is not in use. The keydrop usually pivots above the keyhole and drops down over the ESCUTCHEON.

keyed joint. The POINTING of a mortar joint so that it is concave between bricks. *See* BRICKWORK diagram. *See also* KEY.

keyhole saw. *See* PAD SAW.

keying-in. The insertion of new BONDS into an existing brick wall. The process is used to start a new wall.

key plan. *See* LOCATION PLAN.

key plate. *See* ESCUTCHEON.

keystone (key block). The wedge-shaped stone or brick inserted at the top of an ARCH. It locks the other members of the arch into place.

kg. *abbr.* KILOGRAM.

kick board. A TOE BOARD.

kicker (kicking piece). A block of wood or concrete that takes the thrust of another member.

kicking piece. *See* KICKER.

kick plate (kicking plate). A metal plate attached to the bottom of a door or workbench to prevent damage during use.

kiln. (1) A heated chamber used to bake cement, bricks and lime at high temperatures. (2) A chamber in which warm air is used to dry the moisture from timber during SEASONING.

kiln-dried. *adj.* Describing timber that has been seasoned to a suitable MOISTURE CONTENT by drying with warm air in a KILN. *Compare* AIR-DRIED.

kilo-. A standard prefix that can be placed in front of any SI unit to multiply the unit by 1000. The abbreviated form is 'k' as in kN for kilonewton. *See also* GIGA-, MEGA-, MICRO-, MILLI-, TERA-.

kilogram (kg). The basic unit of mass in the SI system of metric units. *See also* KILO-.

kilonewton (kN). Measure of FORCE. 1 kilonewton = 1000 newtons. *See also* KILO-, NEWTON.

kilowatt (kW). Measure of POWER. 1 kilowatt = 1000 watts. *See also* KILO-, WATT.

kilowatt hour (kWh). An alternative unit of energy still in use for electricity and heating. 1 kilowatt hour = 3.6 megajoules. *Compare* JOULE.

kinetic energy. A form of energy that a body has because of its motion. *Compare* POTENTIAL ENERGY. Moving fluids also possess kinetic energy. *See also* BERNOULLI'S THEOREM.

king bolt. A vertical steel rod used in place of a timber KING POST.

king closer. A brick that is cut as a CLOSER to fill an opening in a course larger than a half-brick. A diagonal corner is cut off the brick from the centre of one end to the centre of one side.

king pile. Long PILES driven in the centre of a trench to support timbers running from both sides of the excavation.

king post (*USA*: joggle post). A vertical timber in the centre of a roof TRUSS. It runs down from the ridge to the centre of the TIE BEAM. *See* ROOF TRUSS diagram.

king post truss. A triangular roof TRUSS made from two rafters joined together at the ridge and linked by a tie beam at their feet. A vertical king post links the tie beam to the ridge and additional struts may run from the base of the king to the rafters. This type of truss is no longer common. *See* ROOF TRUSS diagram.

kip. (*USA*) A kilopound. 1 kip = 1000 pound weight.

kiss marks. Marks left on bricks that have touched one another while being fired in the KILN.

kite winder. The middle WINDER step in a STAIR where it turns through 90°. The step has a shape like a traditional flying kite. *See* STAIRS diagram.

kN. *abbr.* KILONEWTON.

knapped flint. A flint stone that is broken across the middle to show a dark face.

knapping hammer. A hammer used to shape stones.

knee. (1) An upwards curve on a handrail. (2) An ELBOW in a pipe.

knee brace. A structural member placed

across the inside of an angle in a framework. It is used to increase stiffness especially between a wall and a roof, or in a PORTAL FRAME.

kneeler. A stone, with a sloping top, that is set in a gable wall, just below the GABLE COPING.

knee wall. *See* ATTIC WALL.

knife-edge load. A theoretical load on a structure that is assumed to act in a line across the width of the structure.

knife filler. A type of FILLING paste used for decoration. It has a consistency that allows it to be applied by a broad-bladed filling knife.

knobbing. The rough dressing of stones at a quarry by the removal of projections.

knocked down. *adj.* Describing building components that are prepared and delivered to a building site in a form ready for assembly.

knockings. Small stone chips knocked off during hammering or chiselling.

knocking up. The process of mixing and making workable a batch of concrete, mortar, paint or plaster.

knot. Hard distortions in the grain of wood made where the base of a branch grew from the main stem of the tree. *See also* DEAD KNOT, LIVE KNOT.

knot area ratio (KAR). The ratio of the area of all KNOTS in a particular cross-section of wood compared to the total area. The ratio is used for the visual grading of STRESS-GRADED TIMBER.

knot brush. A thick type of paint brush with fibres grouped in round or oval knots.

knotting. A coating of sealant applied to KNOTS before painting new wood. It prevents resin bleeding out and affecting the paint.

knuckle. The parts of a HINGE that surround the pin and form the pivot.

knuckle joint. The joint between the two slopes on one side of a MANSARD ROOF.

kraft paper. A strong type of brown paper used as a BUILDING PAPER.

kVA. *abbr.* Kilovolt-amperes. A measure of apparent 'power' used in an ALTERNATING CURRENT circuit. The apparent power may be different to the active power measured in WATTS. *See also* POWER FACTOR.

k-value. *See* COEFFICIENT OF THERMAL CONDUCTIVITY.

kW. *abbr.* KILOWATT.

kWh. *abbr.* KILOWATT HOUR.

L

label. A type of LINTEL or DRIPSTONE that projects above a door or window.

labour. Work that is done on a material but paid for separately from it. Labour-only work usually forms separate ITEMS in the BILL OF QUANTITIES.

labour constant. The amount of labour needed for a specified amount of work, such as 1 square metre of tiling.

labourer. A general construction worker who has not been formally trained in any particular TRADE.

laced valley. In a tile or slate roof, a valley formed by interlacing tiles across a VALLEY BOARD. The junction is sharper than a SWEPT VALLEY.

lacing. Light timber or metal members that are used as a brace between separate struts. The system forms a composite beam or a stronger structure.

lacing course. A COURSE of dressed stone or brick built into a wall of RANDOM RUBBLE.

lacquer. A surface finish produced by applying many coats of a natural VARNISH that is carried in a volatile solvent. Lacquer was traditionally used to give a glossy finish to furniture. Modern lacquers are often based on NITROCELLULOSE paints and used for coating metals.

ladder. A set of stairs or rungs that can be moved. *See also* EXTENDING LADDER, STEP LADDER.

ladder scaffold. A light form of SCAFFOLD made from ladders braced together. It is often used for painting and decorating.

lag. *vb.* To wrap water pipes or tanks with THERMAL INSULATION to reduce heat loss and protect against freezing.

lagbolt. (*USA*) *See* COACH SCREW.

lagging. (1) Material used to LAG. (2) Temporary boards used to support an arch during construction. (3) A lining used in a tunnel as protection from falling stones and soil.

lagscrew. *See* COACH SCREW.

laitance. A scum of fine particles that can form on the surface of fresh mortar and concrete. It is usually caused by too much water in the mix or by over-working. It leaves a layer of weak concrete which should be cut away.

lake. A traditional form of PIGMENT made by precipitating a dye on to a mineral base such as aluminium hydroxide or China clay.

lake asphalt. A relatively pure form of natural BITUMEN that is almost viscous when quarried from the earth. Lake Asphalt, Trinidad is the main source.

lally column. (*USA*) A hollow column used to support beams. It is sometimes filled with concrete.

lamella roof. A type of roof structure made from pre-formed members that are connected to give a large clear span.

lamina. A layer or plate that is assumed to have negligible thickness.

laminar flow (streamline flow). An orderly form of fluid flow in which the particles of liquid or gas move in straight lines at fixed distances from the wall. In general, laminar flow occurs at low velocities. *Compare* TURBULENT FLOW, CRITICAL VELOCITY.

laminate. (1) *vb.* To compress and bond together layers of different materials to make a new material which may be termed a 'laminate'. (2) The building-up of a structural member from small pieces of timber glued together. (3) A LAMINATED PLASTIC.

laminated beam. A BEAM formed by gluing together layers of timber that run parallel to the axis of the beam. The method is used to make long structural beams and it allows curved shapes.

laminated glass. A type of sheet glass that includes layers of plastic bonded between separate layers of glass. If the glass is broken the sheet is held together by the plastic. *See also* SAFETY GLASS.

laminated joint. *See* COMBED JOINT.

laminated plastic. A sheet, or other shape, that is formed by bonding together layers of paper or fabric and impregnating them with a synthetic resin. The use of a THERMOSETTING polymer such as MELAMINE FORMALDEHYDE produces 'formica' and similar materials that are used for durable WORKTOPS and linings. *See also* MFC.

laminboard. A composite wooden board of high quality made from core strips glued between plywood sheets. The width of the strips is less than those used in BLOCKBOARD.

lamp. A device that generates artificial light. A lamp is usually housed in a LUMINAIRE. *See also* DISCHARGE LAMP, FILAMENT LAMP.

lamp black. A black PIGMENT of carbon that is left as a deposit by the incomplete combustion of HYDROCARBONS such as oil. *See also* CARBON BLACK.

lamp cord. (*USA*) *See* FLEX.

lamson tube. A system of tubes and containers used to send documents around a building. The containers are driven by pneumatic pressure.

land drain. *See* FIELD DRAIN.

landing. A platform at the junction between two flights of stairs. *See also* HALF LANDING, QUARTER LANDING. *See* STAIRS diagram.

landscape. An ARCHITECT who specializes in the surroundings of a building or other structure. The choice and arrangement of trees and shrubs is often a feature of landscape architecture.

land surveyor. A person who specializes in the SURVEY of land and the making of maps (cartography). *See also* CADASTRAL SURVEY, GEODESY.

lantern light. A framed structure that protrudes up from a roof and contains glazing to light the room below. *See also* ROOFLIGHT.

lap. The amount by which the length of one component, such as tile, overlaps an adjacent one. *See also* GAUGE.

lapis lazuli. A stone, containing silicates, that is ground to form the traditional pigment of 'ultramarine'.

lap joint (lapped joint). A joint in which the timbers are overlapped and connected by a bolt, nail or other connector.

lapped tenons. A timber joint in which two TENONS enter a mortise from opposite ends and overlap inside the joint.

lap siding. *See* CLAPBOARD.

larch. A SOFTWOOD tree of the genus *Larix*. It gives hard strong timber.

large-panel construction. A form of INDUSTRIALIZED BUILDING that makes use of large, preformed concrete panels, usually walls.

laser. A beam of visible or infra-red light that, because all the components are in phase, remains sharply defined over a distance. The construction uses of laser beams include alignment of tunnels, levelling and ELECTRONIC DISTANCE MEASUREMENT.

latch (latch bolt). (1) A simple type of door

fastener in which a bar slips over a hook. *See also* THUMB LATCH. (2) The angled tongue that protrudes from a door lock such as a MORTISE LOCK or CYLINDER LOCK. A spring pushes the latch out unless it is withdrawn by turning a key or knob.

latent heat. The heat energy absorbed or released from a substance during a change of state without a change in temperature. Water, for example, has a latent heat of fusion for the change between ice and liquid, and a latent heat of vaporization for the change between liquid and steam. *See also* ENTHALPY, SENSIBLE HEAT.

lateral-force design. The structural design of a building so that the structure can withstand horizontal forces, such as those generated by wind or earthquake.

lateral support. A horizontal support for a wall or column in order to prevent sideways movement.

latex. Originally a natural RUBBER from trees. More usually a synthetic POLYMER that is dispersed in water. The uses of latex include special adhesives, emulsion paints and waterproof coatings.

lath. A strip or slat of wood traditionally used in LATHING.

lathe. A machine that turns pieces of wood or metal while they are given circular shapes by cutters fixed on the lathe. *See also* TURNING.

lath hammer. A traditional plasterer's hammer that has an axe edge for cutting LATHS.

lathing. Any base that provides a key for a coat of PLASTER. It may be a series of timber LATHS or METAL LATHING.

latitude. (1) The position of a point on the Earth's surface measured as the angular distance north or south from the equator. *See also* LONGITUDE. (2) The distance north or south from a reference line that runs east and west.

lattice (lattice beam, lattice girder, open-web girder). An open structure like a beam in which the main members are joined by diagonal braces.

lattice window. A window with small panes of glass, often diamond-shaped, that are set in metal CAMES. A LEADED LIGHT is a common type.

lavatory basin. *See* WASHBASIN.

lavatory pan. *See* WC.

lay. The type of twist used in the construction of a rope.

lay bar. A horizontal GLAZING BAR.

lay board. On a pitched roof, a board fixed to the rafters as a base for the rafters of a subsidiary roof.

layer board. A board that forms the base of a BOX GUTTER.

laying off. The final light brushing of a wet paint surface to eliminate brush or roller marks.

lay light. A horizontal LIGHT fixed in a ceiling.

layout. A drawing that shows the general arrangement of construction or plant.

lay panel. A panel, such as one set in a door, that runs horizontally.

leach (leaching). The process of passing water, or other solvents, through a material in order to remove salts or other dissolved substances.

lead. (1) A soft pure metal element (chemical symbol Pb) with high density and blue-grey appearance. Its ability to be worked flat made it a traditional material for roofing and PLUMBING work. Compounds of lead have also been the basis of many paint pigments. *See* WHITE LEAD. Lead is now relatively expensive for construction and other materials are usually substituted. Uncombined lead is harmful to health if continually absorbed into the body. *See also* LEAD-FREE PAINT, PLUMBOSOLVENCY, RED LEAD. (2) A lead CAME in a LEADED LIGHT window. (3) An electrical CONDUCTOR such as the connection to an appliance.

lead burning. The process of welding lead together without SOLDER.

lead-capped nail. *See* LEAD-HEAD NAIL.

lead cesspool. A RAINWATER HEAD made of lead.

lead damp course. A traditional form of DAMP-PROOF COURSE made from lead sheet. It is too expensive for modern use.

lead dot. *See* DOT.

lead drier. A DRIER made from organic salts of LEAD.

leaded light. A type of LIGHT for a window made up from small panes of glass fixed between CAMES of lead.

leader. (1) In surveying, the person who holds the leading edge of the tape or chain. (2) (*USA*) *See* DOWNPIPE.

leader head. *See* RAINWATER HEAD.

lead flashing. FLASHING that uses LEAD sheet.

lead flat. A type of flat roof covered with flexible sheets of LEAD. The joints between sheets are folded and lapped over wooden ROLLS.

lead-free paint. Paint that does not contain the potentially harmful compounds of LEAD PAINT.

lead-head nail (lead-capped nail, lead-head). A nail with a lead cap around the head. When used to fix roofing materials the lead squashes flat and makes a watertight seal between the roofing and the nail.

lead joint. A SPIGOT-AND-SOCKET JOINT between iron pipes in which the gap is filled with molten lead.

lead line (sounding line). A cord graduated with metre marks and used to take soundings of water depth. A piece of lead is used to weight the line.

lead paint. Paint that contains PIGMENTS based on compounds of LEAD. Compounds such as WHITE LEAD are harmful if absorbed by the body through contact with the paint surface or by breathing the dust. *See* LEAD-FREE PAINT.

lead plug. A small piece of lead, usually cylindrical, driven into a hole in brick or stone work. It acts as a tight fixing for a screw or other fastening.

lead slate. A specially-made piece of FLASHING that fits around a pipe where it passes through a roof.

lead tack. A type of TINGLE.

lead wedge (bat). A short length of lead scrap that is folded into a tapering shape and used to fix FLASHING into a groove such as a RAGLET.

lead wool. Lead cut into fine strands and used for CAULKING around joints in cast iron pipes.

leaf. (1) One part of a compound door or window, such as a folding door or a REVOLVING DOOR. *See also* OPENING LEAF, STANDING LEAF. (2) One of the skins of a CAVITY WALL.

leafing. The action of certain paint PIGMENTS, such as those based on aluminium powder, that float to the surface of the coating and form a protective layer.

lean lime. A form of LIME, such as HYDRAULIC LIME, that contains impurities and is less workable than non-hydraulic lime.

lean mix. A mix of CONCRETE or MORTAR that contains a lower than normal proportion of cement or lime. The opposite of a RICH MIX. *See also* WATER-CEMENT RATIO.

lean mortar. Mortar mixed with a low proportion of cement. *Compare* FAT MORTAR.

lean-to roof (half-span roof). A roof with a single slope, the top of which usually joins the wall of a higher building. *Compare* MONO-PITCH ROOF.

lease. A legal contract and document by

which a person is given use of a property for a specified period, usually for the payment of rent.

least count. The smallest measurement made with the vernier of a surveying instrument.

least squares. A method used to find the value or curve that best fits a set of measurements. The method assumes that the sum of the squares of the deviations of the experimentally obtained values from the optimum values should be a minimum.

Le Chatelier method. A standard test for the SOUNDNESS of PORTLAND CEMENT. A sample of cement paste is placed in a mould, heated in boiling water for a set period and the expansion then measured.

ledge (wale). One of the horizontal timbers on the back of a BOARDED DOOR. *See also* FRAMED, LEDGED AND BRACED DOOR. *See* DOOR diagram.

ledged and braced door. A type of BATTEN door with vertical boards attached to horizontal LEDGES that are diagonally braced. It does not have the outer frame of a FRAMED, LEDGED AND BRACED DOOR. *See* DOOR diagram.

ledged door. *See* BOARDED DOOR.

ledger. (1) A horizontal timber support fixed to a wall. (2) A horizontal pole in SCAFFOLDING.

leeward. The direction towards which the wind blows.

left-handed. *See* HANDED.

leg. In land surveying, one of the measured lines in a TRAVERSE.

L_{eq} index. *See* EQUIVALENT CONTINUOUS SOUND LEVEL.

let in. To cut out an area in one member ready to receive another, as in a HOUSED JOINT.

letter plate. A rectangular plate that is usually set in the main entrance door. The plate contains a slot through which mail can be delivered.

level. (1) A surveying instrument that uses a telescope device to establish horizontal lines. *See also* AUTOMATIC LEVEL, DUMPY LEVEL, TILTING LEVEL. *Compare* TRANSIT. (2) *See* SPIRIT LEVEL.

levelling. (1) The process of using a LEVEL instrument to measure the altitude or relative level of various points on a site. (2) The general process of filling or reducing a surface to make it smooth and horizontal.

levelling compound (self-levelling finish). Various types of compound laid as a finish on a rough floor surface. The material is viscous enough to set itself into a level surface.

levelling rod. *See* LEVELLING STAFF.

levelling rule. A long straight edge containing a SPIRIT LEVEL. It is used to keep the surface of SCREEDS at a uniform level.

levelling staff (levelling rod). A long graduated pole used for survey work. It is held vertically while being sighted through the telescope of a LEVEL or a THEODOLITE. *See also* SELF-LEVELLING STAFF, TARGET ROD.

level survey. A SURVEY of land made by a series of level readings related to a BENCH MARK.

level transit. (*USA*) *See* TRANSIT.

lever. A rigid bar that can rotate on a fixed support called the fulcrum. When a force is applied to the long end of a lever, like a PINCH BAR, then the force is increased at the short end of the lever.

lever board. A wooden LOUVRE.

lever handle. A type of door handle that operates by turning a horizontal bar on a pivot.

lever lock. A type of door lock in which the key must turn several internal levers to release the bolt.

lewis (lewis bolt, rag bolt). A metal bolt or

shackle that is permanently embedded into a large block of concrete or stone.

LHPC. *abbr.* LOW-HEAT PORTLAND CEMENT.

LIB. *abbr.* LOAD-INDICATING BOLT.

lichen. A flat type of plant growth that adheres to external stone and brickwork.

life-cycle cost. A method of calculating the relative costs of different installations by taking into account the initial price and running costs for the life of each object.

lift (elevator). A moving platform, cage or 'car' that carries goods or passengers from one level to another in a tall building. The lift cage travels on vertical rails in a shaft and is usually raised by cables wound by an electrical motor. *See also* HYDRAULIC LIFT.

lift latch. (*USA*) THUMB LATCH.

lift-off butt. A form of BUTT HINGE where raising the door lifts half of the hinge off the hinge pin.

lift shaft (lift well). The vertical opening that runs through a building for the movement of a LIFT.

lift slab construction. A method of erecting tall structures by constructing each floor slab at ground level and then raising them to position on a central core.

lift well. *See* LIFT SHAFT.

ligger. A flexible wooden stick used at the ridge of a THATCH roof.

light. (1) The part of a WINDOW that contains glass and is bounded by the window frame, or by MULLIONS and TRANSOMS. The light may be divided into several panes by GLAZING BARS. *See also* DEAD LIGHT, OPENING LIGHT. (2) Electromagnetic radiation that can be detected by human vision. *See also* INFRA RED, ULTRA VIOLET.

light alloys. ALLOYS that are usually based on ALUMINIUM and MAGNESIUM. They can be used to make lightweight structural components with high strength.

light box. An enclosure built into a wall or ceiling to conceal one or more LAMPS behind translucent glass.

light-gauge copper tube. Copper pipe in various sizes such as those used for domestic plumbing. It is generally joined by CAPILLARY JOINTS or COMPRESSION JOINTS rather than by screwed joints.

lighting. (1) The equipment or process used to supply an area with ILLUMINATION. (2) The process of starting the flame in a fire or boiler.

lightning conductor. A system that reduces the risk of tall buildings being damaged by lightning strikes. A combination of upright pointed rods and thick copper leads are fixed to the upper edges of the building. They are connected to EARTH via a lead running down the outside of the building.

lightning shake. A break across wood fibres caused by compression failure.

light output ratio (LOR). A method of classifying light fittings by the proportion of total light emitted into the upper and lower hemispheres. The LOR may be subdivided into the downward light output ratio (DLOR) and the upward light output ratio (ULOR). *Compare* BRITISH ZONAL SYSTEM.

lightweight aggregate. A low-density type of AGGREGATE that gives a LIGHTWEIGHT CONCRETE. Materials used as lightweight aggregate include CLINKER, volcanic PUMICE, foamed slag or PULVERIZED FUEL ASH.

lightweight concrete. A low-density form of concrete made with LIGHTWEIGHT AGGREGATES or by adding air to the mix. *See* AERATED CONCRETE. Construction with lightweight concrete is used to reduce the load of a structure and to improve THERMAL INSULATION. *Compare* DENSE CONCRETE.

light well. An unroofed shaft in the middle of a building. It gives light and air to windows that open on to the shaft.

lignin. A material that makes up about one third the mass of wood. It contains resins that line and strengthen the cell walls. *See also* CELLULOSE.

limb. *See* LOWER PLATE.

lime. (1) A building material based on CALCIUM OXIDE. This 'quicklime' is made by burning chalk or other forms of calcium carbonate in a kiln. *See also* HIGH-CALCIUM LIME, HYDRAULIC LIME, SLAKED LIME. (2) A term sometimes used for general compounds based on calcium. (3) A HARDWOOD with fine grain and yellowish-brown colour.

lime concrete. An early form of concrete that used LIME as the binder instead of PORTLAND CEMENT.

limed oak. A decorative treatment that gives a 'blonded' effect to OAK and highlights the grain of the wood. The wood is coated with LIME which is allowed to 'pickle' and then brushed off.

lime mortar. A form of MORTAR made from lime and sand, without cement. *Compare* GAUGED MORTAR.

lime paste. *See* SLAKED LIME.

lime plaster. A mixture of LIME and sand. *See also* UNGAUGED LIME PLASTER.

lime putty. A plastic material made by keeping SLAKED LIME soaked in water for days.

lime-soda process. *See* SODA-LIME PROCESS.

limestone. A natural form of CALCIUM CARBONATE.

lime tallow wash. A form of LIMEWASH containing tallow as a binder. It can be used on bituminous roof surfaces.

limewash. A thin, milky type of paint made by soaking SLAKED LIME in excess water. Binders and other additives are used to improve the properties of limewash and it is a traditional coating for brickwork, plaster and stonework. It discourages the growth of lichen and mould but needs frequent renewal.

limit state. The condition at which a structure starts to become unfit for use. This state may happen at the yield point of a metal component or may be caused by corrosion, instability and other reasons.

limit state design. A modern method of structural design that accepts that variations in materials, construction and loads make it impossible to guarantee against failure. The aim is to provide an acceptable probability that the structure will perform satisfactorily during its intended life. *See* CHARACTERISTIC STRENGTH, FACTOR OF SAFETY.

lincrusta. A wall covering with a texture or pattern that stands out in low relief.

line. A length of cord used to set out levels and alignments during building, especially for brickwork.

linear equation. A mathematical expression that gives a straight line when plotted.

linear programming. A mathematical method used to find the optimum use of resources. The method assumes that the various inputs, outputs and costs of an operation are expressed as LINEAR EQUATIONS.

linear regression. A statistical method of finding the straight line that best fits a scatter of data points on a graph. For example, if certain costs and profits are known to have a linear relationship then the linear regression line can be used to make profit predictions.

line drop. A VOLTAGE DROP in an electrical supply line. It is caused by the resistance of the conductor.

line level. A small SPIRIT LEVEL that can be suspended on a LINE used in bricklaying.

linen tape. A measuring tape made of glazed linen cloth and used for rough measuring and setting out. It is light and convenient but not as accurate as a steel tape.

line of collimation. *See* COLLIMATION LINE.

line of thrust. In a structure such as retaining wall, the direction of the single force that is the resultant of the multiple forces acting on the structure.

line pin. A steel pin or nail inserted in

mortar joints at the end of a brick wall and used to fix one end of a bricklayer's LINE.

liner. *See* LINING FITCH.

lining. (1) Sheet material such as plasterboard that is used to cover partitions and ceilings. (2) A surround of thin wood fixed to the inside of a DOOR FRAME. (3) A layer of concrete or other material used to seal the bed of a canal or inside of a tunnel.

lining fitch (liner, lining tool). A narrow paint brush with a slanted edge. It is used to paint narrow lines when decorating.

lining paper. A plain paper pasted on to walls and ceilings as a base for WALLPAPER or PAINT. It can be used to cover defects in the surface.

lining plate. A metal strip used to secure the edge of FLEXIBLE-METAL roofing.

lining tool. *See* LINING FITCH.

link. A standard length of a surveyor's CHAIN.

link dormer. A large form of DORMER window that joins two parts of a roof. It may incorporate a chimney and have LIGHTS in the sides of the dormer.

linoleum (lino). A durable floor covering made from linseed oil compressed on to canvas and supplied as a continuous roll or as tiles. Its traditional use in kitchens and bathrooms is now mainly replaced by vinyl sheeting.

linseed oil. A natural DRYING OIL used in traditional paints and varnishes to form the final film. It is a vegetable oil extracted from the seeds of the flax plant. *See also* RAW LINSEED OIL.

lintel (lintol). A beam that carries the load over an opening such as a door or window. It is often made of precast concrete or steel profile. *Compare* BOOT LINTEL.

lintol. *See* LINTEL.

lip. A section that projects from the STRIKING PLATE of a door lock. The lip is angled to guide and depress the latch of the lock.

lipping (edging strip). A strip of wood that protects the edge of veneered board or blockboard, such as that used in a flush door.

liquefaction. (1) The general process when a substance changes from gas to liquid form. (2) A sand that is saturated with water and liable to flow as a fluid.

liquefied petroleum gas (LPG). BUTANE or PROPANE gas that has been liquefied under pressure and is kept stored in strong containers. It is used as a fuel for some heating appliances.

liquidated damages. *See* DAMAGES.

liquidation. The formal ending of a company and its activities.

liquidator. A person appointed to supervise the LIQUIDATION of a company.

liquidity. The ability of a company or individual to convert assets for cash if necessary.

liquid limit. The MOISTURE CONTENT of CLAY at the point where it changes from the plastic state to the liquid state.

liter. (*USA*) *See* LITRE.

litharge. Lead monoxide, PbO. A reddish-coloured material used in the manufacture of paints and glass.

lithopone. An opaque white PIGMENT often used in interior paints. It is made from barium sulphate ($BaSO_4$) and zinc sulphide (ZnS).

litmus. A vegetable dye that changes colour to red in the presence of an ACID and blue for an ALKALI.

litre. A metric unit of volume. 1 cubic metre contains 1000 litres.

live edge. Paintwork that remains wet enough at the edge to be blended with a new paint without showing an overlap.

live knot (intergrown knot). A KNOT that is

firm because the fibres are intergrown with the surrounding timber. *Compare* DEAD KNOT.

live load (imposed load, superimposed load). A load that is put on a structure but is not part of the structure itself. Replaceable loads may include the weight of people, furniture, vehicles and snow. *See also* MOVING LOAD. *Compare* DEAD LOAD.

live wire. A CONDUCTOR that is connected to an electrical source and can supply a current if a circuit is completed back to the source.

lm. *abbr.* LUMEN.

load. A FORCE imposed on a structure or part of a structure. It may be the weight of the structure itself. *See also* DEAD LOAD, LIVE LOAD.

load-bearing. *adj.* Describing a structure, such as a wall, that supports a LOAD in addition to its own weight and any wind forces. A lightweight partition is not usually load bearing. *Compare* NON-BEARING, SELF-SUPPORTING WALL.

load-extension curve. The result of a tensile test on a sample of metal, as produced by a TENSOMETER. The applied load is plotted against the elongation of the sample and the curve is similar to the STRESS-STRAIN CURVE.

load factor. The ratio of the maximum load that causes the failure of a component or structure to the design load imposed on it. *Compare* FACTOR OF SAFETY.

load-indicating bolt (LIB). A form of HIGH-TENSILE FRICTION-GRIP BOLT. The bolt indicates its state of tension by the compression of small projections on the underside of the head.

loading coat (loading slab). A slab of concrete laid over asphalt TANKING in order to prevent movement by water pressure beneath the tanking.

loam. A mixture of sand, clay and vegetable matter such as HUMUS. Loam made with chopped straw is a traditional material for making plaster and bricks.

lobby. An enclosed space, with two or more doors, that acts as an entrance to a building. The lobby may act as an AIR LOCK as well as a reception area.

local attraction. Deviation of a magnetic compass needle from magnetic north caused by a metal object close to the compass.

local authority. (*UK*) The local city council, district council, county council or similar organization that controls the quality of building by means of the current BUILDING REGULATIONS or planning regulations.

location plan (key plan). A form of BLOCK PLAN that shows the layout of buildings and other features on a building site.

lock. A device that fastens a door by pushing a bolt into a keeper. The bolt is operated by a key. *See also* CYLINDER LOCK, MORTISE LOCK, RIM LOCK. *Compare* LATCH.

lock block. A block of wood in a FLUSH DOOR into which a LOCK can be fixed.

locking bar. A door fastener similar to a HASP AND STAPLE. A long hasp or bar fits over the staple and also protects the securing screws.

locking stile. The STILE on the edge of a door that carries the LOCK.

lock joint. A seam made in FLEXIBLE-METAL roofing.

lock nut. (1) A NUT fitted to a pipe, conduit or other device in a way that prevents movement or loosening. (2) A second NUT on a bolt that is tightened down on top of a first nut to prevent loosening.

lock rail. The central RAIL across a door on which the lock is usually fixed.

lock saw. *See* PAD SAW.

lockshield valve. One of the two valves on a panel RADIATOR that allow the panel to be isolated for removal or repair. The lockshield valve is adjusted to a suitable flowrate which gives the radiator a suitable share of the hot water. *See* HEATING diagram.

loess. SILT that is blown and deposited by the wind.

loft. The space within a pitched roof that is usually entered through a trap door in the ceiling. The loft may contain the CISTERN for the cold-water supply.

loft ladder (*USA*: disappearing stairs). A special ladder fixed on the topside of the trap door into a LOFT space. As the trap door opens it causes the ladder to unfold and extend downward.

log. A length of tree trunk after it had been cut and trimmed.

logarithmic scale. A set of numbers obtained from a mathematical expression that contains a logarithm. Logarithmic values are compressed and do not increase in an even manner. *See also* SOUND LEVEL.

London stock. A distinctive yellow BRICK made from London clay. London has many buildings constructed of such bricks but they are no longer the local STOCK (most common) brick.

long column. A COLUMN that, when overloaded, fails by buckling rather than by crushing.

long dummy. A tool used by plumbers to straighten kinks in lead pipes.

long float. A large type of FLOAT that needs two people to operate it.

longhorn beetle. *See* HOUSE LONGHORN BEETLE.

longitude. The position of a point on the Earth's surface measured as the angular distance east or west of a prime meridian taken as zero degrees. *See also* LATITUDE.

longitudinal section. The shape or the drawing of an object when cut through along its length. *Compare* CROSS SECTION.

long oil. A VARNISH containing a high proportion of oil and a low proportion of RESIN. *Compare* SHORT OIL.

lookout. (*USA*) A wooden bracket that supports the overhanging edge of a roof.

looping in. A system of electrical wiring used for lighting. One conductor is connected to the lamp fitting while the other conductor is run to the switch and back to the lamp.

loop vent. A VENTILATION PIPE that continues above the highest branch of the SOIL STACK.

loose butt hinge. *See* LIFT-OFF HINGE.

loose-fill insulation. A THERMAL INSULATION material that is supplied and installed in the form of small separate particles. Examples include expanded mineral chips and expanded plastic beads. The insulation can be placed loose between ceiling rafters and inside cavity walls.

loose knot. A KNOT that is not intergrown with the surrounding wood and therefore can be easily knocked out. *Compare* LIVE KNOT.

loose-pin hinge. A BUTT HINGE from which the pin can be withdrawn to allow the door to be unhung without unscrewing the hinges.

LOR. *abbr.* LIGHT OUTPUT RATIO.

loss of head. In a fluid, the loss of energy or HEAD caused by HYDRAULIC FRICTION.

lost-head nail (set-head nail). A wire NAIL with a narrow tapered head that can easily be punched below the surface of the wood.

lot. (*USA*) One of the portions of land into which a town is divided by legal boundaries.

loudness. The sensation in human hearing that varies with the energy of a sound. The ear is not equally sensitive at all frequencies and so loudness depends on the frequency as well as the amplitude of the sound wave.

loudness level. A scale of equal LOUDNESS that makes allowance for the changes in sensitivity of the ear that occur with frequency. The unit is the PHON.

louver. *See* LOUVRE.

louvre (louver). A ventilation inlet that is protected from the rain by horizontal slats slanted downward. In installations such as windows, the slats are pivoted at the ends and can be closed together.

louvre window. A type of WINDOW formed from overlapping horizontal glass slats that can be rotated to admit air. *Compare* VENETIAN BLIND.

low carbon steel. A form of plain CARBON STEEL that has a carbon content less than 0.15 per cent. Low carbon steel is soft, ductile and has a low tensile strength. Its uses include concrete reinforcing bars. *Compare* HIGH CARBON STEEL, MILD STEEL.

lower plate (limb). A rotating plate on a THEODOLITE that is graduated in degrees of angle.

low-heat Portland cement (LHPC). A form of PORTLAND cement that gives off less heat during HYDRATION than ordinary Portland cement. It is used for massive structures, such as dams, where uneven temperature gradients in the concrete might cause cracking.

low level flushing cistern. A FLUSHING CISTERN that is mounted on or just above the WC pan, rather than mounted at a height on the wall. *See also* SIPHONIC WC.

low pressure mercury vapour lamp. *See* MERCURY LAMP.

low pressure sodium vapour lamp. *See* SODIUM LAMP.

low-pressure system. A hot water or central heating system in which the water circulates by natural CONVECTION CURRENTS in the pipes rather than by a pump.

LPG. *abbr.* LIQUEFIED PETROLEUM GAS.

lug. (1) A tag that projects from a pipe or frame and is used to secure the component. (2) A metal terminal fixed on the end of a cable to make a good electrical connection.

lug sill. A window or door SILL that is longer than the opening so that the ends can be built into the wall. *Compare* SLIP SILL.

lumber. (*USA*) Wood that has been sawn or processed for use in construction. *See also* TIMBER.

lumber core. (*USA*) *See* COREBOARD.

lumen (lm). The SI unit of LUMINOUS FLUX. One lumen is defined as the flux emitted into one STERADIAN by a point source of one CANDELA.

lumen method. A commonly-used technique of lighting design that is valid if the fittings are mounted overhead in a regular pattern.

luminaire. A light fitting that holds or contains a LAMP. *See also* BRITISH ZONAL SYSTEM, LIGHT OUTPUT RATIO.

luminance. A measure of the brightness of an area of a light source, or a reflecting surface. Luminance may be measured by CANDELAS per square metre, or by the APOSTILB.

luminescence. The process by which certain substances absorb energy by radiation and emit light. *See also* FLUORESCENT LAMP, FLUORESCENT PAINT.

luminous efficacy. The efficiency or ability of a LAMP to convert electrical energy to light energy. It is usually expressed in LUMENS per watt.

luminous flux. The rate of flow of light energy. The SI unit of flux is the LUMEN.

luminous intensity. The power of a light source, or illuminated surface, to emit light in a particular direction. The SI unit is the CANDELA.

luminous paint. *See* FLUORESCENT PAINT.

lump. (*UK*) A collective term for self-employed building workers and their income tax status.

lump hammer. *See* CLUB HAMMER.

lump-sum contract. A simple form of FIXED-PRICE CONTRACT that is usually based on drawings and specification without a BILL OF QUANTITIES. It is suitable for small buildings and alteration work.

lux (lx). The SI unit of ILLUMINANCE. 1 lux = 1 LUMEN per square metre.

lux meter. *See* LIGHT METER.

lx. *abbr.* LUX.

lyctus powder-post beetle. A BEETLE, about 5 mm long, that is destructive to timber, especially the sapwood of hardwood. It spends much of its life cycle as a larva that bores tunnels and feeds within the wood. It leaves FLIGHT HOLES of 1–2 mm in diameter.

M

M. *abbr.* MEGA-.

m. *abbr.* MILLI-.

macadam. A road surface formed from broken stone of uniform size compacted by rolling. The interlocking between aggregate particles gives strength to the roadbase. The aggregate may also be bound together by a cement and coated by BITUMEN or tar. *See also* TARMACADAM.

macerator. A device that uses liquid and mechanical action to soften and separate solid SEWAGE waste. Macerators are attached to some WCs to allow the use of smaller waste pipes.

machine screw. A metalwork fixing like a BOLT but with a head that can receive a screwdriver.

made ground. Ground that has been raised in level by FILL brought from elsewhere.

magazine boiler. A boiler that takes coal or other solid fuel from an attached bunker.

magenta. A crimson red colour. *See* SUBTRACTIVE COLOUR.

magnesia. Magnesium oxide, MgO. *See also* MAGNESITE.

magnesite. A natural form of MAGNESIUM CARBONATE. When heated it forms magnesia (MgO) and it is used in fire-resistant materials.

magnesite flooring. A form of JOINTLESS FLOORING formed with 'oxychloride cement' which is made from magnesium oxide, magnesium chloride and fillers

magnesium. A pure metal element (chemical symbol Mg) which is soft and light. It can be combined with aluminium and zinc to produce LIGHT ALLOYS of high strength. Magnesium corrodes in preference to nearby metals and is used in SACRIFICIAL PROTECTION.

magnesium alloy. *See* LIGHT ALLOYS.

magnesium carbonate ($MgCO_3$). The main chemical component of MAGNESITE.

magnesium-oxychloride cement. The material used in MAGNESITE FLOORING.

magnesium sulphate ($MgSO_4$). The main chemical compound in EPSOM SALTS.

magnetic bearing. In surveying, the horizontal angle between a survey line and the direction of magnetic north pole. *Compare* BEARING.

magnetic field. A region in which a magnetic force of attraction or repulsion can be detected, such as with a magnetic compass needle.

magnetic flux. A measure of the strength of a MAGNETIC FIELD over a given area.

magnetic north pole. The focus of the Earth's natural magnetic attraction and the place to which magnetic compass needles point. The magnetic pole is situated in Canada but moves position and does not coincide with the true geographical north pole. *See also* DECLINATION.

magnetic variation. (1) *See* DECLINATION. (2) Variations in the position and effect of the MAGNETIC NORTH POLE.

magnetite (Fe_3O_4). One of the forms of

iron oxide found in iron ore. *Compare* HAEMATITE.

mahogany. A HARDWOOD tree which give a durable reddish-brown timber. Mahogany was originally obtained from the *Meliaceae* family but various tropical trees are now also used for 'mahogany'. The wood is traditionally used for furniture and can be highly polished. *See also* SAPELE, UTILE.

main. A large pipe that brings water or gas to a building. *See also* GRAVITY MAIN.

main beam (primary beam). A BEAM that transfers its load on to a wall or column, rather than to another beam. *Compare* SECONDARY BEAM.

main contractor. A CONTRACTOR who is responsible for most of the work on a site.

mainframe computer. A general-purpose computer with the high speed and large storage capacity needed for tasks like national databases of information. It is larger than a MINICOMPUTER or MICROCOMPUTER.

maintenance. The routine repair and renewal of a building needed to maintain the utility, structural soundness and value of the building. *See also* RENOVATION, REFURBISHMENT.

maintenance factor. In lighting calculations such as the LUMEN METHOD, an allowance for the loss of light output caused by the build-up of dust and the deterioration of the lamp. A factor of 0.8 is often used.

maintenance management. The organization of MAINTENANCE by appropriate schedules, inspections, decisions and execution of necessary works.

maintenance period. *See* DEFECTS LIABILITY PERIOD.

maisonette. (1) (*UK*) A self-contained FLAT on two levels. (2) (*USA*) A DUPLEX APARTMENT.

make good. (1) To make a repair that produces an effect as good as new. *See also* RENOVATION. (2) To make repairs to an existing construction or to decoration that has been affected by new work nearby.

male thread. A THREAD on the outside of a pipe or tube; it can be screwed to a matching FEMALE THREAD.

mall (maul). A heavy MALLET.

malleability. The ability of some materials, especially metals, to be beaten into thin sheets. Malleability is one aspect of a general property of TOUGHNESS. *See also* DUCTILITY.

mallet. A tool like a HAMMER with a large wooden or rubber head.

management contract. A CONTRACT in which the role of the main contractor is as a manager of sub-contractors rather than as a builder.

mandrel. (1) A cylindrical piece of wood used to shape metal such as lead pipe. (2) A shaft that holds an object to be shaped on a lathe.

manganese. A pure metal element (chemical symbol Mn) contained in most steels.

manganese steel. STEEL that contains more than about 1 per cent of MANGANESE. It is tougher than ordinary steels.

manhole. A chamber that allows access from the ground to a drain, sewer or other underground services. Deep manholes are made of brick or concrete and may have ladder rungs built in. *See also* BENCHING, INSPECTION CHAMBER, SLIPPER.

manhole cover. A removable plate fitted over the entrance to a MANHOLE or INSPECTION CHAMBER. Manhole covers for foul drains are sealed by bedding themselves into a groove in the outside frame.

manifold. A pipe or chamber with a number of inlets or outlets attached.

manipulative joint. A type of COMPRESSION JOINT in which the ends of the pipes are opened out slightly. *Compare* NON-MANIPULATIVE JOINT.

manometer. An instrument that measures

pressure by comparing the levels of liquid in a U-shaped glass tube. The liquid may be water, oil or mercury. *See also* U-GAUGE.

mansard roof (French roof, *USA*: gambrel roof). A pitched roof where each face has two slopes. The lower slope is steeper than the upper one. *See* ROOF diagram.

marble. A METAMORPHIC ROCK formed from limestone. The surface can be highly polished and usually contains random veins and other patterns of colouring.

marbling. The use of paint to reproduce the patterns of MARBLE, or other stone, as decoration.

margin. (1) The exposed area of each TILE or SLATE. The length of the margin is the same as that of the GAUGE. (2) In the production of concrete, an extra strength added to the CHARACTERISTIC STRENGTH to allow for normal variations in the production process. (3) The exposed face of a STILE or RAIL in a door. (4) In a stair, the distance between the top of the CLOSE STRING and the line of the NOSINGS. *See* STAIRS diagram.

margin light. A narrow pane of glass at the edge of a SASH window.

margin of safety. *See* FACTOR OF SAFETY.

margin towel. A narrow rectangular trowel for working with plaster in a narrow width.

marine glue. A general term for a strong and durable type of waterproof ADHESIVE used in boat building and other exposed construction. A marine glue may be an EPOXY RESIN in TWO-PACK form.

marine ply. PLYWOOD that has been made with MARINE GLUE to make it weatherproof.

marked face. The working FACE of timber.

marking gauge (joiner's gauge). A tool used by joiners to mark parallel lines on the face of a piece of wood. *See also* FACE PLATE.

marl. A soil or rock containing significant amounts of LIME.

marquetry. The process of decorating the surface of timber by INLAYING with fine wood, ivory or other material.

marsh gas. A natural gas, mainly METHANE, that can be formed by the decay of vegetable matter. Marsh gas can be an explosive and poisonous hazard in mines and sewers.

martensite. A hard form of steel produced when the metal is quickly cooled in the process of HARDENING. *Compare* AUSTENITE.

mash hammer. *See* CLUB HAMMER.

masking. The use of tape, paper or other means to protect an adjacent surface from paint and to produce a sharp edge.

masking noise. Background NOISE that occurs at the same time as another sound and obscures that other sound.

mason. A person who works with stone.

masonry. (1) The craft of working with stone. (2) Construction based on stone, brick, concrete and similar materials derived from rock or earth.

masonry cement. A form of HYDRAULIC CEMENT that makes a MORTAR suitable for working with stone. It is based on Portland cement with plasticizers and other additives.

masonry nail. A hardened steel nail that can be driven by a hammer into brick, concrete or stonework.

masonry paint. A thick durable paint, such as CEMENT PAINT, that is designed for the exterior surfaces of buildings. Some types contain sand or fibres

masonry tools. *See* MASONRY TOOLS diagram.

masonry unit. A single brick, block, stone or tile used in building.

mason's mitre. A corner of solid stone that is carved to give the appearance of a MITRE JOINT.

mason's putty. A LIME PUTTY mixed with

stone dust and Portland cement. It is used in pointing ASHLAR stonework.

mason's scaffold. A SCAFFOLD that supports itself and is not fixed to the wall where it would damage the surface.

mason's stop. A MASON'S MITRE.

mass concrete. Concrete without REINFORCEMENT. It may be used in structures such as foundations that rely on gravity for stability. *Compare* REINFORCED CONCRETE.

mass law. A principle of SOUND INSULATION by which heavyweight construction transmits less sound than lightweight construction. The SOUND REDUCTION INDEX of a single-leaf partition increases uniformly with mass per unit area.

mass retaining wall. A RETAINING WALL that resists sideways forces by the action of its weight rather than by structural means.

mastic. A waterproof, adhesive filler that remains permanently flexible. It can be used as a SEALANT in external joints in buildings and for glazing.

mastic asphalt. An ASPHALT that contains a high proportion of binder so that the material is spreadable when heated and semi-solid at normal temperatures. Mastic asphalt is dense, waterproof and durable.

mastic asphalt surfacing. A WEARING COURSE for roads or roofs made from MASTIC ASPHALT to which sand or chips may be added.

mat. (1) A RAFT, made of concrete or steel, on which a structure rests. (2) A mesh of reinforcement in a slab of concrete.

matchboarding (matched boards). Boards that are shaped at the edges, such as TONGUED AND GROOVED, so that they fit closely together when used for lining. *See also* VEE JOINT.

matched boards. *See* MATCHBOARDING.

matched floor. A floor made from MATCHBOARDING.

matching. (1) The use of MATCHBOARD. (2) The arrangement of materials, such as wallpapers and VENEERS, so that colours or patterns join together in suitable ways.

materials science. The scientific study of the properties and interactions of the substances from which things are made.

matrix. (1) In general, items laid out in a regular grid pattern. (2) A material in which particles of another material are embedded. In concrete, for example, the aggregate is held in a matrix of cement paste.

matt. A uniform surface finish, usually on paint, that absorbs or scatters light and appears non-shiny. *Compare* EGGSHELL, GLOSS.

mattock. A tool shaped like a PICK AXE with a broad blade at one end. It is used for digging and removing roots in hard ground.

mattress. (1) A platform of BLINDING or other concrete laid on the ground as a base for PLANT. (2) A large flexible mat of materials, such as concrete pieces woven together with ropes, that is sunk on to a river or seabed to prevent scour.

maturing. The process of improving some materials, such as the soaking of LIME PUTTY and the AGEING of varnish.

mat well. A depression in the floor just inside an external door so that a floormat can be fitted flush with the floor.

maul. *See* MALL.

maximum demand. The highest instantaneous value of electrical power drawn by a consumer.

MCB. *abbr.* MINIATURE CIRCUIT BREAKER.

MDF. *abbr.* Medium density FIBREBOARD.

mean. In statistics, the arithmetic mean or 'average' of a set of values is the sum of the values (ignoring negative signs) divided by the number of values. *Compare* MEDIAN, MODE.

mean depth. *See* HYDRAULIC MEAN DEPTH.

mean radiant temperature. The average RADIANT TEMPERATURE caused by surrounding surfaces.

means of escape. A route from any point in a building by which people can leave safely in case of fire.

measure and value contract. A CONTRACT where the contractor receives a BILL OF QUANTITIES as well as drawings to use for the TENDER.

measurement. The analysis of the working drawings of a building and the production of details for the amounts of the various materials and work needed. In the United Kingdom this task is performed by a QUANTITY SURVEYOR. *See also* BILL OF QUANTITIES, STANDARD METHOD OF MEASUREMENT.

measuring frame. *See* BATCH BOX.

measuring weir. A WEIR built in a river to help measure the flowrate.

mechanical advantage. The ratio of the lift or other force provided by a machine to the force applied to the machine.

mechanical bond. A BOND formed between two surfaces or components by some form of interlocking between indentations.

mechanical engineering. The design, production and operation of machinery.

mechanical engineering services. The equipment and materials concerned with the heating, ventilation and air conditioning of a building.

mechanical saw. A powered saw such as a BAND SAW, CIRCULAR SAW or JIG SAW.

mechanical ventilation. The use of an electrical FAN or other device provide VENTILATION of an internal space rather than the use of NATURAL VENTILATION.

median. In statistics, the value in a sample of numbers such that exactly half the numbers have a value greater than the median. It is only the same as the MEAN if the sample is symmetrical.

medium. A general term for the liquid component of PAINT that dries to act as the BINDER for the PIGMENTS. *See also* VEHICLE.

medullary ray. *See* RAY.

meeting rail (check rail). A RAIL on the frame of a SASH window that meets another such rail when the window is closed. *See* WINDOW diagram.

meeting stile. On a folding door, a closing STILE that meets in the middle with another such stile when the door is closed.

mega-. A standard prefix that can be placed in front of any SI unit to multiply the unit by 1,000,000. The abbreviated form is 'M' as in MJ for megajoule. *See also* GIGA-, KILO-, MICRO-, MILLI-, TERA-.

melamine-faced chipboard (MFC). CHIPBOARD on to which MELAMINE LAMINATE is bonded as a protective coating. MFC is used for furniture units, especially in kitchens.

melamine formaldehyde. A SYNTHETIC RESIN that has THERMOSETTING properties. It is used for adhesives and for LAMINATED PLASTICS.

melamine laminate. A popular form of LAMINATED PLASTIC. *See also* MELAMINE-FACED CHIPBOARD.

melamine-surfaced chipboard. *See* MELAMINE-FACED CHIPBOARD.

member. A component of a structure such as a column or beam. A FRAME is formed by a number of members fixed together.

membrane. A continuous thin skin or covering of material such as plastic sheeting used in a DAMP-PROOF COURSE or VAPOUR BARRIER.

membrane theory. A theory of structural design for thin shells which are assumed to behave like the walls of a bubble.

mending plate. A flat metal plate that is

screwed to either side of broken woodwork to act as a repair.

meniscus. The curved surface of a liquid that is in contact with the edge of a container. The effect is particularly noticeable in a narrow glass tube.

mensuration. The measurement of length and the calculation of areas and volumes based on such measurements.

Merchalli scale. A measurement of earthquake shocks on a scale of 1 to 12. It is based on the effects upon people and buildings rather than upon the energy of the source as in the RICHTER SCALE.

mercury. A silvery-white metal element (chemical symbol Hg) that is a liquid at room temperature. 'Quicksilver' is an old name for mercury. Its liquid properties make it useful for thermometers, barometers and other instruments. Unconfined mercury vapour or mercury is a health hazard.

mercury vapour lamp. A DISCHARGE LAMP that uses MERCURY vapour. The colour quality of most mercury discharge lamps is 'corrected' by phosphors, such as in the FLUORESCENT LAMP.

Merulius lacrymans. *See* DRY ROT.

mesh. (1) Steel bars joined together in a web as reinforcement for concrete. (2) EXPANDED METAL sheet. (3) A cloth woven from fine wire and used to screen sand or water. *See also* SIEVE.

metal-casement putty. A form of PUTTY or other GLAZING COMPOUND that can be used on metal or sealed surfaces of hardwood, concrete or stone.

metal coating. A thin film or coat of metal such as nickel or zinc that is applied to the surface of metal objects to protect them from corroding. *See also* CHROMIUM PLATING, GALVANIZING, SHERADIZING.

metal lathing. EXPANDED METAL used as a LATHING for plaster. It is used where a surface is too irregular to provide a normal key for the plaster.

metallic paint. A paint containing small flakes of a metal such as aluminium, bronze or zinc. The paint can be used as a sealer or a rust-inhibiting primer.

metal-sheathed mineral insulated cable. Electrical CABLE made from copper tube in which the conductors are surrounded by mineral insulation such as MAGNESIA. The fire resistant nature of the cable makes it useful or compulsory for some fire protection equipment.

metal trim. Pressed metal edges and trims such as ARCHITRAVES and SKIRTINGS. They are fixed in place before plastering and included in the plaster surface.

metal valley. A VALLEY GUTTER that is lined with FLEXIBLE METAL.

metamorphic rocks. Rocks formed from other rocks that have been subjected to combinations of high temperature and pressure. Marble, for example, is a metamorphosed version of limestone. *Compare* IGNEOUS ROCKS, SEDIMENTARY ROCKS.

methane (CH_4). The simplest HYDROCARBON that exists naturally in the rocks. It is a major ingredient of NATURAL GAS and can be a hazard when tunnelling and mining. *See also* SEWAGE GAS.

methylated spirit. An industrial version of pure 'alcohol', or ethanol, that is used as a solvent for SHELLAC and for cleaning. It is made unsuitable for drinking by the addition of methanol and dye.

metre (m). The basic unit of length in the SI system of metric units.

metric ton. *See* TONNE.

mezzanine. An intermediate level or storey, often between the ground and first floors.

MFC. *abbr.* MELAMINE-FACED CHIPBOARD.

mica. One of several MINERALS that can be cleaved into thin lustrous flakes. Mica is a good electrical insulator and is also used as an EXTENDER in paints. *See also* VERMICULITE.

mica-flap valve. A hinged sheet of mica used as a non-return valve to admit air into a drain. *See also* AIR-ADMITTANCE VALVE.

micro-. A standard prefix that can be placed in front of any SI unit to divide the unit by 1,000,000. The abbreviated form is 'μ' as in μA for milliamp. *See also* GIGA-, KILO-, MEGA-, MILLI-, TERA-.

micro-bore system. A central heating system that pumps water to and from the radiators by pipes of very small diameter, such as 6 mm.

microcomputer. A computer whose electronic processing units are miniaturized within microchips. The uses of microcomputers include desktop business computers and control units for machinery. *Compare* MINICOMPUTER, MAINFRAME COMPUTER.

micrometer. An instrument that measures thickness or angle with high accuracy.

micron. One millionth of a METRE.

microptic theodolite. A lightweight and compact form of THEODOLITE.

microstrainer. A MESH of fine stainless steel used in the initial FILTRATION of water supplies.

mid-feather. A leaf of brickwork that separates two flues inside a chimney.

migration of plasticizer. An increase in brittleness of a plastic material, such as floor tiles, caused when some plasticizer is absorbed by a neighbouring material such as concrete.

mild steel. A form of CARBON STEEL that has a carbon content of 0.15–0.25 per cent. Mild steel is relatively ductile and its uses include pipes and joists. *Compare* LOW CARBON STEEL, HIGH CARBON STEEL.

milkiness. A defect in varnish or lacquer that appears as a whitish film in the transparent surface. It can be caused by working in damp conditions.

milk of lime. SLAKED LIME in water.

milled lead. LEAD that is rolled into sheets.

mill finish. The surface of ALUMINIUM components that are allowed to weather naturally without treatment such as ANODIZING.

milli-. A standard prefix that can be placed in front of any SI unit to divide the unit by 1000. The abbreviated form is 'm' as in mA for milliamp. *See also* GIGA-, KILO-, MEGA-, MILLI-, TERA-.

mill scale. A layer of black iron oxide formed on the surface of steel during rolling. *See also* PICKLING.

millwork. (*USA*) Items of woodwork, such as doors or windows, that are prefabricated in a planing mill or woodworking shop.

mineral. A material that is found naturally in the ground, but the term excludes ORGANIC compounds.

mineral fibre. A general term for fibres made from inorganic MINERALS such as glass.

mineral-insulated cable. *See* METAL-SHEATHED MINERAL-INSULATED CABLE.

mineral spirit. *See* WHITE SPIRIT.

mineral-surfaced bitumen felt. A form of bitumen ROOFING FELT that is used as a final layer in BUILT-UP ROOFING. Fine particles of slate or other stone are embedded in the upper surface to give protection.

mineral turpentine. *See* WHITE SPIRIT.

mineral wool (rock wool). A general term for various open-structured fibrous materials made from MINERAL FIBRES or slag. Mineral wool is used for THERMAL INSULATION and as a resilient layer.

miniature circuit breaker (MCB). A compact form of CIRCUIT BREAKER.

minibore system. *See* MICRO-BORE SYSTEM.

minicomputer. A comparatively cheap computer that has less ability than a MAINFRAME COMPUTER but is more powerful than most MICROCOMPUTERS.

minimum cement content. A lower limit to the quantity of cement specified in a DESIGNED MIX of concrete.

minus sight. In surveying, a FORE SIGHT.

mirror screw. A type of screw used to fix a mirror. A shiny dome fits over the head of the screw and matches the mirror.

mission tiles. (*USA*) *See* ITALIAN TILES.

mist coat. A very thin coat of paint, usually sprayed on to a surface.

mitre (*USA*: miter). A joint formed between two members with similar cross-section so that the join cuts the angle in two. A right-angled mitre is formed by cutting each of the two faces at 45° to its axis. *See* TIMBER JOINTS diagram.

mitre block (mitre box). A device used to help in the cutting of MITRE joints. One form of mitre block is an L-shaped block of wood containing slots at 45° to guide cutting.

mitre brad. A CORRUGATED FASTENER.

mitred cap. In a staircase, a NEWEL cap into which the handrail is mitred.

mitre dovetail (secret dovetail). A form of DOVETAIL joint in which the joining pins are hidden.

mitred valley. *See* CLOSE-CUT VALLEY.

mitre joint. *See* MITRE.

mitre knee. A MITRE made where a horizontal part of a STAIR HANDRAIL meets a sloping part.

mitre saw. *See* TENON SAW.

mitring machine. *See* TRIMMING MACHINE.

mix. (1) The proportions of ingredients in a batch of CONCRETE, MORTAR or PLASTER. (2) A batch of concrete, mortar or plaster that has been mixed ready for use.

mixed drainage system. *See* COMBINED SYSTEM.

mixing valve. A valve that combines supplies of hot and cold water. The outlet temperature may be controlled manually or by THERMOSTAT.

mix proportions. *See* MIX.

MJ. *abbr.* Megajoule. 1 MJ = 1,000,000 joules. *See also* JOULE and MEGA-.

moat drain. A drain laid along the side of a building to intercept the natural drainage of the ground and to protect the building.

MOATS. *abbr.* (*UK*) Methods of Assessment and Test. A publication of the BRITISH BOARD OF AGREMENT.

mode. In statistics, the value in a sample of numbers that occurs most frequently. *Compare* MEAN, MEDIAN.

modular brick. A type of BRICK, of various standard sizes and proportions, that fits measurements used in metric design.

modular component. Part of a building whose dimensions are designed to coordinate with other components of a MODULAR SYSTEM.

modular masonry unit. A BRICK or BUILDING BLOCK that has the same dimensions in each direction, including joints.

modular ratio. In a composite material, such as reinforced concrete, the ratio of the ELASTIC MODULUS of the reinforcement to the elastic modulus of the masonry.

modular system. A form of DIMENSIONAL COORDINATION in which the components of a building are based on the size of a standard MODULE on a PLANNING GRID.

module. (1) A unit of size chosen as the basic unit of a MODULAR SYSTEM. It is used by itself or in simple multiples. (2) A component in prefabricated construction.

modulus of elasticity. *See* ELASTIC MODULUS.

modulus of rigidity. *See* SHEAR MODULUS.

modulus of rupture. The STRESS at which a brittle material like mass concrete fails.

Mohr's hardness scale. A relative measure of the property of HARDNESS of a material. The scale ranges from 1 for talc to 10 for diamond.

moisture barrier. A general term that can mean a DAMP-PROOF MEMBRANE against liquid water or a VAPOUR BARRIER against water vapour.

moisture content. In general, a measurement of the water contained in a material. (1) The moisture content of timber is given as the percentage weight of water in a sample compared to the final dry weight. The result can be greater than 100 per cent for green timber. Timber used in construction has moisture contents in the range 15–22 per cent. *See also* FIBRE SATURATION POINT, WARP. (2) The moisture content of soil is given as the weight of water in a sample compared to the weight of solids. (3) A measure of ABSOLUTE HUMIDITY.

moisture equilibrium. The state of a material, such as wood, when it no longer increases or decreases in MOISTURE CONTENT for given conditions of surrounding temperature and humidity.

moisture gradient. The continuous change in MOISTURE CONTENT between different parts of a material, such as the inside and outside of a brick wall.

moisture meter. An instrument that measures the MOISTURE CONTENT of materials such as wood and masonry. The electrical properties of resistance or capacitance depend on moisture content and can be used by the meter.

moisture movement. The change in size of a building component with natural variations of MOISTURE CONTENT. Materials such as wood, concrete and brick expand with increased moisture and contract with drying.

mold. (*USA*) *See* MOULD.

mole. *See* BREAKWATER.

mole drain. A temporary drain formed by drawing a steel cylinder through the ground by tractor. It is used in heavy soils like clay.

moler brick. A lightweight brick or block made from DIATOMITE.

moment of force (moment). A measure of the turning effect of a FORCE about a particular point. It is equal to the size of the force multiplied by the perpendicular distance from the point to the line of action of the force.

moment of inertia (I, second moment of area). A measure of the resistance of a body to rotation. It is calculated as the sum of the products of the area or mass of each element in the body and the square of their distance from the axis of rotation. *See also* RADIUS OF GYRATION.

moment of resistance. A measure of the resistance inside a beam to forces that try to bend the beam. It can be calculated as the highest BENDING MOMENT that the beam can carry without exceeding the design STRESS of the material.

momentum. A measure of the ability of a body to maintain its motion. It is calculated as the product of the mass of a body and its velocity.

monitor roof (monitor). A flat or pitched roof with a long built-in rooflight. The glazing is usually vertical.

Monk bond (flying bond, Yorkshire bond). In brickwork, a modified form of FLEMISH BOND that has a sequence of two stretchers and one header repeated in each course.

monkey tail. A downward curve at the end of a handrail.

monkey-tail bolt. *See* EXTENSION BOLT.

monochromatic light. LIGHT that contains radiation of a single frequency, or single colour. Most colour is reproduced by a mixture of frequencies as in ADDITIVE COLOUR or SUBTRACTIVE COLOUR.

monolithic. Part of a structure that is formed as a single block without joints.

monolithic screed. A floor SCREED that is laid on to a fresh concrete slab so that is well-bonded to the concrete.

monomer. A basic molecule from which a POLYMER is made.

mono-pitch roof (penthouse roof). A pitched roof with a single slope. *Compare* LEAN-TO ROOF.

monumental mason. A MASON who does decorative work in stone, such as shapes and lettering.

mopboard. (*USA*) *See* SKIRTING BOARD.

mortar. A mixture used to lay bricks, blocks and stones and to form the BOND between them. Various types of CEMENT MORTAR are common. *See also* GAUGED MORTAR, LIME MORTAR, MASONRY MORTAR. *See* BRICKWORK diagram.

mortar board. *See* HAWK.

mortar-cube test. A standardized test for the strength of CEMENT. A special mortar is hardened in standard moulds to make cubes which are then tested for CRUSHING STRENGTH.

mortgage. An agreement in which a property, such as a house, is used as the security for a loan. The money is usually loaned for the purchase of the mortgaged property.

mortise (mortice). A slot in wood or stone that is made to receive and grip another item, such as a TENON. *See also* OPEN MORTISE.

mortise-and-tenon joint. A joint made between parts of a framework that are usually at right angles to one another. A MORTISE is cut into one member to fit a TENON protruding from the other. *See also* HAUNCHED TENON. *See* TIMBER JOINTS diagram.

mortise chisel. A CHISEL with a thickened blade which can be struck by a MALLET and used to make MORTISES.

mortise joint. *See* MORTISE-AND-TENON JOINT.

mortise latch. A type of LATCH that is set in a MORTISE cut into the STILE of a door. *Compare* RIM LATCH.

mortise lock. A LOCK set in a MORTISE cut into the edge of a door. *Compare* RIM LOCK.

mortising machine (square chisel). A form of power tool or machine that cuts MORTISES in timber.

mosaic. (1) A pattern or picture made up from small pieces of coloured stone or glass bedded into cement on a wall, floor or ceiling. *See also* TESSELLATED, TESSERA. (2) A composite map or photograph made by joining views of adjacent areas.

motorized valve. A valve that is operated by an electric motor and used to open or close the circulation of water in a heating system. *See* HEATING diagram.

mottler. A flat type of paint brush used when MOTTLING.

mottling. A form of GRAINING that reproduces the highlights and shades of a natural wood surface.

mould. (1) A container or structure used to shape materials like plaster or concrete while they set. FORMWORK is a type of mould used on site. (2) A growth of various type of FUNGI that feed on organic materials often present in paper, pastes and oils. In moist conditions, such a damp wall, mould growth usually shows as a green stain.

moulded. A description of a material that has been subject to the process of MOULDING. Moulded plywood for example is bent into a curve.

moulding. (1) The process of shaping a piece of material in a MOULD. (2) A continuous projection, groove or other shape in a surface. It may be for decoration or used to lead water clear of a surface. *See also* CORNICE, OGEE, OVOLO, SCOTIA. *See* MOULDINGS diagram.

moulding machine. A machine designed to cut a MOULDING into timber or stone.

moulding plane. A type of hand PLANE used to cut MOULDINGS in wood.

mould oil. *See* RELEASE AGENT.

mouse. In computer-aided design (*see* CAD), a hand-held device that is attached to the computer and which generates shapes on the VDU screen. The direction of the cursor on screen follows the movement of the mouse on a desk.

movement joint (expansion joint). A flexible joint between slabs of material, such as concrete, that allows movement of the slabs without damage. The movement may be expansion or contraction caused by temperature changes, drying, and subsidence.

moving load. A type of LIVE LOAD, such as a vehicle, that moves about on a structure.

moving stair. *See* ESCALATOR.

ms. *abbr.* Millisecond. *See also* MILLI-.

MS. *abbr.* MILD STEEL.

muck. A general term for waste rock, earth or rubbish that needs to be removed from a site.

mullion. An intermediate vertical member of the window frame that divides a window into separate LIGHTS. The lights may be further divided by GLAZING BARS. *Compare* TRANSOM.

multiple glazing. Windows that are constructed with more than one layer of glass. The purpose is usually to improve thermal or sound INSULATION. *See also* DOUBLE GLAZING.

multi-ply. A general name for PLYWOOD with more than three layers.

multi-point water heater. A form of INSTANTANEOUS WATER HEATER that supplies more than one tap.

multi-unit wall. (*USA*) *See* CAVITY WALL.

multi-view drawings. *See* ORTHOGRAPHIC PROJECTION.

Munsell system. A method for classifying the colour of surfaces in terms of the three factors of HUE, CHROMA and greyness value. *Compare* COLOUR COORDINATES.

muntin. (1) In a framed construction, such as a door, an internal member that separates two panels. *See* DOOR diagram. (2) (*USA*) A GLAZING BAR or MULLION.

Muntz metal. A type of BRASS used for castings. It generally contains three parts of copper and two parts of zinc.

mural. A decoration made directly on a wall, often by painting.

MW. *abbr.* Megawatt. 1 MW = 1,000,000 watts. *See also* MEGA- AND WATT.

mycelium. The body of a FUNGUS that contains a mass of HYPHAE filaments.

N

N. *abbr.* NEWTON.

Na. Chemical symbol for SODIUM.

nail. A simple mechanical fastener made from a metal spike with a point at one end and a head at the other end. It is used to join components, usually wooden, by driving the nail with a HAMMER. *See also* ANNULAR NAIL, CLOUT NAIL, LOST-HEAD NAIL, WIRE NAIL.

nailable. A material in which NAILS can be used to make fastenings.

nail float. *See* DEVIL FLOAT.

nailing ground. *See* COMMON GROUND.

nail plate. *See* CONNECTOR.

nail punch. A short steel tool with a blunt point that is used with a hammer to drive the head of a NAIL below the surface of wood.

naphtha. A volatile and highly-flammable liquid made from petroleum. It is used in some preservative paints for it properties of penetration and of killing FUNGI.

nappe. The sheet of water that flows over the top of a DAM or WEIR.

narrow-ringed timber. Timber that has grown slowly so that the ANNUAL RINGS are close together.

natural asphalt. A mixture of BITUMEN and mineral material that occurs naturally in some rock formations. *See also* ASPHALT.

natural bed. Stonework that is laid so that its planes are orientated in the same direction that they had in the ground. *Compare* FACE-BEDDED.

natural cement. A natural clay or limestone that, when burnt, makes a HYDRAULIC CEMENT.

natural circulation. The movement of air or water by natural CONVECTION rather than by mechanical means.

natural frequency. A characteristic frequency of an object at which it will tend to vibrate if it is disturbed. *See also* RESONANCE.

natural gas. A gas that comes from natural deposits in the earth, usually near oil. It consists of HYDROCARBONS such as METHANE. *Compare* TOWN GAS.

natural light. Light obtained from the sky. *See also* DAYLIGHT FACTOR. *Compare* ARTIFICIAL LIGHT.

natural resin. One of various natural POLYMERS used in traditional paints. They are usually obtained from plant sources such as tung tree seeds or from insects such as SHELLAC.

natural sand. SAND that is produced by the natural disintegration of rock.

natural seasoning. *See* AIR DRIED.

natural stone. Stone that has been quarried from the ground. *Compare* CAST STONE.

natural ventilation. The VENTILATION of a room or building by windows and other INFILTRATION without the use of fans and other MECHANICAL VENTILATION.

nautical mile. A unit of length especially used in navigation. 1 nautical mile = 1852 metres (6076.1 feet).

navvy. A traditional term for a labourer who specializes in excavation work.

NBC. *abbr.* (*USA*) National Building Code.

NBS. *abbr.* (*USA*). National Bureau of Standards. *See also* STANDARDS ORGANIZATIONS.

NC. *abbr.* NOISE CRITERION.

NDT. *abbr.* NON-DESTRUCTIVE TESTING.

neat size. The size of wood or joinery after cutting and planing.

NEC. *abbr.* (*USA*) National Electrical Code.

necking (neck). A moulding between the top of a COLUMN and the CAPITAL.

needle. (1) A horizontal beam passed through a wall to act as a temporary support. *See also* ACROW, PROP. (2) A horizontal beam in a FLYING SHORE.

needle scaffold. A type of SCAFFOLD hung from NEEDLES driven into the wall.

needle shower. A body SHOWER with water jets from the sides as well as from the top.

neon. An inert gas that, when used in a discharge tube, gives out a red light. *See also* DISCHARGE LAMP, NOBLE GASES.

neoprene. A form of synthetic rubber based on the polymer polychloroprene.

nest of saws. Different saw blades that can be used in the same frame or handle.

net floor area. The usable floor space of a building. It excludes the area occupied by partitions, stairs, columns and similar spaces.

network. *See* NETWORK ANALYSIS.

network analysis (programme evaluation and review technique). A graphical management aid that represents the stages of a complex project by lines or circles linked together by their time scales and dependencies. *See also* ARROW DIAGRAM, CRITICAL PATH METHOD, CRITICAL ACTIVITY, FLOAT TIME, GANTT DIAGRAM, PRECEDENCE DIAGRAM.

neutral axis (neutral plane). In a structural member that is being bent, the line or plane where there is zero stress.

neutralizing. The treatment of a plaster, concrete or mortar surface before it is painted. The aim of the treatment is to prevent free LIME disrupting the paint.

neutral plane. *See* NEUTRAL AXIS.

newel (newel post). (1) In a flight of STAIRS, an upright post that supports the handrail. *See* STAIRS diagram. (2) The central column that supports a spiral stair.

newel cap. A wooden cap fixed to the top of a NEWEL POST.

newel drop. A decorative feature in a stair where a NEWEL POST projects down through the ceiling line.

newton (N). The SI unit of FORCE.

new town. (*UK*) A town that has been planned and promoted by a government-sponsored authority.

NFB. *abbr.* Narrow flange beam (used in structural steelwork).

NFPA. *abbr.* (*USA*) National Fire Protection Association.

NHBC. *abbr.* (*USA*) National House Building Council.

nib. (1) On a roof TILE, a projection underneath which hooks over the tiling BATTEN. (2) The top edge of a vertical sheet of asphalt where it fits into a slot in the wall. (3) Small particles of foreign matter, such as grit, that project out of a paint film and causes it to feel rough or 'bitty'.

NIBS. *abbr.* (*USA*) National Institute of Building Science.

nickel. A pure metal element (chemical symbol Ni) that resembles iron. It resists corrosion and is used for alloys such as nickel steel.

nicker. *See* CENTRE BIT.

nicol prism. A slice of naturally-occurring mineral that produces polarized light. *See* POLARISCOPE.

night latch (night lock). Any latch, like a CYLINDER LOCK, that is operated by key from the outside and by knob or handle on the inside.

night vent. A small opening, usually in a window, that provides a constant low level of ventilation in rooms such as bedrooms.

nippers (crampon). Two curved levers that are hinged together like a pair of scissors and close together when lifted. They are used from a crane for grabbing and lifting large blocks of stone.

nipple. (1) A small valve in the pipework of a heating system that can be used to release air from the system. (2) A short piece of pipe with a taper thread on the outside of both ends, used to join pipes. (3) A small entrance tube that is used to inject grease into a machine.

nitrocellulose (cellulose nitrate). Various compounds made by treating CELLULOSE with nitric and other acids. It is used for plastics, explosives and CELLULOSE PAINT.

NLMA. *abbr.* (*USA*) National Lumber Manufacturers' Association.

NNI. *abbr.* NOISE AND NUMBER INDEX.

noble gases (inert gases, rare gases). The elements helium, neon, argon, krypton, xenon and radon. They are chemically extremely unreactive and generally exist in the uncombined state.

noble metals. Metals such as gold, silver and platinum that do not readily oxidize or react with acids. They are low in the ELECTROCHEMICAL SERIES.

node. (1) A junction on a NETWORK ANALYSIS diagram. *See also* ACTIVITY. (2) A point in a framework, such as a TRUSS, where two or more members meet.

no-fines concrete. Concrete made without fine AGGREGATE such as sand. It has large pores that resist the CAPILLARY movement of water.

nog. *See* FIXING BRICK.

nogging (dwang, intertie, noggin). A series of short horizontal timbers fixed between STUDS or RAFTERS to stiffen a timber frame. *See* WALL diagram.

noise. Sound that is not wanted at the source. This environmental definition ignores the quality of sound. *See also* TRAFFIC NOISE LEVEL, EQUIVALENT CONTINUOUS SOUND LEVEL.

noise absorption. *See* SOUND ABSORPTION.

noise and number index (NNI). A method used to predict and measure annoyance caused by NOISE around airports. It combines the average PERCEIVED NOISE LEVEL of the aircraft with the number of aircraft heard in a given period.

noise control. *See* SOUND INSULATION, SOUND ABSORPTION.

noise criterion (NC). A method of assessing NOISE produced by heating and ventilating equipment. The NC value is found by comparing a recorded noise with standard NC curves of sound levels plotted against frequency.

noise dose. *See* EQUIVALENT CONTINUOUS SOUND LEVEL.

noise exposure. *See* EQUIVALENT CONTINUOUS SOUND LEVEL, NOISE AND NUMBER INDEX, TRAFFIC NOISE LEVEL.

noise insulation. *See* SOUND INSULATION.

noise rating (NR). A method of assessing NOISE produced by industrial plant. The NR value is found by comparing a recorded noise with standard NR curves of sound levels plotted against frequency.

noise reduction coefficient (NRC). For a

particular building material, the numerical average of the values of SOUND ABSORPTION measured over standard frequency ranges.

noise spectrum. A display of the sound frequencies present in a NOISE. It is made up from SOUND LEVELS measured over a series of frequency bands.

nominal mix. A method of specifying CONCRETE in which the quantities of ingredients for the mix are given by volume, rather than weight. *Compare* DESIGNED MIX, STANDARD MIX.

nominal size of timber. The size of UNWROUGHT TIMBER after it is sawn but before it is planed or worked. The DRESSED SIZE or finished size is about 6 mm less in each direction than the nominal size.

nominated sub-contractor. A supplier of work or materials who is specified in the BILL OF QUANTITIES and must be used by the main contractor.

nominated supplier. A supplier of materials who is specified in the BILL OF QUANTITIES and must be used by the main contractor.

nomogram. A single diagrammatic presentation of the relations between several variables. The value of an unknown can be read from a simple construction, usually a straight line, between two other variables.

non-bearing. *adj.* Describing a structural member, such as a wall, that supports no load apart from its own weight. *Compare* LOAD BEARING, SELF-SUPPORTING WALL.

non-combustible (incombustible). A material that does not take part in the chemical process of COMBUSTION or burning. The properties of a non-combustible material may still change in a fire. *See* FIRE RESISTANCE.

non-critical activity. In CRITICAL PATH analysis, an activity that is not part of the 'critical path'.

non-destructive testing (NDT). Various methods of testing the quality of materials or components, such as concrete, without causing them damage. Techniques include the use of electrical properties, X-rays, RADIOACTIVITY, ULTRASONIC waves, strain and other mechanical properties. *See also* SCHMIDT HAMMER.

non-ferrous metal. A metal other than IRON or an ALLOY that is not based on iron.

non-flammable. *adj.* Not easily FLAMMABLE. It is often dangerous to assume that materials have such properties; the concept of FIRE RESISTANCE is preferred.

non-hydraulic lime. A form of lime that hardens by combining with carbon dioxide. *See also* HIGH-CALCIUM LIME.

non-manipulative joint. A COMPRESSION JOINT between pipes that requires no work to the pipes except cutting. *Compare* MANIPULATIVE JOINT.

Norfolk latch. *See* THUMB LATCH.

normal distribution. *See* GAUSSIAN DISTRIBUTION.

normalizing. (1) A HEAT TREATMENT in which steel is heated and slowly cooled. This softens the metal and makes it less brittle. (2) The adjustment of SOUND INSULATION measurements to allow for the acoustics of the measurement room.

Norman brick. A BRICK with flatter proportions than modern sizes. In the USA the nominal dimensions are 12 by 4 by 2 2/3 inches.

north-light roof. A factory roof made from a series of pitches that have alternate steep and gentle slopes. The steeper slopes are glazed and, in the northern hemisphere, face north so as to avoid direct sunlight.

nose. (1) A blunt overhang such as a ledge or NOSING. (2) The lower part of the shutting STILE of a door or CASEMENT WINDOW.

nosing. A rounded overhang to the edge of a stair TREAD, window sill, flat roof or similar component. *See* STAIRS diagram.

nosing line. A sloping line that would join the NOSINGS of a stair. *See also* MARGIN.

notch. A groove cut into a component so that another component can be fitted.

notched weir. A WEIR used to measure flowrate in a river.

notching. A join made between two timbers by cutting a NOTCH in one or both of them.

novation. In a LEASE or other legal document, the substitution of a new obligation for an old obligation by mutual agreement.

novelty siding. *See* GERMAN SIDING.

noys. A measurement of aircraft noise that can be converted to PERCEIVED NOISE LEVEL.

NPL. *abbr.* (*USA*) National Physical Laboratory.

NR. *abbr.* NOISE RATING.

NRC. *abbr.* NOISE REDUCTION COEFFICIENT.

N-truss. *See* PRATT TRUSS.

NTS. *abbr.* Not to scale (used on drawings).

nut. A square or hexagonal metal collar that contains an internal thread designed to couple with the shaft of a BOLT. *See also* LOCKNUT.

nylon. A general term for a class of plastics made of polyamides. It is a THERMOPLASTIC material with high strength and toughness.

NZS. *abbr.* New Zealand Standard.

O

oak. Various types of HARDWOOD trees belonging to the genus *Quercus*. The wood is valued for the look of its open-pored grain which is sometimes imitated by GRAINING or highlighted as in LIMED OAK. The hardness and durability of the timber has made it useful for traditional house construction and for joinery such as doors and windows.

oakum. A fibrous material produced from hemp or the strands of old rope. It was traditionally used for CAULKING the seams of some buildings or pipes.

oblique butt joint. In joinery, a BUTT JOINT that is made at some angle other than a right-angle.

OBM. *abbr.* ORDNANCE BENCH MARK.

obscured glass (translucent glass). GLASS treated so that it allows light to pass but cannot be seen though. The surface of the glass is made irregular by moulding or by roughening with sand-blasting. *See also* OPAL GLASS.

obsidian. A dark volcanic rock with glass-like qualities.

ochre. A yellow-brown coloured PIGMENT made from sand containing iron oxide.

octave (octave band). In sound, the range of frequencies between any one frequency and double that frequency.

octave band analyser. An instrument that can analyse the frequencies present in a particular sound.

OD. *abbr.* (1) ORDNANCE DATUM. (2) Outside diameter.

ODM. *abbr.* OPTICAL DISTANCE MEASUREMENT.

oedometer. An instrument used to measure the CONSOLIDATION of a soil sample under test.

offset. (1) A ledge produced by a change of thickness in a wall or the joint between two displaced lengths of parallel pipe. (2) In surveying, a horizontal distance measured at right angles to a survey line.

offset hinge. *See* CLEANING HINGE.

offset screwdriver. A screwdriver with a bend in its shaft so that it can turn screws at right angles to its length.

off-white. A colour, such as 'cream', that is mainly white with a little darker pigment added.

ogee. The outline of a moulding or other edge that makes two curves, such as in the letter S. *See* MOULDINGS diagram.

O/H. *abbr.* Overhead.

ohm. The unit of ELECTRICAL RESISTANCE. It is defined as the POTENTIAL DIFFERENCE in volts divided by the CURRENT in amperes. *See also* OHM'S LAW.

Ohm's law. The principle that, for electrical conductors at constant temperature, the CURRENT flowing is directly proportional to the POTENTIAL DIFFERENCE across the conductor. The constant of proportionality is the RESISTANCE of the conductor measured in OHMS.

oil paint. A paint in which the BINDER is an

oil that forms a film by oxidation or polymerization. *Compare* EMULSION.

oil stain. A thin paint containing a little PIGMENT that may used for colouring wood floors.

oilstone. A HONE.

oilstone slip. A small HONE with a curved edge that is used for sharpening tools.

oleo-resin. (1) A form of natural RESIN taken from pine trees and used to make TURPENTINE. (2) In paint, a mixture of a DRYING OIL and a hardening RESIN that produces a high GLOSS.

olive. *See* GLAND.

one-hour wall. (*USA*) A wall with a FIRE RESISTANCE of at least one hour.

one-part (one-pack). *adj.* Descring an adhesive or sealant that is supplied and used from a single container. *Compare* TWO-PACK.

one-pipe system. (1) A system of drainage from a building that combines the waste and the SOIL outputs. *Compare* TWO-PIPE SYSTEM. (2) A heating system in which each radiator is connected to a single circulating pipe. *Compare* TWO-PIPE SYSTEM.

opacity. The COVERING CAPACITY of a paint.

opal glass. A form of whitish OBSCURED GLASS made by adding minerals to the molten glass.

OPC. *abbr.* Ordinary Portland cement. *See* PORTLAND CEMENT.

open assembly time. The time allowed between the application of GLUE to surfaces and the assembly of the components.

open caisson. A type of CAISSON that is open at both top and bottom.

open cell. A cellular material in which the adjacent cells are interconnected. *Compare* CLOSED CELL.

open channel. A channel that contains air above the surface of the water. The channel may be covered and the water flows by gravity rather than pressure.

open circuit. An electrical circuit that is not complete. Some equipment, such as security alarms, are run with an open circuit and activate if the circuit is closed.

open cornice. (*USA*) *See* OPEN EAVES.

open defect. Any timber defect, such as a KNOT or CHECK, that shows as a hole or split in the surface.

open-divided scale. A type of SCALE RULE with markings arranged so that only the divisions at each end are fully subdivided into the smallest units.

open eaves. EAVES of a roof in which the undersides of the rafters are left exposed rather than covered with a SOFFIT BOARD.

open fire. A fire that burns in a FIREPLACE rather than in an enclosed burner.

open floor. A suspended wooden floor with the JOISTS exposed on the underside because there is no ceiling beneath.

open grained. *See* WIDE-RINGED TIMBER.

opening leaf. The LEAF in a folding door that is hinged and opens. *Compare* STANDING LEAF.

opening light. A window LIGHT that can be opened, unlike a DEAD LIGHT.

open joint. An unsealed gap between panels in a RAIN SCREEN CLADDING. Water that passes through the join is collected and drained by trays or channels.

open mortise (slot mortise). A MORTISE that is open on three sides and makes a slot completely through a member.

open-newel stair. A stair without NEWELS, like a GEOMETRIC STAIR.

open plan. A layout for the interior of a building that features large spaces and few fixed partitions.

open-riser stairs. A stairway where the TREADS are supported without RISERS between them.

open roof. A pitched roof without a ceiling so that the rafters can be seen.

open slating. SLATES or TILES that are laid on a roof with a gap between those in the same COURSE.

open stair. (*USA*) A stair that has no wall at one or both sides.

open string. *See* CUT STRING.

open-tank treatment. A method of timber treatment where the PRESERVATIVE chemicals are applied by simple immersion of the timber, rather than by the use of a PRESSURE TANK.

open tender. A TENDER that is advertised among a wide range of contractors rather than limited to pre-selected contractors.

open traverse. A surveying TRAVERSE in which the final line is not joined to the starting point.

open valley. A roof VALLEY where the tiles or slates are arranged so that the gutter at the bottom of the valley can be seen. *Compare* SECRET GUTTER.

open-vented system. A water-heating circuit with an open expansion pipe at the top. *Compare* SEALED SYSTEM.

open-web girder. *See* LATTICE.

open-well stair. A STAIR in which the FLIGHTS are arranged around an open area.

operational research. A systematic study of business or industrial operations whose results are used to help make management decisions. *See also* NETWORK ANALYSIS.

optical distance measurement (ODM). In surveying, the use of traditional sighting instruments such as a THEODOLITE. *Compare* ELECTROMAGNETIC DISTANCE MEASUREMENT.

optical-reading theodolite. A THEODOLITE with horizontal and vertical scales engraved on glass. Prisms transmit an image of the scale into an eyepiece next to the telescope.

optical square. A hand-held surveying instrument used to fix points at right angles to a line such as a survey line. Mirrors or prisms are used to view two points at the same time.

optical thermometer. *See* PYROMETER.

optimization. The process of choosing the most efficient balance between various requirements that may conflict.

orange peel (pock marking). A paint defect where a spray-painted surface dries with a rough surface like an orange. It is caused by various incorrect techniques of spray painting.

orbital sander. A sanding tool where an electric motor quickly moves an abrasive pad in a circular motion. The tool is used to produce a smooth sanded finish on timber. *Compare* BELT SANDER.

ordinary Portland cement (OPC). *See* PORTLAND CEMENT.

ordinate. A straight line on a graph or map that is parallel to an axis and can be used to locate a particular point.

Ordnance Bench Mark (OBM). In the United Kingdom, a BENCH MARK established by the ORDNANCE SURVEY. The height of this bench mark is accurately taken from the ORDNANCE DATUM.

Ordnance Datum (OD). The main DATUM used by the ORDNANCE SURVEY to establish heights in Great Britain. It is based on the mean value of sea level readings at Newlyn in Cornwall.

Ordnance Survey. In the United Kingdom, the organization and work of the official survey and mapping of the country.

ore. A naturally-occurring aggregate, such as rocks, from which metals or other substances are extracted.

Oregon pine (Oregon fir). Other names for DOUGLAS FIR.

organic. A description of the many compounds that are based on carbon rather than on MINERALS. Such compounds were originally obtained from living 'organic' sources.

oriel window. A type of window that projects from an upper storey and is carried on CORBELS.

orientation. The position of a building in relation to its surroundings.

origin. On a graph, the point where the two axes intersect and where the scales usually start at zero.

O-ring joint. A type of water-tight joint made between pipes such as drains. A flexible ring is inserted in the socket of one pipe and allows some movement in the joint.

orthogonal. *adj.* In general, describing drawings or axes that involve right angles. *See* ORTHOGRAPHIC PROJECTION.

orthographic projection (multi-view drawings). A drawing PROJECTION that shows solid objects, such as a building, by a set of PLANS, SECTIONS and ELEVATIONS drawn on the same flat plane. Connections are drawn between lines that exist in all the different viewpoints. Various conventions exist such as 'first angle' (or European) projection. The projection is commonly used to produce WORKING DRAWINGS. *Compare* AXONOMETRIC PROJECTION, ISOMETRIC PROJECTION.

orthotropic. *adj.* Describing substances that, like wood, have different physical properties in different directions. *Compare* ISOTROPIC.

osier. *See* WITHY.

osmosis. The movement of water or other solvent through a porous skin or membrane. The direction of the movement is in one direction, towards the more concentrated solution.

osmotic pressure. The pressure difference set up between two sides of a membrane by the process of OSMOTIC PRESSURE.

outbuilding (outhouse). A smaller building, such as a garden shed, that is part of the main building but not attached to it.

outer string (side string). The STRING of a stair that is furthest from a wall. *Compare* WALL STRING. *See* STAIRS diagram.

outfall. The place where a sewer or drainage channel discharges into the sea, river or watercourse.

outhouse. *See* OUTBUILDING.

outlet. (1) An opening in a gutter or other drain. (2) An electric SOCKET OUTLET.

outline planning permission. (*UK*) A provisional approval for a certain size and type of building subject to detailed plans and formalities. *See* PLANNING PERMISSION.

outrigger. A beam or structure that projects from a building and carries a structure such as a FLYING SCAFFOLD.

outside glazing. Glazing where the glass is inserted into the frames and secured from the outside. *Compare* INSIDE GLAZING.

out-turn cost. The actual final cost of a project.

oven-dry timber. Timber dried in a ventilated oven at 103°C.

overcast sky. For design purposes, a sky that is assumed to have continuous layer of cloud. *See also* STANDARD SKY.

overcloak. At a seam in FLEXIBLE-METAL ROOFING, the part of an upper sheet that overlaps the lower sheet (the UNDERCLOAK).

overflow pipe. A drainage pipe installed above the normal water level in a storage tank or cistern. If a BALL VALVE fails the pipe allows excess water to flow safely away. *Compare* WARNING PIPE.

overflow warning pipe. *See* WARNING PIPE.

overgrainer. A thin type of paint brush used when GRAINING.

overhand work. External bricks that are

laid in a wall from the inside of the building rather than from a scaffold on the outside.

overhang. In general, part of a structure such as roof or floor that projects beyond the wall that carries it.

overhanging eaves. EAVES of a roof that project over the wall.

overhead door (up-and-over door). A type of door that opens by lifting up and sliding horizontally at the head of the door.

overheads (overhead expenses). The fixed costs of running an organization that have to be paid although they are not the result of any particular material or labour. The charge for the overheads is usually spread over the items in a CONTRACT.

oversailing course. A STRING COURSE of brick or stone that helps form the projection of a CORBEL. *See* ROOF diagram.

oversite (solum). The ground area within the containing walls of a building, after the removal of topsoil and vegetation.

oversite concrete. A layer of concrete that is laid over the ground beneath a suspended timber floor to keep the construction dry.

overtime. Extra payment for labour that is engaged outside normal working hours, such as on a weekend. *See also* DOUBLE TIME.

overtone. A FREQUENCY of sound that is equal to the FUNDAMENTAL FREQUENCY of a sound multiplied by a whole number. Most sounds contain overtones or harmonics that accompany the fundamental and give it a particular quality or 'timbre'.

ovolo. A MOULDING that includes an outward curve in its profile. It often makes a quarter of a circle. *Compare* SCOTIA. *See* MOULDINGS diagram.

oxidation. (1) A chemical process that usually involves the combination of a substance with OXYGEN to form an oxide. The simple RUSTING of iron is an example of oxidation. (2) A process in the drying of some paints that causes them to harden by chemical reaction of the resins rather than by simple EVAPORATION.

oxide. A compound formed by the joining of a chemical element with oxygen, usually by the process of OXIDATION. Most metal ORES are oxides of the metal.

oxter piece. A vertical member in ASHLERING.

oxy-acetylene flame. A high-temperature flame that is used for welding and cutting metal. It is produced by burning a combination of oxygen and acetylene gas from pressurized cylinders.

oxychloride cement. *See* MAGNESITE FLOORING.

oxygen. A colourless gaseous element (chemical symbol O) that exists in the atmosphere. It is highly reactive: most of the oxygen on Earth is combined with other elements in the rocks. *See also* ORE, OXIDATION.

ozone. A form of oxygen gas containing three atoms to a molecule instead of two. Ozone can be formed by an electric discharge such as a spark and it is chemically very active.

ozone treatment. One of several methods of DISINFECTION used in the treatment of water supplies to ensure that a safe quality is maintained in the distribution system. Air is dried, subjected to a high voltage and the OZONE produced is injected into the water. *Compare* CHLORINATION.

P

package deal (design-and-build contract). An all-in contract for a large project in which a CONTRACTOR takes responsibility for all aspects of design and construction.

packing. In general, material that is used to fill gaps between parts of an assembly to make them fit tightly.

pad. *See* PADSTONE.

paddle mixer. A bin containing two rotating shafts used for mixing MORTAR.

pad foundation. An isolated slab of concrete that acts as a foundation for a single load.

pad saw (compass saw, keyhole saw, lock saw). A hand SAW with a thin tapered blade used for cutting sharp curves. *See* WOODWORKING TOOLS diagram.

padstone (cushion, pad). A stone or concrete block placed under a heavy load, such as at the end of a steel beam, in order to distribute the forces. *See also* BASE PLATE.

paint. A liquid coating that is applied to various surfaces and components. It dries as a hard film that gives protection and decoration. The main components of paint are PIGMENTS carried in a liquid MEDIUM. *See also* DRYING OIL, MEDIUM, THINNER, VEHICLE. *Compare* VARNISH.

paint can (paint kettle). A container used to contain PAINT ready for application by brush.

painter's putty. GLAZIER'S PUTTY used to fill holes in surfaces before they are painted.

paint harling. A rough surface coating, like ROUGH CAST, formed from paint-covered stone chippings thrown on to a thick sticky paint surface.

paint kettle. *See* PAINT CAN.

paint stripper (paint remover). A liquid that is coated over an old paint film to soften the paint and allow removal by scraping.

paint system. A series of coats of different PAINTS that have different purposes. A typical three-coat system consists of PRIMER, UNDERCOAT and top coat.

pale (picket). An upright board or stake in a FENCE or PALISADE.

palette knife. A knife with a thin flexible blade that is used to mix and match PAINT, often on a 'palette board'.

palisade. An enclosure formed from a FENCE made up of vertical PALES attached to RAILS or RIBANDS.

pallet. (1) *See* FIXING FILLET. (2) A movable platform that can be lifted by a fork-lift truck. It is used to move and store materials.

pan-and-roll tiles. (*USA*) *See* ITALIAN TILES.

pane. A sheet of glass cut to fill the space between the GLAZING BARS of a window and form a LIGHT.

panel. (1) An area of sheet material that fills the space inside a frame. (2) A wall of brick or other material that fills the space between the frame of a structure. (3) A span or bay of a concrete slab.

panel absorber. A sheet of material, such

as a suspended ceiling, that provides SOUND ABSORPTION at low frequencies. *Compare* CAVITY ABSORBER.

panel door (panelled door). A door constructed as a framework of STILES and RAILS. The spaces between the framework are filled by panels of wood or glass. *See* DOOR diagram.

panel heater. A device for heating rooms by means of a warm flat panel installed flush with the other room surfaces. The heating is usually by ELECTRIC PANEL HEATERS, and sometimes by hot water pipes.

paneling. (*USA*) *See* PANELLING.

panelled door. *See* PANEL DOOR.

panelling. (1) Materials used to make PANELS. (2) An area of construction formed from PANELS.

panel pin. A thin NAIL with a small head used to fix joinery. It can be driven beneath the surface and hidden.

panel saw. A type of CROSS-CUT SAW used to cut BOARD material.

panel wall. In a FRAME BUILDING, an external wall that only carries its own weight. It may be made up of WALL PANELS.

panic bolt (panic bar). A door bolt used on an emergency exit in a public building. The bolt is opened by pushing on a horizontal bar across the door.

pantile. A traditional form of SINGLE-LAP tile with a cross-section in the shape of a shallow letter S. When laid on a roof the curved shape of each tile fits and overlaps the neighbouring tile in the same COURSE.

pantograph. (1) A device made of interconnected rods used to copy one drawing on to another, with or without a change of scale. (2) A hinged metal framework on the roof of an electric locomotive. It collects the power from the overhead supply wire.

paperhanging. The process of applying WALLPAPER or similar materials to cover surfaces such as walls and ceilings.

PAR. *abbr.* (1) *See* PAR LAMP. (2) Planed all round. *See* WROUGHT TIMBER.

parabola. An open curved shape made by cutting vertically through a cone. In structures it is the shape made by the BENDING MOMENT distribution in a uniformly loaded beam.

paraffin. (1) A waxy HYDROCARBON from the alkane series of compounds formerly known as the 'paraffin' series. (2) (*UK*) Kerosine oil.

parallel coping. A form of COPING with a uniform section rather than a WEATHERING slope. It is used for the top of sloping structures such as a GABLE wall.

parallel gutter. *See* BOX GUTTER.

parallel-motion equipment. A system fitted to a DRAWING BOARD that allows a straight edge to move over the board and stay parallel to the edges of the board.

parallel thread. A type of screw THREAD that keeps the same diameter and is used for connectors such as bolts. *Compare* TAPER THREAD.

Paraná pine. A fine-textured and strong form of SOFTWOOD.

parapet. *See* PARAPET WALL.

parapet gutter. *See* BOX GUTTER.

parapet wall (parapet). A low wall around the edge of a roof or balcony. It may also be the part of a wall that projects above the roof and is therefore exposed on all sides.

parenchyma. The type of wood tissue that stores food and forms the medullary RAYS.

parget (parge). A mixture used in PARGETTING. It is usually made from cement mortar or COARSE STUFF.

pargetting. (1) The process of RENDERING the inside of a brick FLUE so as to seal it. (2) A traditional form of plasterwork on the outside of some British buildings. It is usually decorated in repeating patterns made by pressing a mould into the plaster while it is soft.

paring chisel. A long CHISEL with a thin blade designed to be worked by hand and not struck with a MALLET.

PAR lamp. A FILAMENT LAMP in the form of a spotlamp which contains an internal reflector.

parliament hinge. *See* H-HINGE.

parpen, parpend. *See* PERPEND STONE.

parquet (parquet floor). A polished wooden floor surface made from thin hardwood blocks fixed onto a subfloor in a geometrical pattern. *See also* PLYWOOD PARQUET.

parquet strip. A wooden floor surface formed from hardwood boards that are TONGUED-AND-GROOVED and SECRET NAILED to the subfloor.

partially-fixed joint. A structural joint that does not transfer the full BENDING MOMENT from one member to another.

partially-separate system. A drainage system in which the rainwater from the roof and yard of a building joins the foul water but other rainwater is drained separately. *Compare* COMBINED SYSTEM, SEPARATE SYSTEM.

particle board. *See* CHIPBOARD.

parting agent. *See* RELEASE AGENT.

parting bead. A narrow vertical strip of wood separating the two sashes in a SASH WINDOW.

parting slip. A narrow strip of wood that separates the two weights inside the frame of SASH WINDOW.

partition. An internal wall that separates rooms. It is usually one-storey and not loadbearing. It may be made from brick, blocks or a lined framework. *See also* STUD PARTITION.

partition head (partition plate). The top horizontal member of a STUD PARTITION.

partnership. A legal agreement between two or more persons to run a joint business and share in liabilities and profits.

parts per million (ppm). A measure of concentration expressed as the number of parts of one substance distributed in one million parts of the other substance.

party fence. A FENCE that separates two properties and is often the joint responsibility of both owners.

party wall. A boundary wall that separates two buildings and is shared by their structures. It requires high standards of fire resistance and sound insulation.

pascal. The SI unit of PRESSURE, which is expressed as force per unit area. 1 pascal = 1 NEWTON per square metre.

passings. The LAP or amount by which one sheet of FLASHING overlaps the next sheet.

passive earth pressure. The resistance to horizontal forces offered by a vertical face of earth.

passive solar construction. A building that makes use of the sun's energy by means of fundamental design and materials rather than by special equipment. Heat storage in heavyweight walls is an example. *See also* TROMBE WALL.

paste. (1) In general, any soft mixture that can be worked and moulded. (2) An adhesive coating used to fix decorative materials such as wallpaper. Traditional pastes are based on cellulose; modern pastes use synthetic polymers such as acrylic.

patch. An INSERT of veneer used to patch the surface of plywood.

patent glazing. A system of GLAZING that fixes the glass by dry methods rather than by the use of putty. Most systems use special metal glazing bars to support the glass and to make the seal.

paternoster. A form of LIFT in which a series of suspended open compartments move in an endless chain, like the a fairground wheel. Goods or passengers must load and unload while the compartment moves slowly past the floor without stopping.

patina. A protective film of oxide or other compounds that form on a metallic surface when it is exposed to the air. Copper, for example, slowly acquires a green coating of 'verdigris'.

patterned glass. Glass that is smooth on one side and has a pattern or texture on the other side. *See also* OBSCURE GLASS.

pattern staining. On ceilings, a natural deposit of dirt or dust that forms a pattern revealing the hidden structure behind the surface. It is a result of poor and uneven thermal insulation.

pattress. The base of a light fitting or wall switch.

Pavelock. (*USA*) Trade name for a system of interlocking PAVERS.

pavement. (1) (*UK*) A path or roadside footway with a durable surface such as asphalt or concrete. (2) The general construction of a road. *See also* BASECOURSE, MACADAM, WEARING COURSE.

pavement light. A LIGHT constructed from GLASS BLOCKS set into the pavement to give light to a basement below.

pavement prism. One of the blocks of glass that make up a PAVEMENT LIGHT.

paver. (1) A small slab of hard brick or concrete laid with others as the surface of a PAVEMENT. (2) A large movable concrete mixer used in road-making.

pavilion. (1) A decorative building in a garden. (2) A temporary structure such as used for exhibitions.

payback time. For a project such as energy saving, the time taken for the saving in running costs to match the initial cost of the project.

Pb. Chemical symbol for LEAD.

pc. *abbr.* PRIME COST.

PCA. *abbr.* (*USA*) Portland Cement Association.

pd. *abbr.* POTENTIAL DIFFERENCE.

PE. *abbr.* POLYETHYLENE (especially in the description of plastic pipes).

peak value. In ALTERNATING CURRENT electricity, the maximum value of current or voltage during a cycle.

pean. *See* PEEN.

pearlite. *See* PERLITE.

peat. A type of SOIL consisting mainly of decayed vegetable matter. It is acidic in nature and has a high moisture content.

pebble dash. A protective coating for external walls. It is formed by throwing small stones against RENDERING before the render mix is set. *See also* ROUGH CAST.

pebble walling. A type of wall made from rounded pebble stones, such as those from a beach.

pedestal. The base that supports a COLUMN or other load.

pedestal washbasin. A WASHBASIN supported from the floor by a hollow column that is usually made of matching porcelain.

pediment. An ornamental triangular wall, similar to a low GABLE, which is usually built above a portico, door or window.

peeler log. A log used for cutting thin sheets of veneer.

peeling. A defect that causes a PAINT film to detach and lift away from the underlying surface.

peen (pean, peine). On a HAMMER head, the opposite end to the striking face. It may be in the shape of a ball or wedge.

peen hammer. A hammer without a flat head. Instead there are two cutting PEENS that are used for working stone. *See* WOODWORKING TOOLS diagram.

peg. In general, a short rod of metal or wood that is used to fasten components like tiles.

pegboard. A BOARD with a surface pattern of small holes into which pegs can be inserted.

peggies. Small SLATES of random sizes.

peg stay. A type of casement stay that holds a window open by resting on a PEG that engages holes in the stay.

peine. *See* PEEN.

pellet. A small round piece of wood cut to cover the head of a countersunk screw.

pelmet. A long strip of board or cloth set above a window in order to hide the curtain rail and other fittings.

pelton wheel. A type of TURBINE that is driven by impact of water on cups or buckets mounted on the edge of a wheel.

pencilling. The process of painting the mortar joints in brickwork to highlight their pattern.

pendant. A light fitting or ornament that is suspended from a ceiling or roof.

pendulum saw. A type of CIRCULAR SAW that is suspended above a log and is lowered to cut across the log.

penetration test. A test made on site to assess the load-bearing of a soil. To make the test, a cone-shaped 'penetrometer' is forced into the ground at a known pressure.

penetrometer. *See* PENETRATION TEST.

penstock. (1) A pipe or conduit that supplies water to drive a TURBINE. The penstocks in a hydro-electric station often run down the outside of the dam. (2) A gate that controls the flow of liquid in a sewer pipe or channel.

pentachlorophenol. A timber PRESERVATIVE.

penthouse. A dwelling such as an apartment built on the flat roof of a large building. It usually has special features such as private terraces and views.

penthouse roof. *See* MONOPITCH ROOF.

penultimate certificate. The document that authorizes payment to the CONTRACTOR when work is almost finished. It consists of the total value of labour and materials less amounts for any outstanding work. *See also* FINAL CERTIFICATE, INTERIM CERTIFICATE.

perceived noise level. A method used to assess the NOISE of aircraft. It takes account of annoying higher frequencies in engine noise that are not adequately measured by the A-SCALE. The noise is rated in 'noys' and converted to units of perceived noise decibels PNdB. *Compare* NOISE AND NUMBER INDEX.

percentage of reinforcement. The total sectional area of reinforcement in a concrete beam compared to total area of the concrete.

percolation. The movement of water or vapour through fine pores of rock or soil.

perfect frame. A frame structure that is stable when loaded from any direction. All members of the frame are needed for stability, unlike a REDUNDANT FRAME.

perforated brick. A BRICK with a number of holes that run from top to bottom. These perforations reduce the weight and allow easier firing in the kiln.

performance specification. A detailed description of the minimum standards expected of a material or component. It usually refers to standard tests defined by one of the STANDARDS ORGANIZATIONS.

pergola. A horizontal framework supported on posts. It is used to carry climbing plants and to provide a walkway beneath.

periodic table. An arrangement of the chemical ELEMENTS in order of increasing atomic number or mass and also in groups with related properties.

perlite (pearlite). A natural volcanic glass found mainly in North America. Its expanded form is used a lightweight AGGREGATE for concrete or plaster.

perlite plaster. A type of gypsum plaster made with PERLITE as aggregate instead of sand. It forms a lightweight plaster with good thermal insulation.

perm. (*USA*) Vapor permeance. A measure of VAPOUR DIFFUSIVITY.

permafrost. SUBSOIL that is permanently frozen.

permanent hardness. A form of HARD WATER in which the hardness can not be removed by boiling, unlike TEMPORARY HARDNESS. Permanent hardness is caused by the presence of dissolved salts such as calcium sulphate (gypsum) and magnesium sulphate (epsom salts).

permanent set. A deformation of a material or structure that remains if the deforming load is removed. It includes CREEP and yield.

permanent shuttering. FORMWORK for concrete that remains in place as part of the structure after the concrete has set.

permanent supplementary artificial lighting of interiors (PSALI). A system of combined lighting where parts of an interior are lit for the whole time by artificial light that is designed to balance and blend with the daylight.

permanent supplementary lighting. *See* PERMANENT SUPPLEMENTARY ARTIFICIAL LIGHTING OF INTERIORS.

permanent threshold shift (PTS). A permanent loss of hearing caused by a prolonged exposure to noise. *Compare* TEMPORARY THRESHOLD SHIFT.

permeability. The rate at which a material allows a liquid or gas to move through its structure. The opposite of IMPERMEABILITY.

permeameter. An instrument used to measure the PERMEABILITY of a soil sample.

permissible stress. The ULTIMATE STRESS of a material divided by the FACTOR OF SAFETY.

permissible working load. The maximum routine load that should be placed on a structure. It is less than the ULTIMATE LOAD.

perpend. (1) CROSS JOINT. A vertical joint in brickwork or masonry. *Compare* BED JOINT. (2) A corner of brickwork that has an accurate vertical and is used as a guide for the rest of the wall. (3) *See* PERPEND STONE.

perpend stone (parpen, parpend). A stone that passes completely through a wall and presents a smooth face to both sides.

perspective drawing. A two-dimensional drawing that shows the three-dimensional appearance and relationships of objects as seen from a particular view point. *See also* ISOMETRIC PROJECTION, ORTHOGRAPHIC PROJECTION.

Perspex. A tough, lightweight transparent PLASTIC material sometimes used as a substitute for glass. It an ACRYLIC type of POLYMER which, being THERMOPLASTIC, has relatively low FIRE RESISTANCE.

PERT. *abbr.* Programme evaluation and review technique. *See* CRITICAL PATH ANALYSIS.

pet cock. A small valve used to release unwanted air in the pipes or pump of a hot water system.

petrifying liquid. A type of THINNER that dilutes an EMULSION PAINT with BINDER rather than with water. It is used to thin paints for exterior use without over-diluting them.

petrography. The branch of PETROLOGY that deals with the classification of rocks.

petrol-intercepting chamber (petrol trap). A type of drain TRAP that receives water from garage floors and filling-station forecourts before connecting to the sewer. It contains three ventilated chambers in which petrol or oil float and are kept in the trap.

petrology. The study of the origin and structure of rocks. *See also* PETROGRAPHY.

petrol trap. *See* PETROL-INTERCEPTING CHAMBER.

pewter. An ALLOY based on TIN with some LEAD and perhaps other metals.

PF. *abbr.* PHENOL FORMALDEHYDE RESIN.

PFA. *abbr.* PULVERIZED FUEL ASH.

pH value. A measure of the acidity or alkalinity of a solution, based on the concentration of hydrogen ions present. The scale ranges from 0 for a strong ACID to 14 for a strong BASE. A neutral solution has a pH of 7.

phase change. The change of state when a material changes from solid to liquid, or back again.

phenol. Carbolic acid. A carbon compound used as disinfectant and in the making of plastics.

phenol-formaldehyde resin (phenolic resin). A class of SYNTHETIC RESIN with THERMOSETTING properties. Phenol-formaldehyde resins resist moisture and are used in LAMINATED PLASTIC materials and in glues. *See also* BAKELITE.

phenolic resin. *See* PHENOL-FORMALDEHYDE RESIN.

Phillips screw. A type of SCREW with crossed slots in the head that needs a special screwdriver. *See also* RECESSED-HEAD SCREW, POZIDRIVE SCREW.

phon. The unit of LOUDNESS LEVEL. The phon is defined as being equal to the DECIBEL at a frequency of 1000 Hz. *Compare* SONE.

phosphating. A process in which metal surfaces are treated with phosphoric acid before a surface coating is applied.

phosphor bronze. A tough ALLOY of copper with tin and phosphorous.

phosphorescent paint. Paint containing pigments that absorb energy from visible or ultra-violet light and re-emit the energy as light when the source of radiation is removed. *Compare* FLUORESCENT PAINT.

photocell. A device that produces an electric current or voltage when exposed to light or other electromagnetic radiation.

photoelasticity. A change in the optical properties of transparent materials when they are stressed. The effect can be used to examine the effects of loads on plastic models of structures. *See also* POLARISCOPE.

photoelectric control. A system that uses a PHOTOCELL to switch electric lights on or off automatically in accordance with the levels of surrounding light.

photogrammetry. The process of making maps from photographs, which are usually taken from the air.

photometer. An instrument that detects light and can be calibrated to measure ILLUMINANCE LEVELS.

photon. A single unit or quantum of ELECTROMAGNETIC ENERGY, such as light.

photosynthesis. The process by which plants such as trees make organic material from carbon dioxide and water by the use of light energy absorbed by chlorophyll.

photovoltaic cell. A type of PHOTOCELL.

piano hinge. A long continuous hinge, such as found on the lid of a piano. It may be used for joining long thin sections of materials.

pick. A hand tool for digging and breaking up ground surfaces. It has a long head with two sharp points mounted at right angles to the handle.

pick axe. A form of PICK with a chisel edge at one end of the head and a point at the other end.

picket. (1) *See* PALE. (2) *See* RANGE POLE.

pick hammer. A tool used to draw or drive nails into roof SLATES.

picking. The marking of a stone surface with many small holes made by striking it with a steel point.

picking up. In painting, the blending of a LIVE EDGE of paint with a new coat.

pickling. (1) A process that removes RUST or MILL SCALE from the surface of steel. The metal is dipped in acid solutions, then washed and dried before painting. (2) The process of stripping or removing paint by the use of an alkali, such as sodium hydroxide (caustic soda).

pictorial projection. A method of drawing that gives a perspective or simulated view of an object. Angular perspective with a 'vanishing point' is commonly used for external views of buildings and parallel perspective tends to be used for internal views.

picture rail. A continuous rail of plain or moulded wood fixed to the walls of some traditional rooms at a suitable height for hanging pictures. *Compare* DADO.

pieced timber. A section of timber from which a defective piece has been cut and replaced with a good piece. The replacement piece is usually cut and fitted with a DOVETAIL shape.

piece work. Labour that is paid at a rate that varies with the amount of work completed.

piend. In Scotland, a HIP.

pier. (1) A BUTTRESS or other thickening of a wall that is bonded to the brickwork in order to stiffen the structure. (2) The brickwork in a wall between openings such as doors and windows. (3) The pillar or other support for an arch. (4) A deck built out over the sea and used for loading ships or for recreation.

piezoelectric effect. (1) The production of an electric current or POTENTIAL DIFFERENCE by the application of stress to certain crystals. (2) The production of stress in certain crystals by the application of a POTENTIAL DIFFERENCE.

piezometer. An instrument that uses the PIEZOELECTRIC EFFECT to measure pressure or stress in materials and structures.

pig iron. *See* CAST IRON.

pigment. In paint, finely-ground and insoluble particles that are dispersed through the MEDIUM to give colour and COVERING POWER when the paint film dries. *Compare* BINDER.

pilaster. A rectangular COLUMN or PIER that is attached to a wall.

pile. A long post of concrete, steel or timber that is driven deep into the ground as a support for a structure above. Piles are commonly used in weak ground or for very heavy structures. *See also* BEARING PILE, BORED PILE, DRIVEN PILE, FRICTION PILE, SHEET PILE, SCREW PILE.

pile cap. (1) A metal cap or other cover that protects a PILE as it is driven into the ground. *See also* DOLLY. (2) A block of concrete built around the tops of a group of piles to distribute the load upon them.

pile driver. A machine that uses a powered hammer or heavy weights to drive a PILE into the ground.

pile shoe. A cast-iron point attached to the end of a PILE that has to be driven through hard ground.

piling. (1) The process of driving PILES into the ground. *See also* WATER-JETTING. (2) An unevenness in a wet paint film caused when the paint dries too quickly.

pillar. A vertical COLUMN or PIER.

pillar tap. A water tap such as that used on baths and basins. It has a vertical feed pipe that passes through a hole in the bath or basin.

pilot hole. A small-diameter hole drilled through a material as a guide for drilling a larger hole.

piloti. A series of columns or piers that rise from the ground and support a structure above.

pilot light. In a gas appliance, a small flame that always burns and is used to ignite the main flame.

pilot nail. A temporary nail used to hold timber in place while the main nails are driven, such as in FORMWORK.

pin. (1) A thin wire nail, such as a PANEL PIN. (2) The TENON inserted into a dovetail joint. (3) Concrete or mortar that contains enough water to set and harden but which remains stiff and granular. It may be rammed into places such as the gap between UNDERPINNING piles and the building above. (4) *vb.* To complete the gap between a pile or other foundation and the structure above. DRY MIX concrete may be used, as in UNDERPINNING.

pinch bar. *See* WRECKING BAR.

pinch rod. A length of timber used to check the width of gaps such as those in a door or window opening. *See also* STOREY ROD.

pine. Various type of SOFTWOOD trees that are widely-used for structural timber.

pinholing. Small rough craters on a paint surface that allow moisture to pass. The defect may be caused by the presence of air bubbles on the brush or by grease on the surface.

pin joint. In a structure, a hinged joint between members.

pink primer. A PRIMER for painting woodwork. Traditional primer has a pink colour caused by the presence of red lead PIGMENTS that are no longer in use.

pinning. In Scotland, a stone of different colour or texture set in a wall to give contrast.

pin rail. A wooden batten fixed to a wall and used to fasten hold coat hooks.

pintle. The PIN in a hinge.

pipe. A long hollow tube made of various metals, plastics, concrete and ceramics. They are used to carry water, sewage and gases.

pipe cutter. A tool that cuts metal pipes by the use of hard cutting discs that are turned against the pipe.

pipe drill. A circular-shaped CHISEL used to cut holes in brickwork.

pipe fitter. A person trained to install pipework for water, gas or other services.

pipe hook. A type of FASTENER with a spiked end that is driven into the mortar or timber of a wall. The outer end is a curved hook in which a pipe can be supported.

pipe sleeve. *See* EXPANSION SLEEVE.

pipe tongs. *See* FOOTPRINTS.

pipe wrench (Stillson). A heavy type of WRENCH with serrated jaws. It is used to grip and turn pipes.

pisé de terre. A wall or floor made of rammed earth or COB.

pitch. (1) The slope of a roof or stair. It may be expressed as the ratio of the height to the span. (2) The distance between adjacent objects or lines such as nails, screw threads, or stair NOSINGS. (3) A dark solid material that is left after the distillation of TAR. (4) The FREQUENCY of a sound.

pitch board. A triangular-shaped template or pattern used to set out the STRINGS or other features of a STAIR.

pitched roof. A roof that has one or more surfaces sloping at a significant angle. *Compare* FLAT ROOF.

pitch-faced stone. Stone whose face has been worked in the quarry with a PITCHING TOOL.

pitch fibre. A sheet material formed from a woven fibre that is impregnated with PITCH or BITUMEN. Its uses include pitch fibre pipes for drainage.

pitching piece. In a stair, a horizontal timber that supports the joists of a LANDING and the CARRIAGES of the stair.

pitching tool. A form of HAMMER-HEADED CHISEL with a thick broad edge. It is used to work the face of PITCHED-FACED STONE.

pitchmastic. A material made from PITCH and AGGREGATE that is fluid when hot. It is used as a top coating for JOINTLESS FLOORING.

pitch pine. Various types of wood surfaces,

such as PINE, that feature bright colouring and highlighted grain.

pitch pocket. In SOFTWOOD timber, a gap between GROWTH RINGS that contains resin.

pith. The centre of a log of wood.

pitot tube. An instrument used to measure the total pressure and the velocity of a moving liquid or gas. A narrow tube attached to a MANOMETER is inserted into the stream with the opening facing the current.

pitting. (1) The formation of small holes and craters in the surface of a metal because of corrosion or atmospheric attack. (2) The BLOWING of plaster.

pivot. In general, a point about which something turns or swings. A door may sometimes swing on pivots rather than HINGES.

placing. The process of positioning fresh concrete in its final position for setting and hardening.

plain ashlar. Building stone that has been given a smooth surface.

plain concrete. Concrete that does not use REINFORCEMENT to carry loads or bending forces. It may contain some metal to help reduce shrinkage and cracking.

plain-sawn timber. *See* FLAT-SAWN TIMBER.

plain tile. The usual form of roofing TILE whose surface is flat with a slight outward camber. At the head of the tile there are two NIBS and two nail holes. *Compare* EAVES TILE.

plan. A drawing that shows the layout of a building or object in a horizontal plane. *Compare* SECTION.

plane. Various types of tool for reducing the thickness of wood or for smoothing wooden surfaces. In general, the wood is cut by a sharp 'plane iron' blade that protrudes a little from the metal surface of the 'plane body'. *See also* BENCH PLANE, JACK PLANE, MOULDING PLANE, RABBET PLANE, SMOOTHING PLANE, UNIVERSAL PLANE.

plane angle. A two-dimensional angle measured on a flat surface. *See also* RADIAN. *Compare* SOLID ANGLE.

plane frame. A structural frame, such as a TRUSS, in which all the members are in the same vertical plane. *Compare* SPACE FRAME.

plane iron. The cutting blade in a PLANE. *See also* CAP IRON.

plane of saturation. *See* WATER TABLE.

planer. *See* PLANING MACHINE.

plane stock. The body of a PLANE that holds the PLANE IRON blade.

plane surveying. Land surveying of areas where the curvature of the earth can be ignored without significant error. *Compare* GEODESY.

plane table. A drawing board fixed on a tripod with a swivel head. When used in the field the drawing surface can be turned and oriented with features in the surrounding countryside.

planimeter. A drawing instrument that measures an area on a plan when a moving arm traces the perimeter.

planing machine (planer). A machine with a blade that is used to reduce the thickness of wood and to smooth its surfaces. It may be hand-held.

planing mill. (*USA*) A sawmill where timber is also planed and made into floor boards and weatherboards.

plank. (1) A flat length of sawn SOFTWOOD of 50–100 mm thick and over 300 mm wide. *Compare* DEAL. (2) (*USA*) LUMBER more than 25 mm thick that is laid horizontally like a floorboard.

plank-on-edge-floor. (*USA*) A type of solid floor made from many JOISTS touching one another. The floor surface is fixed directly to these joists without decking.

planning grid. A drawing paper prepared with a network of equally-spaced horizontal and vertical lines. It can be an aid in designing the layout of a building especially if a MODULAR SYSTEM is used.

planning permission. In the United Kingdom and other countries, the formal permission needed from a public authority before development of land or change of use of buildings. *See also* OUTLINE PLANNING PERMISSION.

planometric projection. A view that shows an object in PLAN but with oblique parallel lines from the corners to indicate the front, side and thickness.

planoscope. A levelling instrument set up on a site so that it can be viewed while working and used as a check on ground level. The variations of image seen in the planoscope tells the observer when the height is correct. *Compare* DUMPY LEVEL.

plant. (1) In general, the large items of temporary equipment used on site during construction. (2) The mechanical equipment installed in a building for heating, ventilation and for lifts.

planted. A strip of decorative moulding such as a cove that is attached to a surface rather than cut from the surface.

plaster. A material that is coated onto walls and ceilings while wet to form a smooth hard surface when it dries. It may be made from GYPSUM PLASTER, PORTLAND CEMENT or LIME PUTTY.

plaster base. A surface that takes a coating of PLASTER. Examples include brickwork, LATHING, insulating board and other linings.

plaster bead. *See* ANGLE BEAD.

plasterboard. A common name for GYPSUM PLASTERBOARD.

plasterboard nail. A round-headed nail used to fix PLASTERBOARD to STUDS or JOISTS. The nail is usually GALVANIZED or otherwise treated against corrosion

plasterbox (*USA*: cut-in box). A metal or plastic box inserted in a wall to contain an electrical switch or socket.

plaster dab. One of several small lumps of GYPSUM PLASTER used to fix sheets of plasterboard to a brick or block wall.

plasterer. A person trained and skilled in the use of PLASTER.

plasterer's float. *See* FLOAT.

plasterer's lath hammer. *See* LATH HAMMER.

plasterer's putty. *See* LIME PUTTY.

plaster of Paris. *See* HEMIHYDRATE PLASTER.

plaster slab. A precast BUILDING BLOCK made of GYPSUM PLASTER.

plaster stop. *See* ANGLE BEAD.

plastic. (1) *adj.* Having the property of PLASTICITY. (2) A particular class of materials. *See* PLASTICS.

plastic design. A method of structural design for steel or reinforced-concrete frames. The aim is to involve many PLASTIC HINGES that can transfer loads among themselves.

plastic hinge. In structural design, a point that is assumed to be stressed to the YIELD POINT and move slightly. *See also* PLASTIC DESIGN.

plasticity. (1) The property of a material such as a 'plastic' that allows it to keeps a new shape after it has been deformed or moulded. *Compare* ELASTICITY. (2) The ability of mortar or plaster to be easily worked and spread.

plasticizer. (1) An ADMIXTURE that increases the workability of concrete or mortar. (3) A material that is added to a PLASTIC material or a PAINT during manufacture to help keep the final product or paint film flexible.

plastic laminate. *See* LAMINATED PLASTIC.

plastic modulus. A value used in the PLASTIC DESIGN of steel structures.

plastic paint. A type of PAINT that can be moulded or shaped after it has been applied.

plastics. In general, materials and products that, at some stage in their manufacture, show PLASTICITY and are easily moulded. Most plastics are POLYMERS. *See also* THERMOPLASTIC, THERMOSETTING.

plastic wood. Various types of paste made from cellulose and resins in volatile solvents. They are used to make small repairs to wood and other surfaces.

plastisol. A protective coating for metals made from POLYVINYL CHLORIDE.

plat. (*USA*) A plan that shows the boundaries and ownership of land.

plate. (1) A horizontal timber member that is usually supported along its length, such as in a SOLE PLATE. (2) Steel that is thicker than about 3 mm.

plate cut. *See* BIRDSMOUTH JOINT.

plate floor. A reinforced-concrete floor with flat underside. *Compare* HOLLOW TILE FLOOR.

plate girder. A GIRDER made up of web and flange plates welded together.

plate glass (polished plate glass). A high-quality form of GLASS that is polished to give smooth parallel sides. *Compare* FLOAT GLASS, SHEET GLASS.

platform frame. A type of timber-frame construction in which the floor platforms are built over the full thickness of the walls. The walls for the next storey are separately built on this platform. *Compare* BALLOON FRAME.

platform roof. *See* FLAT ROOF.

plenum system. A type of AIR-CONDITIONING system in which the rooms are kept at a pressure above atmospheric pressure by blowing in clean air. Used air is taken out by other ducts or allowed to escape through gaps.

pliers. A hand tool that pivots like scissors to grip or cut objects within its jaws.

plinth. (1) A course of stones, bricks or other material that runs along the base of a wall and projects from it. (2) A base plate that runs along the bottom of a piece of furniture. (3) A square-shaped base for a column or pedestal.

plot. (1) To draw a map or graph from measurements. (2) In computer-aided design (*see* CAD), to produce a drawing on a mechanical plotter that moves its pens under the control of the CAD computer. (3) An area of land for building.

plotter. An instrument that produces drawings or maps by a mechanical method. In a CAD plotter, the pens are guided over the paper by signals from the computer.

plough (*USA*: plow). (1) A type of PLANE used to cut grooves in wood at a constant distance from the edge. (2) *vb.* To cut a groove.

ploughed and tongued joint. *See* FEATHERED JOINT.

plough strip. A strip of wood that has been produced by a PLOUGH.

plow. (*USA*) *See* PLOUGH.

plug. (1) (wall plug) A small piece of wood or other material driven into a hole to help grip a screw or nail. (2) A connector on the end of an electrical cable with protruding contacts that fit into a SOCKET OUTLET. (3) A fitting that screws into a pipe to close it.

plug-centre bit. A form of drill BIT that widens an existing hole.

plug cock. A simple form of valve that opens and closes when the handle of the plug is turned through 90°. The liquid flows through a hole drilled in the plug.

plugging. The process of drilling holes in a masonry to receive PLUGS into which screws and nails are fixed.

plugging chisel (plugging drill). A hardened steel bar with a point that is used to make holes for PLUGS. The chisel is held in the hand and struck with a hammer. *See also* STAR DRILL.

plumb. Vertical or to make vertical.

plumb bob (plummet). The weight that hangs on the end of a PLUMB LINE to show the vertical direction.

plumb cut. The vertical cut in a BIRDSMOUTH JOINT where the foot of a rafter fits over a wall plate.

plumber. A person trained and qualified to work with PLUMBING.

plumbing. (1) The installation or operation of sanitary services, water supplies and heating in and around a building. (2) The use of a PLUMB LINE to check the vertical direction.

plumb level. A type of SPIRIT LEVEL designed to check whether a surface is vertical.

plumb line. A cord on which is a hung a PLUMB BOB weight. The line is suspended free of obstruction to show the vertical direction.

plumbosolvency. The ability of some water to dissolve small amounts of LEAD from water pipes. SOFT WATER or ACIDIC water can acquire amounts of dissolved lead that are possibly toxic.

plumb rule. A straight-edged board used to check whether a surface is vertical. A PLUMB LINE hangs from the top of the board.

plummet. *See* PLUMB BOB.

plus sight. (*USA*) *See* BACK SIGHT.

ply. (1) A thin sheet or VENEER of wood, or other material. It makes up one of the layers on PLYWOOD or LAMINATE. (2) One of the sheets of ROOFING FELT that makes up a roof cover.

plymetal. PLYWOOD that is faced with sheet metal on one or both sides.

plywood. A composite sheet of timber made from VENEERS glued together. An odd numbers of layers give a BALANCED CONSTRUCTION and the grains of adjacent layers run in different directions to give extra strength and stability. *See also* MARINE PLY.

plywood parquet. A form of PARQUET floor made from squares or tiles of plywood with a top veneer of hardwood.

PNdB. *abbr.* Perceived noise decibel. *See* PERCEIVED NOISE LEVEL.

pneumatic caisson. A CAISSON in which compressed air is used to keep out water.

pneumatic mortar. *See* GUNITE.

pneumatic structure. *See* AIR-SUPPORTED STRUCTURE.

pneumatic tools. Various tools or plant that is driven by a supply of compressed air. The method has advantage for percussive drills and for safety in wet places.

pneumatic water supply. A water supply with a closed cistern in the basement from which water is forced by compressed air.

pocket. (1) An opening made in a wall to insert a beam. (2) In the side of a SASH WINDOW frame, a hole to provide access to the sash weight.

pocket rot. A form of timber decay that spreads from small areas.

pock marking. *See* ORANGE PEEL.

podger. A large SPANNER used in erecting steel-work. It has a pointed handle used to align holes drilled in plates that are to be bolted together.

point. (1) An outlet on an electrical supply or a gas supply. (2) The sharp end of a SAWTOOTH.

pointing. (1) The process of scraping mortar out of joints in brickwork and pressing in new mortar. *See also* RECESSED POINTING. *Compare* JOINTING. (2) The use of mortar to fill the joints between RIDGE or HIP TILES on a roof.

point load (concentrated load). A structural load that bears on a relatively small area, such as beneath a column. *Compare* UNIFORMLY DISTRIBUTED LOAD.

point of contraflexure. *See* CONTRAFLEXURE.

Poisson's ratio. An elastic material under STRESS experiences STRAIN in directions perpendicular to the direction of the force. The ratio is that of the strain in the cross direction to the strain in the longitudinal direction. *See also* ELASTIC MODULUS.

poker vibrator. A vibrating tool that is inserted into fresh concrete to consolidate it. *See also* VIBRATOR.

polar coordinates. A system of COORDINATES that defines a position on a plane by means of a length and the angle between the length line and a reference line.

polar curve. A graphical method that shows the directional qualities of light emitted from a LAMP or LUMINAIRE. The output is plotted onto polar co-ordinate paper with the light source at the centre.

polariscope. Optical equipment that uses POLARIZED LIGHT to display the patterns of internal stress in models of structures. *See also* PHOTOELASTICITY.

polarized light. Light or other ELECTROMAGNETIC RADIATION whose vibrations are restricted to certain directions. *See also* POLARISCOPE.

pole plate. A horizontal beam at the foot of a pitched timber roof. It rests across the tie beams of the roof TRUSS and supports the feet of the common RAFTERS.

pole wall. (*USA*) A type of log cabin wall made from vertical poles.

poling wall. One of a number of vertical boards used to support the sides of an excavation. *See also* WALING.

polished plate glass. *See* PLATE GLASS.

polished work. Building stone, such as marble, that has been worked with abrasives to give a smooth polished surface.

poll. (1) The striking face of a HAMMER. (2) The process of making KNAPPED FLINT.

poll adze. A type of ADZE with a blunt head in addition to the cutting edge.

pollarding. The regular process of cutting back the top shoots of a tree to encourage growth of the trunk.

polyamide. *See* NYLON.

polycarbonate. A THERMOPLASTIC type of POLYMER used to produce tough, transparent plastic materials that are sometimes used as a substitute for glass.

polyester. A large class of synthetic resins made as POLYMERS with THERMOPLASTIC or THERMOSETTING properties. Polyesters are used to make plastics, fibres, adhesives and paints (such as ALKYDS).

polyethylene (polythene). A THERMOPLASTIC type of POLYMER made from ethylene. The density of the polymer is varied to make a wide range of plastic materials that can be flexible or rigid.

polygon. A closed plane figure made of three or more straight lines that meet.

polygon of forces. In STRUCTURAL DESIGN, the principal that if forces acting at a point are in equilibrium then they can be graphically represented by a closed POLYGON with one side for each force. *See also* TRIANGLE OF FORCES.

polyisocyanurate. A THERMOSETTING type of POLYMER made from ISOCYANATE. In foamed form it is used for thermal INSULATION.

polymer. A compound with a long chain-like structure made from the union of many smaller MONOMER molecules. Some natural compounds such as CELLULOSE are classed as polymers but most polymers are artificially produced by POLYMERIZATION. Most modern PLASTIC materials are synthetic polymers. *See also* CO-POLYMER, RESIN.

polymerization. The joining of MONOMERS to make POLYMERS in a chemical process that usually involves the use of heat and pressure. *See also* CONDENSATION.

polypropylene. A THERMOPLASTIC type of POLYMER based on propylene. It is used to produce a variety of plastic materials which range from fibres to rigid articles.

polystyrene. A THERMOPLASTIC type of POLYMER based on the benzene ring molecule. It is used to produce a variety of plastic materials which include insulation foams and water tanks.

polytetrafluoroethylene (PTFE). A durable THERMOPLASTIC type of POLYMER with a waxy texture and a low coefficient of friction. Its uses include the jointing of pipes, bearings, and 'non-stick' coatings. Teflon is the trade name for one type of PTFE.

polythene. A common name for POLYETHYLENE.

polythene film. A flexible, waterproof sheet material based on POLYETHYLENE. Its uses in construction include temporary protection from weather, DAMP COURSES and VAPOUR BARRIERS.

polyurethane. A THERMOSETTING class of POLYMERS made from ISOCYANATE. They are used for adhesives, paints, insulating foams, as well as for rigid articles.

polyvinyl acetate (PVA). A POLYMER based on the vinyl acetate MONOMER. As a synthetic resin it is used for EMULSION PAINTS, sealants and adhesives.

polyvinyl chloride (PVC). A THERMOPLASTIC type of POLYMER based on the vinyl chloride MONOMER. Depending on the amount of plasticizer used, PVC products have varying properties: they may be flexible such as in sheets, or rigid such as in unplasticized (uPVC) pipes.

pony wall. *See* SLEEPER WALL.

poor lime. LIME that contains impurities.

popping. *See* BLOWING.

population. The total number of items in a group that is the subject of statistical examination.

porcelain. A fine CERAMIC material that is almost translucent. It is made from a mixture of KAOLIN and certain minerals containing silica which are fired at high temperature. In addition to traditional tableware, porcelain is used to make high quality fittings for bathrooms.

pores. (1) In general, small openings in the surface of a material. (2) In timber, small round holes in the structure of HARDWOOD through which sap passes.

porosity. A measure of the porous nature of a soil. It is the ratio of the volume of voids in a soil to the total volume of the sample.

porous. *adj.* Describing the property of a material with a significant number of PORES that can allow the passage of liquids and gases.

porphyry. A hard reddish rock containing large crystals of minerals.

portal. (1) A large entrance or gateway. (2) A tunnel opening.

portal frame. A wall and roof frame with a double pitch that gives a large clear span suitable for factory and warehouse spaces. The frame is stiffened by a strong rigid joint between the roof beams and the columns. The loads in a portal frame may be further distributed by pin joints at the apex of the roof and at the bottom of the columns. *See* THREE-HINGED FRAME.

portico. A covered entrance to the doorway of a building.

Portland blast-furnace cement. A cement that is based on PORTLAND CEMENT but also contains up to 65 per cent of BLAST-FURNACE SLAG to give the cement POZZOLANIC qualities.

Portland cement. The type of CEMENT used to make most modern CONCRETE and MORTARS. Portland cement acquired its name because the hardened concrete it makes was thought to look like PORTLAND STONE. Portland cement is made by firing a mixture of limestone (or chalk) and clay in a kiln. The fused products are ground into powder and have GYPSUM added as a setting RETARDER. Most Portland cement is sold as ordinary Portland cement (OPC) but the chemical composition of the cement can be varied to give other types such as LOW-HEAT PORTLAND CEMENT, RAPID HARDENING PORTLAND CEMENT and SULPHATE-RESISTING PORTLAND CEMENT.

Portland stone. A cream-coloured LIMESTONE from Portland on the south coast of England. It weathers well and has been used for facing buildings.

Portsmouth valve. A type of BALL VALVE.

post. A major upright support in a building or frame.

post-tensioned concrete. A type of PRESTRESSED CONCRETE in which the tension is applied to the reinforcement after the concrete is poured and set.

potential difference (pd). A measure of the difference in electrical potential that exists between two points in a CIRCUIT when an electrical current flows. The unit of pd is the VOLT. *Compare* with ELECTROMOTIVE FORCE, which is also measured by the volt.

potential energy. A form of energy that a body or a fluid has because of its height. *Compare* KINETIC ENERGY. *See also* BERNOULLI'S THEOREM.

pot floor. *See* HOLLOW-TILE FLOOR.

pot life. The length of time for which a GLUE can be used after it is mixed or opened.

pot type boiler. A boiler where water is heated in a jacket that surrounds the burner.

pounce powder. (*USA*) A fine powder dusted on to drawing paper to give smoother handling.

powder-post beetle. *See* LYCTUS POWDER-POST BEETLE.

power. (1) The rate of using ENERGY, or of doing work. The SI unit of power is the WATT. *See also* HORSEPOWER. (2) Electrical ENERGY.

power factor. In ALTERNATING CURRENT electrical supplies, a measure of energy lost as heat in a circuit when the voltage and current are not in phase. *See also* KVA.

power float. A type of trowel or FLOAT that uses a power-operated rotary blade to give a smooth finish to floor screeds.

power station. A plant built to produce large amounts of electrical energy for distribution to industry and households. *See also* DYNAMO, HYDRO-ELECTRIC POWER STATION, THERMAL POWER STATION, TURBINE.

pozidrive screw. A type of SCREW with crossed slots in the head that needs a special screwdriver. *See also* RECESSED-HEAD SCREW, PHILLIPS SCREW.

pozzolan (pozzolana). (1) A natural volcanic material from Pozzuoli, Italy which was used as an early form of HYDRAULIC CEMENT. (2) Various other materials, such as PULVERIZED FUEL ASH, have POZZOLANIC qualities.

pozzolanic. *adj.* Describing the ability of a material to combine with lime in the presence of water and produce an HYDRAULIC CEMENT. Pozzolanic materials may be added to PORTLAND CEMENT concrete to give properties such as low heat output and long-term strength. *See also* PORTLAND BLAST-FURNACE CEMENT.

PP. *abbr.* POLYPROPYLENE (especially in the description of plastic pipes).

PPC. *abbr.* PROBABLE PROFIT CONTRIBUTION.

ppm. *abbr.* PARTS PER MILLION.

Pratt truss (N-truss). A TRUSS with vertical struts between different panels. The truss is used for roofs and bridges.

preamble. An introduction to a section of a document, such as a BILL OF QUANTITIES.

pre-boring. The use of a drill to prepare a holes for a nail. The holes, which are less than the diameter of the nails, reduce the risk of the splitting the timber and allow more nails to be driven into a joint.

precast concrete. Concrete units such as columns, beams and lintels that are made separately and fixed into the construction later. *Compare* CAST-IN-SITU.

precast stone. *See* ARTIFICIAL STONE.

precedence diagram (activity-on-node diagram). A form of CRITICAL PATH ANALYSIS in which the activities are represented by circles. The arrows connecting these 'nodes' show the sequence of activities. *Compare* ARROW DIAGRAM.

precipitation. (1) The formation in a liquid of an insoluble substance which can then be separated. (2) The water that falls on the ground as rain, hail, snow or dew.

precision. The degree of fineness with which a measurement is read from an instrument. It does not allow for errors in the instrument.

pre-cure. The initial setting of GLUE before the surfaces are clamped together.

prefabricated building. A description of a building whose walls and other major components are made in a factory and assembled on site.

preliminaries. The main introduction clauses to a document such as a BILL OF QUANTITIES.

preliminary treatment. An initial treatment of SEWAGE that removes large solids, oils, plastics and other materials from the effluent.

pre-mixed. A bagged mixture of all the ingredients needed for PLASTER or CONCRETE, except water.

present value. An accounting idea used in DISCOUNTED CASH FLOW. It is the present value of receiving, for example, £100 at the end of a given number of years with a certain rate of compound interest.

preservative. (1) In general, a chemical treatment that protects a material and prolongs its useful life. (2) Various chemicals applied to timber to protect it from attack by FUNGI such as dry rot or insects such as BEETLES. Chemicals used include copper compounds, CREOSOTE and pentachlorophenol. *See also* TANALIZED TIMBER.

pressed brick. A BRICK shaped in a mould that gives sharp edges and smooth surfaces. *Compare* WIRECUT BRICK.

pressed steel. Sheet steel that is pressed into shape while hot and used to make items like window SILLS.

pressure. (1) The concentration or spread of a force when it acts on a surface. It is measured as the force per unit area. The SI unit of pressure is the PASCAL. (2) In a fluid, such as water, the pressure at a point below the surface depends on the height of fluid above the point. *See* HEAD.

pressure bulb. *See* BULB OF PRESSURE.

pressure gauge. An instrument for measuring pressure. *See* BOURDON GAUGE, MANOMETER.

pressure gun. A tool used to apply sealant, mastic or putty when CAULKING around a joint. The gun is pressurized by pushing a handle, like a grease gun.

pressure head. *See* HEAD.

pressure tank. A sealed tank in which the pressure is increased to impregnate timber with PRESERVATIVE. *Compare* OPEN-TANK TREATMENT.

pressure vessel. In general, a large container designed to resist considerable pressure caused by a fluid such as steam.

pressurized area (pressurized route). An area or route, such as a stairwell, that forms a fire escape from a building. Increased air pressure is used to keep out smoke. The plant that supplies the pressurized air must be protected from failure and it may operate automatically in case of fire.

pressurized structure. *See* AIR-SUPPORTED HOUSE.

pre-stressed concrete (pre-tensioned concrete). Concrete that is stressed by its internal REINFORCEMENT or by pressure from outside. This pre-compression of the concrete gives reduced tensile forces when it is loaded and results in less cracking. A concrete member may be stressed by tightening the reinforcement before the concrete is cast or after the concrete is set as in POST-TENSIONED CONCRETE. *See also* TENDON.

pre-tensioned concrete. *See* PRE-STRESSED CONCRETE.

priced bill. A BILL OF QUANTITIES on which a contractor has added prices. It may be one of the contract documents.

pricking up. The process of scratching the first coat of plaster to make a key for another coat.

prick punch. A pointed hand tool used to make a small hole as a start for a nail or for a drill. *Compare* NAIL PUNCH.

primary beam. *See* MAIN BEAM.

primary circuit (primary system). The pipe system that transfers heat between the boiler and an intermediate device such a HEAT EXCHANGER. *See also* INDIRECT SYSTEM, SECONDARY CIRCUIT. *See* HEATING diagram.

primary colour. *See* ADDITIVE COLOUR and SUBTRACTIVE COLOUR.

primary energy. The total energy contained in a raw FUEL such as coal or oil. Not all of the primary energy, which may be expressed as a CALORIFIC VALUE, can be extracted from a fuel.

primary filtration. In water treatment, an initial stage of course filtration or straining that reduces the load on the SECONDARY FILTRATION. Primary treatment may not be needed if the source of water is relatively clean.

prime cost. A specified sum allowed in a tender or contract for the supply of certain materials such as plumbing.

primer. (1) A type of paint used for a PRIMING COAT. (2) A bituminous coating used to stick ROOFING FELT on to the decking of a roof.

priming coat. The first coat of PAINT applied to a raw surface. The priming coat is usually designed to seal the surface, to help prevent corrosion, and to be a suitable base for the next coat such as the UNDERCOAT. *See also* THREE-COAT WORK.

principal. *See* PRINCIPAL RAFTER.

principal rafter (principal). One of the main RAFTERS in a roof. Principal rafters support the PURLINS on which the COMMON RAFTERS are laid. *See* ROOF TRUSS diagram.

principal stresses. The resultant stresses in two planes that arise when an object is subject to forces from different directions.

prism. A flat-sided solid block of glass that is used to reflect and deviate light within telescopes and other instruments.

prismatic compass. A hand-held compass that uses a prism to give a view of the compass card while an object is sighted.

prismoidal formula. A formula used to predict the volume of earth in an excavation: Volume $= L/6(A_1+A_2+4\times A_m)$ where L is the length of excavation, A_1 and A_2 are the cross-section areas of the ends, and A_m is the area at the midpoint. *Compare* SIMPSON'S RULE.

prism square. *See* OPTICAL SQUARE.

private sewer. A length of SEWER before it joins and becomes part of a public sewer that is the responsibility of a drainage authority.

probability. In statistics, a measurement or estimate of the likelihood that a particular event will occur, expressed on a scale from zero (impossible) to one (certain). *See also* CHARACTERISTIC STRENGTH, PROBABLE PROFIT CONTRIBUTION.

probable error. In statistics, a deviation equal to the STANDARD DEVIATION multiplied by 0.6745.

probable profit contribution (PPC). The use of statistics to predict likely profits from future work that is not yet certain. PPC is obtained by multiplying the anticipated profit by the statistical probability of obtaining that profit.

process engineering. The branch of engineering that deals with industrial processes such as continuous manufacturing.

production drawing. *See* WORKING DRAWING.

productivity. A measure of the output obtained from given amounts of labour or capital. Various comparisons may be used, such as the ratio of the financial turnover to the number of employees.

professional indemnity insurance. An insurance policy taken out by individuals to cover possible legal claims against them for PROFESSIONAL NEGLIGENCE.

professional negligence. A deficiency in agreed standards of skill and care exercised by an individual engaged in design, management or other professional duties. *See also* PROFESSIONAL INDEMNITY INSURANCE.

profile. (1) (*USA*: batterboard). One of a number of horizontal boards fixed in the ground just outside the proposed foundations of a building. The upper edges are set at a DATUM level for the building and marked with nails or saw cuts to show the lines of walls and other features. *See also* RANGE LINE. (2) A type of TEMPLATE for making a moulding.

profile paper. Drawing paper that is ruled in a grid of squares.

programme evaluation and review technique (PERT). *See* CRITICAL PATH ANALYSIS.

progress chart. A wall chart that displays the separate operations of a construction site, their relationships to one another and their progress. *See also* BAR CHART.

progression. The process of making a TRAVERSE.

projecting scaffold. A platform made of scaffolding built out from a wall rather than resting on the ground.

projection. The display of solid objects such as machines and buildings on a plane surface like drawing paper. Imaginary lines from each point on the object are projected onto the plane from a particular viewpoint. *See* AXONOMETRIC PROJECTION, ISOMETRIC PROJECTION, ORTHOGRAPHIC PROJECTION.

prop. A temporary support, especially the upright members of FORMWORK. *See also* ACROW, NEEDLE.

propane. A member of the alkane (paraffin) series of HYDROCARBON compounds, C_3H_8. It is a flammable gas that can be stored in liquefied form and used as a fuel. *See also* LIQUEFIED PETROLEUM GAS.

protected escape route. A route in a building that has extra FIRE RESISTANCE to provide some safety for exit in a fire.

protected opening. A door or other opening through a wall or floor that is part of a system of COMPARTMENTATION against the spread of fire. The opening must be have a shutter or door of appropriate FIRE RESISTANCE and automatically close in case of fire.

protected shaft. A stairway, lift or other shaft that connects different floors of a building that has a system of fire COMPARTMENTATION. The shaft itself is also protected by a system of fire-resisting walls, doors and shutters.

protective coating. Various systems of covering the surface of a material to increase its durability. *See* ANODIZING, GALVANIZING, METAL COATING, SHERADIZING.

Protimeter. (*UK*) Brand name for a type of MOISTURE METER.

provisional sum. A sum of money in the BILL OF QUANTITIES that is set aside to cover the cost of unexpected work.

Prussian blue. An intense blue PIGMENT made from a compound of iron, potassium and cyanide.

pry bar. A tool with a claw used to pull out nails.

PS. *abbr.* POLYSTYRENE (especially in the description of plastic pipes).

PSA. *abbr.* (*UK*) Property Services Agency.

PSALI. *abbr.* PERMANENT SUPPLEMENTARY ARTIFICIAL LIGHTING OF INTERIORS.

psf. *abbr.* Pounds per square foot.

psi. *abbr.* Pounds per square inch.

psychrometer. *See* HYGROMETER.

psychrometric chart. A combined set of graphs that plots the relationships between the different variables used to specify humidity such as DEW-POINT, DRY BULB, RELATIVE HUMIDITY AND VAPOUR PRESSURE.

PTFE. *abbr.* POLYTETRAFLUOROETHYLENE.

P-trap. A type of TRAP with a horizontal outlet that may take the waste from a WC pan. The trap makes a shape like a letter P on its side. *Compare* S-TRAP. *See also* SEAL.

PTS. *abbr.* PERMANENT THRESHOLD SHIFT.

puddle. Wet clay, sometimes mixed with sand, used to waterproof the sides of a pond or canal.

puff pipe. *See* ANTI-SIPHON PIPE.

pugging. (1) A heavyweight material placed in the cavity of a floor or wall to improve the SOUND INSULATION. Materials used include sand and slag wool. *See also* MASS LAW. (2) In paint preparation, the mixing of a PIGMENT with MEDIUM so that the pigment forms a thick paste. *See also* PUG MILL.

pug mill. A machine in which rollers or knives rotate in a pan. It is used to crush and mix clay, mortar or paint.

pull. A handle that opens a drawer or cupboard.

pull box. In a CONDUIT system, a box through which cables can be pulled.

pulley head. The horizontal headboard in the frame of a SASH WINDOW.

pulley stile. One of the vertical boards on each side of the frame of a SASH WINDOW.

pulling. *See* DRAG.

pulling up. The softening of a previous dry coat of paint or varnish when a new coat is applied.

pull-out strength. The WITHDRAWAL LOAD needed to pull out a nail.

pulverized fuel ash (fly ash, PFA). A fine ash left in the burners of power stations after the combustion of powdered coal. The ash may be used as a LIGHTWEIGHT AGGREGATE or a POZZOLAN.

pumice. A lightweight porous rock formed by volcanos. It is used as an abrasive and a lightweight aggregate.

pump. A mechanical device used to raise liquids or drive them through pipes.

punch. (1) A hand tool driven by a hammer that is used to cut holes or to drive home an object. *See* NAIL PUNCH, PRICK PUNCH. (2) A chisel used to cut stone.

punched work. ASHLAR stone faced with rough diagonal marks made by a PUNCH.

puncheon. A short vertical post in the middle of a TRUSS or scaffold.

punching shear. The effect when a loaded column pushes a hole through a base.

punner. A heavy block or wood or metal on a handle that is repeatedly dropped to compact earth or to bed stones.

purlin. One of a series of horizontal roof beams laid across the PRINCIPAL RAFTERS or TRUSSES. It carries the COMMON RAFTERS or other forms of roof decking. *See also* UNDER-PURLIN.

purlin roof. A form of roof in which the PURLINS are supported by cross walls instead of on TRUSSES.

purpose-made brick. A BRICK made to a non-standard shape.

push bar. On an emergency exit door, a horizontal bar that runs across the inside of the door and releases the lock when the bar is pushed.

push plate. A protective plate fixed to a door at a position convenient for pushing.

putlog. A short beam or length of scaffold tube on which SCAFFOLD boards are fixed.

putty. *See* GLAZIER'S PUTTY, LIME PUTTY, MASON'S PUTTY.

putty knife. A knife with broad flexible blade used to apply PUTTY around glazing and to apply STOPPING.

PVA. *abbr.* POLYVINYL ACETATE.

PVA adhesive. Type of ADHESIVE made from POLYVINYL ACETATE. It is used in wood-working and general applications.

PVC. *abbr.* POLYVINYL CHLORIDE.

pyramidal light. Type of ROOF LIGHT whose sloping sides make a pyramid shape.

pyrites. Yellow mineral of iron containing sulphur.

pyrometer. An electrically-operated instrument for measuring high temperatures. It may use the electrical properties of metal junctions or optical properties.

Q

quad. *See* QUARTER ROUND.

quadrant. (1) In general, a quarter circle with an angle of 90° at the centre. (2) A moulding with a cross-section like a quarter circle. (3) A CASEMENT STAY with a curved shape.

quadrant bearing. The conversion of an angle from a WHOLE CIRCLE BEARING to an angle less than 90°. The conversion was needed to use trigonometric tables.

quadrilateral. A flat figure of any shape formed from four straight sides.

quantities. The amounts of materials and work needed to construct a particular building, used in the BILL OF QUANTITIES.

quantity surveying. The preparation of BILLS OF QUANTITIES and other CONTRACT documents. Also the processes of advising, supervising and arbitrating on matters relating to the costs of building.

quantity surveyor. A person professionally qualified in QUANTITY SURVEYING. In the United Kingdom, the RICS is the recognized professional body.

quarrel. A pane of glass in LEADED LIGHTS.

quarry face. Stone that shows its natural face after it has been quarried.

quarry stone bond. Any masonry bond in a RUBBLE WALL.

quarry tile. A burnt clay tile usually used for flooring. The tile surface is waterproof but not glazed like a CERAMIC TILE.

quarter bend. A 90° bend in a pipe. *Compare* ELBOW.

quartered log. A log that has been cut into four quarters ready for CONVERSION by QUARTER SAWN TIMBER.

quartering. See QUARTER SAWN TIMBER.

quarter landing (quarter-space landing). The platform at the turn between two flights of stairs at 90° to one another. *Compare* HALF LANDING. *See* STAIRS diagram.

quarter round (quad). A MOULDING with an outward curve that turns through 90°. It makes up one quarter of a circle. *See* MOULDINGS diagram. *Compare* SCOTIA.

quarter-sawn timber (rift-sawn timber). Timber that is cut lengthwise along its radius so that the GROWTH RINGS are at high angles to the surfaces of the timber. *Compare* FLAT-SAWN TIMBER.

quartile. In statistics, one of the divisions when a sample is divided into four groups with equal frequencies.

quartz. A natural crystalline mineral form of silicon dioxide.

quartz-halogen lamp. *See* TUNGSTEN-HALOGEN LAMP.

quartzite tiles. A type of floor tile quarried from a sandstone rock containing QUARTZ.

queen bolt. A vertical steel bolt used in place of a QUEEN POST.

queen closer. A brick cut in half along its length to act as a CLOSER.

queen post. One of two upright posts each set at about one third span of a QUEEN POST TRUSS.

queen post truss. A triangular roof TRUSS made from two rafters joined together at the ridge and linked by a tie beam at their feet. Two vertical QUEEN POSTS link the tie beam to the ridge. This type of truss is not used much now.

quenching. The process of rapidly cooling a heated metal to improve hardness or toughness.

quetta bond. A brickwork BOND that forms gaps in the middle of a wall which is usually one-and-a-half bricks thick. The outside faces show FLEMISH BOND and the gaps are filled with grout and reinforcement. The bond is similar to RAT-TRAP BOND except that the bricks are laid flat rather than on edge.

quicklime. *See* CALCIUM OXIDE.

quickset level. A type of LEVEL with a system, such as a ball and socket joint, for easy setting-up.

quicksilver. An old name for the metal MERCURY.

quick start circuit. A system of CONTROL GEAR for a DISCHARGE LAMP that causes the lamp to 'strike' and start without the delay needed by some arrangements.

quilt. *See* BLANKET.

quirk. A narrow groove at the edge of a BEAD or MOULDING. *See* MOULDINGS diagram.

quirk bead. A semi-circular MOULDING that is separated from a board by a QUIRK. *See* MOULDINGS diagram. *Compare* BEAD.

quoin (coin). (1) An outer corner of a building. (2) Special bricks or stones set in the corner of a wall. *See also* BLOCK QUOIN.

quoin header. A brick that is a HEADER in the face wall and a STRETCHER in the side wall.

R

rabbet. *See* REBATE.

rabbet plane. A type of PLANE designed to cut REBATES in wood.

raceway. (*USA*) A rectangular DUCT that carries cables through a building.

racking back. In laying brickwork, the process of first building up the corners in steps that rise from the middle of the wall.

radial arm saw. A type of power SAW in which the motor and circular blade are attached to an arm that pivots over the work table.

radial brick (compass brick). A brick that tapers so that it can be laid in curves.

radial circuit. A system of electrical wiring in which each power outlet or appliance has separate cables radiating from the central CONSUMER CONTROL UNIT. *Compare* RING MAIN.

radial shrinkage. The drying shrinkage of timber that occurs at right angles to the ANNUAL RINGS. *Compare* TANGENTIAL SHRINKAGE.

radian. The SI unit of PLANE ANGLE. It is defined as the angle between those two radii of a circle that cut a length of circumference equal in length to the radius. 1 radian = approximately 57.3°.

radiant heating. A form of heating that mainly relies on heat RADIATION or INFRA-RED RADIATION, rather than on CONVECTION.

radiant temperature. An assessment of the component of thermal comfort caused by radiation from surrounding surfaces. It can be taken as equal to the mean temperature of surrounding surfaces or measured by a GLOBE THERMOMETER. *See also* COMFORT TEMPERATURE, ENVIRONMENTAL TEMPERATURE.

radiation. The transfer of heat energy through a space by means of electromagnetic radiation, or INFRA-RED RADIATION. This transfer can take place in a vacuum. *See also* SOLAR RADIATION. *Compare* CONDUCTION, CONVECTION.

radiator. (1) Any surface that emits heat by RADIATION. (2) A high-temperature heating device that gives off heat by RADIATION. (3) A HEATING PANEL, which in fact transfers most of its heat not by radiation but by CONVECTION. *See* HEATING diagram.

radiator key. A small spanner that fits and turns a release nut on a hot-water RADIATOR. It may used to release AIR LOCKS in the system.

radioactivity. The spontaneous disintegration of unstable nuclei in certain atoms accompanied by the emission of alpha or beta particles and gamma rays. *See also* GAMMA RADIOGRAPHY.

radiography. The use of X-RAYS or RADIOACTIVITY for NON-DESTRUCTIVE TESTING.

radius of gyration (k). A value used in the calculation of the SLENDERNESS RATIO. It is calculated as the square root of the MOMENT OF INERTIA after it is divided by the cross-sectional area.

radon. A colourless, radioactive gas given off by some rocks and soil. If it accumulates in poorly ventilated buildings it can be a health hazard.

rafter. A roof beam that slopes from the RIDGE to the EAVES. *See also* COMMON RAFTER, HIP RAFTER, JACK RAFTER, PRINCIPAL RAFTER, TRUSSED RAFTER. *See* ROOF diagram and ROOF TRUSS diagram.

rafter filling. An infilling of brick between rafters at wall plate level.

raft foundation. A FOUNDATION formed by a continuous slab of concrete.

rag bolt. *See* LEWIS BOLT.

ragfelt. (*USA*) Bitumen felt. *See* BITUMEN MATERIALS.

raggle. *See* RAGLET.

raglet (raggle, reglet). A groove cut into stone, brickwork or a mortar joint in order to receive the edge of lead FLASHING.

ragstone. A coarse-grained form of sandstone.

rail. (1) A horizontal member in a framework such as a door. *See also* BOTTOM RAIL, LOCK RAIL. *See* DOOR diagram. (2) A horizontal member in a fence or stair. *See also* HANDRAIL.

railing. A fence constructed from posts and RAILS.

rain gauge. An instrument used to measure the amount of rain that falls at a particular place.

rain index. *See* DRIVING-RAIN INDEX.

rain leader. (*USA*) *See* DOWNPIPE.

rain screen cladding. An outer covering of wall panels that protects a structure from driving rain. Water that passes through OPEN JOINTS is collected and drained by CATCHMENT TRAYS or channels.

rainwater head (hopper head, leader head). An enlarged head at the top of some DOWNPIPES, into which water from the gutter flows.

rainwater pipe. *See* DOWNPIPE.

rainwater shoe. A short bend at the bottom of a DOWNPIPE that directs water clear of the wall surface and often discharges into a GULLY.

raising plate. *See* WALL PLATE.

rake (batter). A uniform slope or angle of inclination.

raked joint. A mortar joint that is cleared of mortar for a depth of about 20 mm before POINTING or plastering.

rake moulding. (*USA*) The moulding at the top of a BARGE BOARD just below the tiles.

raking back. *See* RACKING BACK.

raking bond. *See* DIAGONAL BOND.

raking cornice (raking coping). A CORNICE or COPING with a slope, such as along the top of a GABLE.

raking flashing. A FLASHING between a sloping roof and a stone chimney for example. The flashing is let into a RAGLET groove in the stone. *See* STEPPED FLASHING.

raking out. The process of cleaning mortar out of brickwork joints before POINTING.

raking riser. A stair RISER that slopes inward from the NOSING.

raking shore (inclined shore). A long SHORING beam that slopes between a wall and the ground. *Compare* FLYING SHORE.

ramp. (1) An area of road or pathway that slopes and smoothly joins one level with another. (2) A bend in a HANDRAIL. (3) A short, steep length of drain pipe.

random ashlar. (*USA*) A form of COURSED SQUARE RUBBLE.

random courses. Courses of a wall or roof that vary in depth.

random shingles. SHINGLES with constant length but varying width.

random slates (rustic slates). SLATES of varying width. *See also* DIMINISHING COURSES.

range. (1) In surveying, the alignment of points with the aid of RANGE POLES or a telescope. (2) In statistics, a measure of the dispersion of a sample given by the difference the highest and lowest values. (3) A large stove or cooker that usually burns solid fuel.

range pole (ranging rod, picket). A straight pole of wood or metal, about 2 metres long with one pointed end. They are used to set out points in a survey line.

ranging line. A string stretched between PROFILES to show the face of a wall or other line.

ranging rod. *See* RANGE POLE.

Rankin cycle. The cycle of heat and energy changes in a steam engine.

rank set. A coarse setting of the BACK IRON of a plane.

rapid gravity filter. A form of GRAVITY FILTER that uses a pressure head of 2 or 3 metres to pass water through the filter bed. The filter is less effective than a SLOW SAND FILTER but it is compact and works many times faster.

rapid-hardening Portland cement. A form of PORTLAND CEMENT that is more finely ground and hardens more quickly than ordinary Portland cement. Although more expensive than normal cement, it can allow early dismantling of formwork and may speed construction.

rasp. A type of coarse FILE.

rare gases. *See* NOBLE GASES.

ratchet brace (ratchet drill). A type of BRACE used by carpenters where there is not enough space to turn a handle. It has a wheel and ratchet arrangement that allows turns of the drill.

rate. In ESTIMATING, the unit cost of a particular material or activity that is entered in a BILL OF QUANTITIES.

rate of growth. The number of ANNULAR RINGS in timber measured along a radius. Closely-spaced rings indicate slow growth which can mean more strength in SOFTWOODS.

rat-trap bond. A BRICKWORK bond in which the bricks are laid on edge to form a cheap type of wall one brick thick with gaps left in the middle. The outside faces show a type of FLEMISH BOND and the gaps are filled with grout and reinforcement. *Compare* QUETTA BOND.

Rawlbolt. A type of bolt that anchors a machine or other load into masonry. The bolt engages a lead plug or other liner that is inserted into a drilled hole.

raw linseed oil. LINSEED OIL that has been refined but not heat-treated such as by boiling.

Rawlplug. An insert that fits into drilled hole as a grip for a nail or screw. They now tend to be made of plastic but wood fibre has been used.

raw sienna. *See* SIENNA.

ray (medullary ray). A series of cells that radiate outward from the centre of the tree trunk, across the grain of the wood. The rays may sometimes be seen as ribbons or flecks of grain in wood such as quarter-sawn oak.

RCCB. *abbr.* RESIDUAL CURRENT CIRCUIT BREAKER.

reaction. (1) In general, an equal and opposite force that is set up as a result of a force. (2) In a support such a column or wall, the upward resistance that opposes the downward load.

ready-mixed concrete (RMC). Concrete that is mixed at a separate batching plant, transported to the site in a mixer truck, and delivered ready to be poured. The system is economical for some projects; for sites with restricted space RMC may be necessary.

real estate. (*USA*) Land and buildings on the land.

ream. The process of widening and smoothing a hole.

rebate (rabbet). A continuous step-like recess cut into the edge of a timber, such as in the glazing bar of a window.

rebate plane. *See* RABBET PLANE.

receptacle. (*USA*) An electrical SOCKET OUTLET set in the surface of a wall or floor as an outlet for current. It usually contains slots into which the prongs of a PLUG are inserted. *See also* SHUTTERED SOCKET.

receptor. (*USA*) The collecting tray at the bottom of a shower bath.

recessed-head screw. A SCREW that does not have a normal slot across the head. Examples include the PHILLIPS SCREW and POZIDRIVE SCREW.

recessed joint. A type of POINTING in which the mortar is set back from the face of the brick and makes a shadow. *See* BRICKWORK diagram.

reciprocal levelling. A method of taking LEVELS from two positions in order to eliminate instrumental error. The method is used when the positions are far apart and normal checks cannot be used.

reciprocating drill. A hand drill that is rotated by pushing down on the handle. It is used in joinery for drilling small holes.

reconditioning. A high temperature steam treatment for HARDWOODS that have suffered irregular shrinkage. *See also* COLLAPSE.

reconstructed stone. *See* ARTIFICIAL STONE.

recovery peg (reference peg). A peg fixed in the ground at a point whose position is known in relation to others.

rectangular coordinates. *See* CARTESIAN COORDINATES.

rectangular hollow section (RHS). A structural steel section with a rectangular cross-section.

rectangular tie. A WALL TIE made of wire bent in the shape of a rectangle.

rectifier. A device that converts ALTERNATING CURRENT into DIRECT CURRENT.

red lead. A red PIGMENT made of LEAD oxides. It has been used as an INHIBITING PIGMENT on metal and timber but it can be harmful to health. *See* LEAD.

red oxide. A red PIGMENT made from iron oxides. It was formerly thought to be an INHIBITING PIGMENT.

reduced level. The height of a point above or below a DATUM, as calculated from the survey readings.

reducer. (1) A paint THINNER. (2) A section of pipe that reduces in diameter in the direction of flow.

reducing power. The strength or white PIGMENT and its ability to make another pigment paler.

reduction of levels. In surveying, the calculation of different levels from the readings in a field book.

redundant frame (stiff frame). A frame structure that has more members than are needed for a PERFECT FRAME.

redwood. A large and fast-growing coniferous tree from North America. The timber is very durable but contains much SAPWOOD.

reeded. *adj.* Describing a pattern, such as in glass or cardboard, of half-rounded BEADS.

re-entrant angle. An angle, such as the corner of a building in a courtyard, that points inward. *Compare* SALIENT CORNER.

reference object (reference mark). The first point used when measuring the angle between two distance points.

reference peg. *See* RECOVERY PEG.

refined tar. COAL TAR that has been distilled to remove water and some oils.

reflectance (reflection factor). A measure of the ability of a surface to reflect visible light. It is expressed as the ratio of the LUMINOUS FLUX reflected to the flux incident upon the surface, and it has a maximum value of 1.

reflection. The return of light or sound energy from a surface without change of wavelength. *See also* DIFFUSE REFLECTION.

reflection coefficient. The ratio of sound energy reflected from a building surface compared to the sound energy falling on the surface. *See also* SOUND ABSORPTION COEFFICIENT.

reflection factor. *See* REFLECTANCE.

reflectivity. The fraction of heat RADIATION that is reflected from a surface. *Compare* EMISSIVITY.

reflector. A device or surface on which light or sound is reflected.

refraction. A change in direction of a wave motion, such as light or sound, that occurs at the junction of different materials.

refractory. *adj.* Describing a material that has the ability to resist high temperatures.

refractory lining. The use of refractory bricks or other materials to line a furnace.

refractory mortar. A type of MORTAR suitable for use in boilers and furnaces.

refrigerant. The fluid that circulates in a REFRIGERATION CYCLE. It needs be safe, have a high latent heat, and be easily converted from liquid to vapour and back. FREON is a common refrigerant.

refrigeration cycle. A process in which a REFRIGERANT repeatedly changes state and transfers heat from one place to another. Heat is absorbed from the surroundings by evaporation of the refrigerant and heat is released during liquefaction. The cycle may be used to operate a REFRIGERATOR or a HEAT PUMP.

refrigerator. A device that uses a REFRIGERATION CYCLE to cool a chamber or other area. The EVAPORATOR collects the heat and the CONDENSER coils give off the heat. *See also* ABSORPTION REFRIGERATOR, COMPRESSION REFRIGERATOR, REFRIGERANT.

refurbishment. The process of repairing and refitting an existing building to make it as good as new. *See also* MAINTENANCE, REHABILITATION.

register. (1) A ventilation outlet for a room that also has a DAMPER to vary the flow of air. (2) A DAMPER that controls the draught in a chimney.

reglet. *See* RAGLET.

Regault hygrometer. A form of HYGROMETER that uses a chilled surface to detect the DEW POINT temperature.

regrate. *vb.* To remove a stone surface to make it appear new.

regression. *See* LINEAR REGRESSION.

regular-coursed rubble (*USA*: coursed ashlar). A type of masonry construction in which random blocks of stone are grouped and laid in courses. *See also* RUBBLE WALLING.

rehabilitation. Extensive repairs and modernization designed to upgrade a building to modern standards. *Compare* REFURBISHMENT.

reinforced bitumen felt. A BITUMEN MATERIAL based on hessian as a woven reinforcement.

reinforced concrete. Concrete in which steel rods or mesh are set in order to carry tensile stresses. *See also* PRESTRESSED CONCRETE.

reinforced masonry. Stone or block walling with steel bars or mesh in the cavities to proved extra strength.

reinforced woodwool. Slabs of woodwool that are stiffened by wood battens or steel sections.

reinforcement. The use of rods, mesh or fabric set into other materials, such as concrete, to give extra strength.

reiteration. In surveying, the repeated measurement of an angle in order to increase the precision.

relative density (specific gravity). The DENSITY of a substance divided by the density of water which, in SI units, is taken as 1000 kilograms per cubic metre. *See also* HYDROMETER.

relative humidity (RH). A measure of HUMIDITY that compares the actual amount of moisture in the air with the maximum the air can hold at that temperature. It is calculated as the percentage ratio of VAPOUR PRESSURE divided by SATURATED VAPOUR PRESSURE.

relaxation. A CREEP and loss of tension in a TENDON used for PRE-STRESSED CONCRETE.

relay. An electrical SWITCH that is operated by an electromagnetic SOLENOID coil or other device. A relay allows a low-power primary circuit to control a high-power circuit.

release agent (parting agent). A substance, such as grease, that is applied to FORMWORK to give concrete a smooth finish and discourage sticking.

relief. A pattern that stands out from an otherwise smooth surface.

relieving arch (safety arch). An arch set inside brickwork or stonework above an opening to lessen the load on the LINTEL.

render coat (rough coat). (1) A first coat of a plaster on a wall. (2) A coat of sand and cement used to waterproof an external wall. *See also* BELL CAST, PEBBLE DASH, ROUGH CAST, STUCCO.

rendering. The process of applying a sand-cement RENDER COAT to a wall.

renovations. Repairs that restore a building to its original condition. *See also* REFURBISHMENT.

re-saw. *vb.* To rip saw timber into smaller pieces.

reseal trap. A type of TRAP sometimes fitted below a washbasin or sink where there is a danger of siphonage. A relatively large volume of water helps break the siphon.

resection. In surveying, a method of locating a point away from the base line.

reservoir. A large tank or artificial lake used to store water. *Compare* DAM.

residual current circuit breaker (RCCB). A CIRCUIT BREAKER that monitors the current in a circuit and breaks the circuit if a leak to earth is detected.

residual stress. Internal stresses that remain in materials after a manufacturing process, such as welding of metal.

resilient flooring. A floor covering, such as vinyl tiling, with some elastic properties that help it to absorb IMPACT SOUND.

resin. (1) Various natural substances that are produced by the sap of some plants and trees such as pines. They have a shapeless structure and may exist as viscous liquids or as solids. They have been used as paint BINDERS. *See also* GUM, OLEO-RESIN. (2) Various synthetic substances with a POLYMER structure and properties similar to natural resins. *See also* SYNTHETIC RESIN.

resin-bonded. A material such as plywood that is glued with a SYNTHETIC RESIN.

resistance. (1) *See* ELECTRICAL RESISTANCE. (2) *See* THERMAL RESISTANCE.

resistance moment. *See* MOMENT OF RESISTANCE.

resistivity. *See* THERMAL RESISTIVITY.

resonance. Vibrations that occur in an object at a particular frequency of sound or mechanical motion. It occurs when the NATURAL FREQUENCY of an object coincides with the frequency of any applied vibrations. Resonance produces effects in ACOUSTICS, SOUND INSULATION and on the behaviour of structures in wind and vibration.

resorcinol. A type of SYNTHETIC RESIN used to make a THERMOSETTING glue with good resistance to water.

rest bend. A right-angled bend used to join a vertical waste pipe to a main drain.

restraint fixing. A form of FASTENER that holds a component, such a wall, but does not carry load from the wall.

retaining wall. A free-standing wall designed to hold back earth or other solid material that rises higher on one side of the wall. *See also* MASS RETAINING WALL, SURCHARGE.

retarded hemihydrate plaster. A form of gypsum PLASTER made by adding a retarder to HEMIHYDRATE PLASTER to slow the setting. It is popularly used, under various tradenames, as an undercoat plaster and for finishing GYPSUM PLASTERBOARD.

retarder. An ADMIXTURE that is added to concrete, mortar or plaster in order to lengthen the time taken to set or harden.

re-tempering. The undesirable practice of remixing mortar or plaster that has begun to set.

retention money. A percentage of the payment due to a contractor that is held back until the end of the DEFECTS LIABILITY PERIOD. *See also* PENULTIMATE CERTIFICATE.

reticule (graticule). A glass circle within the telescope of a surveying instrument. The sighting lines such as CROSS HAIRS are engraved on this diaphragm.

return. (1) A change in direction at the end of a wall, usually at right angles. (2) *See* RETURN PIPE.

return pipe (return). A pipe in a hot water system that takes the cooled recirculating water back to the boiler. *See* HEATING diagram. *Compare* FLOW PIPE.

return wall. A short length of wall at right angles to a main wall.

reveal. The part of the JAMB of a door or window that is not covered by the frame and is visible.

reverberation. The continuing presence of sound in a room when the source of sound stops. It is caused by rapid multiple reflections that blend with the direct sound.

reverberation time (reverberation period). A measure of REVERBERATION which is an important factor in the ACOUSTIC quality of a room. It is defined as the time taken for a particular frequency of sound to decrease to one millionth of its original value, a fall of 60 dB. *See also* EYRING FORMULA, SABINE'S FORMULA, STEPHENS AND BATE FORMULA.

reverse. A type of TEMPLATE used to check the shape of a plaster moulding.

revetment. A covering of stone or other material used to protect an earth surface, such as a bank, from the action of weather or water. *See also* RIP-RAP.

revolving door. A door made from up to four leaves that are hung on a central pivot and can turn within a circular enclosure. The door is used as an entrance and exit for public buildings by entering one of the quarters and moving with the revolving leaves. The system provides an AIR LOCK between the building and the outside.

rewirable fuse. A simple FUSE that, after use, can be rethreaded with wire. *Compare* CARTRIDGE FUSE.

Reynold's number. A number that, in a suitable formula, is used to predict whether a fluid will have LAMINAR FLOW or TURBULENT FLOW. For water in a closed pipe the Reynolds number is about 2000. *See also* CRITICAL VELOCITY.

RH. *abbr.* RELATIVE HUMIDITY.

rheology. The study of the flow and change of shape of materials.

rheostat. An electrical RESISTANCE that can be varied in value.

rhone. *See* EAVES GUTTER.

RHPC. *abbr.* RAPID-HARDENING PORTLAND CEMENT.

RHS. *abbr.* RECTANGULAR HOLLOW SECTION.

RI. *abbr.* ROOM INDEX.

RIBA. *abbr.* Royal Institute of British Architects. *See also* ARCHITECT.

riband. (1) A board that runs beneath the FORMWORK of a reinforced concrete beam and provides support during casting of the beam. (2) A flat rail that supports the pails in a PALISADE.

ribbing. A raised pattern of ANNUAL RINGS in the surface of timber that is caused by different shrinkage rates in spring and summer.

ribbon. (1) A strip of slates or tiles set into a wall for decoration. (2) *See* RIBBON BOARD.

ribbon board (ribbon, ribbon strip). In BALLOON FRAMING, a JOIST that is attached to the wall studs and supports the floor joists.

ribbon courses. Alternate courses of long and short roofing slates or tiles.

ribbon rail. *See* CORE RAIL.

ribbon saw. A BAND SAW with a narrow blade.

ribbon strip. *See* RIBBON BOARD.

rich lime. *See* HIGH-CALCIUM LIME.

rich mix (fat mix). A mix of MORTAR or CONCRETE that contains a higher than normal proportion of cement or lime. The opposite of a LEAN MIX. *See also* WATER-CEMENT RATIO.

Richter scale. A logarithmic scale of earthquake shock based on the energy at the source of the earthquake. *Compare* MERCALLI SCALE.

RICS. *abbr.* (*UK*) Royal Institution of Chartered Surveyors. *See also* SURVEYOR.

riddle. A coarse form of SIEVE with large holes.

ride (riding). The effect when a door touches the floor as it opens.

ridge (ridge board, ridge pole). A horizontal board set on edge at the peak of a pitched roof where the RAFTERS meet. *See* ROOF diagram.

ridge capping. The covering on a roof RIDGE which may be made of a variety of materials.

ridge course. The top course of roof tiles or slates that is next to the RIDGE.

ridge pole. *See* RIDGE.

ridge stop. A form of flexible metal FLASHING installed over the junction of a roof ridge and a wall that rises above the ridge.

ridge tile. A ROOF TILE designed to cover a RIDGE. It may be half-round, HOGSBACK or angular in shape and is bedded in cement mortar. *See also* UNDER-RIDGE TILE.

riding. *See* RIDE.

riffler. A type of file used for working concave surfaces in stone.

rift-sawn timber. *See* QUARTER-SAWN TIMBER.

rigger. (1) A trained person who erects and maintains lifting machinery. (2) A person who assembles metal SCAFFOLDING. (3) A long-haired paint brush or LINING FITCH used for painting lines.

right-handed. *See* HANDED.

right of way. The legal right to pass over land owned by another and the path or road used for this right.

rigid arch. A fixed arch without hinges.

rigid damp course. A DAMP-PROOF COURSE made of a rigid material like brick or slate.

rigid frame. A structural frame in which all members are firmly fixed together without HINGES.

rigidity (stiffness). The ability of a material or structure to resist shear and twisting forces. It may be measured by the ELASTIC MODULUS.

rim latch. A type of LATCH mounted on the

surface of the door in a metal box. It is usually set on the opening STILE of the door and is operated by a knob. *Compare* MORTISE LATCH.

rim lock. A LOCK mounted on the surface of the door. It is usually contained in a metal box and operated by a key. *Compare* MORTISE LOCK.

ring course. The COURSE of stones or bricks that is nearest to the EXTRADOS or outer surface of an ARCH.

ring main. A system of electrical wiring in which each power outlet is connected to the mains supply by two cable routes that can share the current load. A number of outlets or appliances are usually wired into the ring that uses less cable and allows simpler connections than RADIAL CIRCUITS for each outlet.

ring shake. A SHAKE that runs along the growth rings in timber. *See also* SHELL SHAKE.

ring tension (hoop stress). The TENSILE STRESS within a circular wall or tank that retains a liquid or solid.

rip. A saw cut made along the grain of the wood. *Compare* CROSS CUT. *See also* SWAGE-SETTING.

ripper. A long, thin-bladed tool that can be inserted under roof SLATES and used to cut the fixing nails during roof repairs.

ripping bar. (*USA*) *See* CROW BAR.

ripple finish. A surface finish with intentional ripples in the paint film.

rip-rap. The large stones used in a REVETMENT.

rip saw. A hand-held SAW intended to cut along the grain of wood. It can have widely-spaced POINTS. *Compare* CROSS-CUT SAW.

rise. (1) The vertical distance between the RIDGE of a roof and the top of the walls that support that roof. (2) The vertical spacing between the TREADS in a stair. *See* STAIRS diagram. (3) The height of an arch measured from the SPRINGING LINE to the inside surface of the INTRADOS.

rise and fall. In surveying, a method of recording and calculating vertical distances form successive rises or falls in height.

rise-and-fall table. A CIRCULAR SAW bench on which timber can be raised and lowered.

rise and run. The pitch or gradient of a member expressed as the vertical height for each unit of horizontal run.

riser. (1) The vertical front of step. (2) A stone that is deeper than one course in SNECKED RUBBLE.

rising-butt hinge. A form of HINGE that, as it is opened, causes a door to rise clear of floor carpet.

rising damp. The general upwards movement of water from the ground though a wall. It is often caused by CAPILLARY ACTION and is usually prevented by a DAMP-PROOF COURSE. *See also* ELECTRO-OSMOSIS.

rising main. A supply of electricity, water or gas that enters a building and passes up through several storeys.

rive. *vb.* To split stone, especially SLATES or SHINGLES.

riven slate. SLATE TILES that are sawn rather than split.

rivet. A round-headed metal bolt that fastens components such as sheets of metal. After passing through both components the seal is made by flattening the shaft of the rivet, sometimes while it is red-hot.

rivet plug. A nylon plug that joins sheets of thin material when it is expanded by inserting a screw.

riving knife. A steel knife that set behind the blade of a CIRCULAR SAW and keeps newly-cut wood apart.

RL. *abbr.* REDUCED LEVEL.

RMC. *abbr.* READY-MIXED CONCRETE.

RMS. *abbr.* ROOT MEAN SQUARE.

road tar. Coal TAR that has been treated and blended to conform to agreed specifications for use as a BINDER in road construction. Road tars have the advantage of good adhesion in damp conditions and are mainly used as surface dressings on roads. *See also* COATED MACADAM, TACK COAT.

rock. A natural AGGREGATE of MINERALS that make up the EARTH'S crust. *See also* IGNEOUS ROCK, METAMORPHIC ROCK, SEDIMENTARY ROCK.

rock anchor (rock bolt, roof bolt). A long bolt inserted and wedged into the roofs of mines and the sides of excavations as a support.

rock asphalt. A natural form of BITUMEN found in a fine-grained limestone rock.

rock bolt. *See* ROCK ANCHOR.

rocking frame. A moving platform that shakes MOULDS when they are filled with concrete. The movement helps to compact the concrete.

Rockwell hardness test. A standard test for HARDNESS in which the depth of penetration of a point, or a steel ball, is measured. *Compare* BRINELL HARDNESS TEST.

rock wool. *See* MINERAL WOOL.

rod. (1) A board on which the dimensions of a joinery assembly, such as stairs, are laid out in full size. (2) A TIE. (3) *See* LEVELLING STAFF.

rodding. The process of cleaning drains or clearing blockages with DRAIN RODS.

rodding eye. An ACCESS EYE in a drain that can be used for RODDING.

roll. (1) A rounded length of wood over which edges of FLEXIBLE-METAL ROOFING are folded in order to join sheets.

rolled-steel joist (RSJ). A structural steel beam with an I-shaped cross-section and tapered FLANGES. RSJs are produced by HOT ROLLING and used in the frames of buildings. *Compare* COLD-ROLLED SECTION, UNIVERSAL BEAM. *See also* FLANGE, WEB.

roller. (1) A revolving drum covered in fabric and mounted on a handle. It is used instead of a brush for applying paint to large regular areas. (2) A small hand roller used to press down the edges of WALLPAPER.

roller blind. A type of internal BLIND that is stored on a spring-operated roller at the top of the window.

roller coating. The process of using a ROLLER to apply paint or other coatings.

roller shutter. A flexible shutter made of horizontal slats that can be wound and stored on a roller mounted above an entrance. They are usually made of metal and used to protect large windows and doorways.

roll roofing. (*USA*) Various roofing materials, like roofing FELT, that are supplied in roll form.

rollock. *See* ROWLOCK.

Roman brick. A narrow brick, about 50 mm deep including a mortar joint.

Roman cement. (1) *See* POZZOLAN. (2) An early type of cement made by burning lumps of MARL found in London clay. *See also* EMINENTLY HYDRAULIC LIME.

Roman tile. A SINGLE-LAP TILE with a roll at one side. *See also* DOUBLE-ROMAN TILE.

rone. *See* EAVES GUTTER.

roof. The top covering that protects a building from the weather. *See* ROOF diagram. *See also* FLAT ROOF, PITCHED ROOF, RAFTER, TILE, SLATE.

roof boards (sarking). Boards attached to the common RAFTERS of a roof and used as a base for roofing materials. *See also* ROOF DECKING.

roof bolt. *See* ROCK BOLT.

roof cladding. The layer of material or elements that make up the weatherproof part of a ROOF. *See also* BUILT-UP ROOFING, SLATES, TILES.

roof decking. Sheets of material, or a similar construction, that cover the roof structure and act as a base for the ROOF CLADDING.

roof guard. *See* SNOW BOARD.

roofing felt. A flexible BITUMEN MATERIAL based on fibre and usually supplied in rolls. It is used as the waterproof layer in some roof claddings. *See also* BUILT-UP ROOFING, MINERAL-SURFACED BITUMEN FELT, SARKING FELT.

roofing nail. A type of nail used to fix components such as slates or tiles on to a roof. The use of aluminium, copper or galvanizing gives the nail some corrosion resistance. *See also* LEAD-HEAD NAIL.

roof ladder. A CAT LADDER used to move over the surface of a roof.

roof light. A fitting in the roof that admits light into a building from the sky. *See also* LANTERN LIGHT, MONITOR ROOF, SKYLIGHT.

rooflight sheet. (1) A transparent sheet of roofing material, such as a THERMOPLASTIC. *See also* FIRE VENT. (2) A sheet of roofing material with an opening into which a ROOF LIGHT is fitted.

roof space. The space between the ceiling of the top floor and the underside of the roof.

roof terminal. The opening of a ventilation pipe that projects above the roof.

roof tile. *See* TILE.

roof truss. A TRUSS used to support a roof. *See* ROOF TRUSS diagram.

room acoustics. *See* ACOUSTICS.

room index (RI). A measure of the proportions of a room combined into a single number that is used for interior lighting calculations. It is calculated as the length multiplied by the width divided by the mounted height of the fitting and divided by the sum of the length and the width. *See also* UTILIZATION FACTOR.

room-sealed boiler. A gas heating device with a BALANCED FLUE.

root. (1) In carpentry, the part of a TENON that widens at the shoulders. (2) The part of a dam where it meets the hillside.

root mean square (root mean square value, RMS). A type of average value that is applied to fluctuating quantities such as ALTERNATING CURRENT electricity and sound pressure. The RMS value is usually equal to the peak value multiplied by the square root of 2.

ropiness. A defect of paintwork in which the brush marks are visible and do not flow out. A common cause of this 'ropy' finish is that the paint film has begun to harden.

rose. (1) A decorative plate fixed on a door around the door handle. (2) An electrical fitting. *See* CEILING ROSE.

rose bit. A type of drill BIT used for COUNTERSINKING.

rose diagram. A method of indicating the direction of wind conditions at a location on a map. Lines radiate from the point and the different lengths of the lines represent wind speeds or other qualities from that direction of the compass.

Rosemary. (*UK*) A common brand of PLAIN TILE used for roofing.

rosin (colophony). A natural type of RESIN formerly used to make VARNISH. *See also* OLEO-RESIN.

rot. A relatively rapid decay of timber caused by the presence of a FUNGUS that feeds on the wood. *See also* DRY ROT, WET ROT.

rotary cutting. A common method of producing VENEER from a log. The soaked log is revolved against a long knife which peels off a thin continuous sheet. *Compare* SLICED VENEER.

rotary veneer. A VENEER produced by rotary cutting.

rotor. The revolving part of a large DYNAMO. *Compare* STATOR.

rotten knot. *See* UNSOUND KNOT.

rough arch. *See* RELIEVING ARCH.

rough ashlar. Stone brought from the quarry before it dressed or worked.

rough bracket. A bracket beneath a STAIR.

rough cast (harl, harling, slap dash). A protective coating for external walls, similar to PEBBLE DASH. It is made with a RENDERING mix containing rough stones.

rough coat. *See* RENDER COAT.

rough cutting. The cutting of common brickwork. *Compare* FAIR CUTTING.

rough floor. (*USA*) *See* SUB-FLOOR.

rough ground. *See* COMMON GROUND.

roughing-in (roughing-out). Doing the first rough work of a trade, such as roughly shaping a piece of wood or installing pipe without connections.

rough string. A CARRIAGE beneath a STAIR.

rough work. Brickwork that is eventually hidden by other brickwork, by plaster or rendering.

round. (1) A property of paint that has a stiff consistency and gives a thick coat. (2) A RUNG of a ladder.

round knot. A KNOT that is cut and revealed across its length. *Compare* SPLAY KNOT.

rout. *vb.* To cut a groove in wood.

router. (1) A tool that cuts (routs) a groove in wood. (2) A PLOUGH type of PLANE.

row house. (*USA*) A house in a row of three or more similar houses that are joined. *See also* TERRACE HOUSE.

rowlock (rollock). A course of bricks laid on edge, such as for the top of a wall. *See also* RAT-TRAP BOND.

rowlock arch. A RELIEVING ARCH built from courses of brick HEADERS laid on edge.

rowlock-back wall. A brick wall with the front face laid in a regular bond and the back face laid with bricks on edge.

rowlock cavity wall. A brick wall built in RAT-TRAP BOND.

RSC. *abbr.* Rolled-steel channel (used in structural steelwork).

RSJ. *abbr.* ROLLED-STEEL JOIST.

RST. *abbr.* Rolled-steel tee (used in structural steelwork).

rubbed brick (rubber). A soft smooth brick without a FROG that can be rubbed to an exact shape. *See also* GAUGED BRICK.

rubbed finish. (1) A surface effect on concrete made by smoothing with an abrasive material such as carborundum. (2) *See* RUBBING DOWN.

rubbed joint. A glued joint made between two narrow pieces of timber. Both surfaces are planed smooth, coated with glue and rubbed together to exclude air and excess glue from the joint.

rubber. (1) An ELASTOMER material that is obtained naturally from the LATEX of certain trees. Rubber materials are made after various treatments and combinations with artificial POLYMERS. (2) *See* GAUGED BRICK, RUBBED BRICK.

rubbing down (rubbing, flatting down). The process of SANDING a painted surface with sandpaper or other abrasive to make it level and form a key for another coat.

rubbing stone. An abrasive stone used for rubbing bricks. *See* RUBBER.

rubble. (1) Broken pieces of brick, concrete plaster and similar material. (2) Stone for walls that has not been smoothed like ASHLARS. It is sometimes roughly squared.

rubble ashlar. A stone wall built with an ASHLAR face and a RUBBLE back.

rubble wall. A stone wall of various types but without the fine joints of ASHLAR construction. *See also* REGULAR-COURSED RUBBLE.

rule. (1) A flat straight edge, often made of steel, that is marked with graduations and used for drawing or for measuring. (2) Various types of straight edge used for working plaster. *See also* FLOATING RULE, JOINT RULE.

rummel. *See* SOAKAWAY.

run. (1) A length of pipework or cables in a building. (2) A BARROW RUN. (3) A paint defect caused when excess paint flows downward and forms a ridge. (4) (*USA*) In a staircase, the horizontal depth of the TREAD that is stepped on. *See also* GOING.

rung (round, stave). (1) One of the horizontal bars that connect the two side posts of a LADDER and act as a step. (2) A bar fixed into a wall to act as a step.

run line. A straight line of paint made with a LINING TOOL and a straight edge.

runner. (1) In general, a timber frame member used to stiffen other members and to transfer loads. (2) One of two horizontal rails on which drawers slide. (3) One of the two side post of a LADDER. (4) A long strip of carpet, such as used in a passageway.

running bond. *See* STRETCHER BOND.

running rule. A STRAIGHT EDGE nailed to an undercoat of plaster to help form a cornice.

run off. The quantity of rain and snow water collected from a CATCHMENT AREA in a given time. It is the rainfall minus natural evaporation and infiltration into the ground. *See also* IMPERMEABILITY FACTOR.

rust. A reddish coating of iron oxides that forms on the surface of iron or steel after exposure to moisture and air. *See also* CORROSION, OXIDATION.

rusticated. *adj.* Describing a decorative effect in stone or brickwork given by rough surfaces and a recessed mortar joints.

rustic brick. A type of FACING BRICK with a textured surface, such as a SAND-FACED BRICK.

rustic joint. A mortar joint that is recessed below the surface of the stone or brick.

rustic slates. *See* RANDOM SLATES.

R-value. The value of THERMAL RESISTANCE for a particular layer in a structure. *Compare* U-VALUE.

rwp. *abbr.* Rainwater pipe. *See* DOWNPIPE.

rybate. A STRETCHER stone.

S

SAAT. *abbr.* (*UK*) Society of Architectural and Associated Technicians.

sabin. A unit of the SOUND ABSORPTION of a surface. It is calculated as the absorption coefficient of a surface multiplied by the area of that surface.

Sabine's formula. A widely-used acoustic formula for predicting REVERBERATION TIME in a room. The time is proportional to the volume of the room and inversely proportional to the total SOUND ABSORPTION of the room surfaces. *Compare* EYRING FORMULA.

sable brush. Various types of paint brush made from fine animal hair.

sacrificial protection. The principle of protecting a metal against CORROSION by attaching another metal that is higher in the GALVANIC SERIES and therefore corrodes first. GALVANIZING a coating of zinc on to iron is an example. *See also* CATHODIC PROTECTION, CORROSION.

saddle. (1) A FLASHING of flexible metal that is cut and fixed beneath tiles or slates at an intersection with other parts of the roof. (2) A fixing that passes around a pipe and is screwed down on each side. (3) A pipe fitting that clamps over a hole in a drain or sewer and makes a new connection. (4) A metal block at the top of a suspension bridge tower over which the cables pass.

saddle-back board. *See* CARPET STRIP.

saddle-back coping. A COPING cap of triangular section with a sharp point.

saddle bar. A horizontal bar that stiffens LEADED LIGHTS.

saddle joint (water joint). A joint where the stones in a cornice form a saddle shape and throw water clear of the joint.

saddle piece. A flexible metal SADDLE.

saddle roof. A pitched roof with two similar slopes and gables at each end.

saddle scaffold (straddle scaffold). A SCAFFOLD that sits over the ridge of a roof and is used to repair chimneys.

saddle stone. The stone that forms the apex of a gable.

safety arch. *See* RELIEVING ARCH.

safety factor. *See* FACTOR OF SAFETY.

safety glass. Various type of glass that are difficult to shatter and which break into small safe pieces rather than into splinters. Examples include TOUGHENED GLASS, WIRED GLASS and the laminated plastic in SAFETY GLASS.

safety lighting. A secondary system of lighting in a public building that helps people to leave the building safely when the general lighting fails. The safety lights must have an independent supply of electricity.

sag correction. In land surveying, a correction applied to tape measurements to allow for the catenary sag of the tape.

sagging. (1) *See* CURTAINING. (2) The formation of a dip in a surface or structure. *Compare* HOGGING.

sailing course. *See* STRING COURSE.

Saint Andrew's cross bond. *See* DUTCH BOND.

salamander. A portable heater used to dry a building during construction.

salient corner. A corner that points outward from a building, as opposed to a RE-ENTRANT corner.

sally. An inward-pointing cut in a timber, such as a BIRDSMOUTH JOINT.

salt. (1) One of a large group of chemical compounds formed when the hydrogen of an acid is replaced by a metal. Many salts exist naturally as minerals in the ground. For example *see* CALCIUM CARBONATE, CALCIUM SULPHATE. *See also* HARDNESS. (2) Sodium chloride or 'common' table salt. *See also* BRINE.

salt glaze. A glaze obtained on the surface of bricks, drainpipes and tiles by the adding common SALT during firing. *See also* VITREOUS.

sand. Fine particles of mineral material formed by the natural wearing of rock or by the crushing of stone. Sand is used as a FINE AGGREGATE.

sandblasting. A method of cleaning or reducing a surface with a high velocity stream of air containing sand.

sand box. A traditional device used to support a post holding formwork such as CENTERING. The post sits in a box of dry sand; to remove the formwork the sand is allowed to flow away from the box.

sand-dry. *adj.* Describing the state of a paint film when it is dry enough for sand not to stick to the surface.

sander. *See* SANDING MACHINE.

sand-faced brick. A type of FACING BRICK with a decorative coat of sand on least two faces. *See also* RUSTICATED.

sand filter. A device that cleans water supplies or sewage by FILTRATION of the liquid through layers of SAND. In the slow sand filter, water descends through the sand bed by gravity and there is also a BIOLOGICAL FILTER effect from the layer of bacteria that develops on the surface of the sand. A rapid sand filter passes the water under pressure and has a simpler mechanical effect. *See also* COARSE FILTER, FILTER BED, GRAVITY FILTER, RAPID GRAVITY FILTER. *Compare* MICRO-STRAINER.

sanding. The use of GLASSPAPER or SANDPAPER to smooth wood surfaces, either by hand or by sanding machine. *See also* RUBBING DOWN of paint surfaces.

sanding machine (sander). A power tool that smooths surfaces with GLASSPAPER or other types of abrasive. *See also* BENCH SANDER, DISC SANDER, ORBITAL SANDER.

sanding sealer. A special first coat of paint that seals the wood and is hard enough to be sanded.

sand-lime brick. *See* CALCIUM SILICATE BRICK.

sandpaper. An abrasive paper containing sharp sand. *See also* GLASSPAPER.

sand pile. A method of COMPACTION for silty soil achieved by pouring sand into boreholes and driving the sand by a ram.

sandstone. A SEDIMENTARY ROCK made from grains of quartz sand cemented together with a durable material like a silicate.

sandwich beam. *See* FLITCH BEAM.

sandwich panel. A composite panel in which hard outer layers such as plywood or plaster contain a core of foamed plastic or other cellular material.

sanitary appliance. In general, a bathroom or kitchen fitting that is connected to a water supply and to drainage.

SANZ. *abbr.* Standards Association of New Zealand.

sapele. A HARDWOOD timber often used as a substitute for MAHOGANY veneer.

saponification. The general process that produces soap when oil or fat is reacted with an alkali. A paint film can be decomposed by the same process when alkalis in cement or other building materials react with oils in the paint to give a soluble soap product.

sap stain. *See* BLUE STAIN.

sapwood. Timber taken from the outer wood of a tree. Sapwood is lighter in colour than HEARTWOOD, more likely to ROT but easier to treat with preservative.

sarking. *See* ROOF BOARDS.

sarking felt. ROOF FELT that is laid beneath tiles or slates, with or without ROOF BOARDS.

sash. One of the sliding LIGHTS in a SASH WINDOW. *See* WINDOW diagram.

sash balance. A spring that operates a SASH WINDOW instead of SASH WEIGHTS and associated mechanisms.

sash chain. A chain used in a SASH WINDOW in place of a SASH CORD.

sash chisel. A type of CHISEL with a narrow blade.

sash cord. The CORD that is attached to the sash of a SASH WINDOW, passes over a pulley and then suspends the SASH WEIGHT.

sash door. A door in which the upper half contains glass.

sash fastener. A locking bolt for a SASH WINDOW. It is fixed on the meeting rail of one sash and engages the meeting rail of the other.

sash pulley. One of the pulleys that carry the SASH CORDS in the frame of a SASH WINDOW.

sash weight. A long metal weight that hangs inside the CASED FRAME of a SASH WINDOW. The weight is used to counterbalance the SASH and allow the window to open easily and stay in position.

sash window (hanging sash). A type of window in which two LIGHTS open vertically by sliding inside the CASED FRAME. The weight of the windows is countered by the SASH WEIGHTS. *See* WINDOW diagram.

satin finish. A semi-MATT finish with a fine-textured sheen on the surface.

saturated air. A sample of air that contains the maximum amount of water vapour it can hold before CONDENSATION begins. The RELATIVE HUMIDITY of saturated air is 100 per cent. The mass of water able to be held as vapour is constant for a given temperature and increases as the air is warmed. *See also* HUMIDITY.

saturated vapour pressure (SVP). The VAPOUR PRESSURE of a sample of SATURATED AIR. The SVP of a fixed sample of air increases as the temperature of the sample increases.

saturation. (1) A property of COLOUR that describes the purity or intensity of a HUE compared to a neutral gray of similar lightness. (2) The condition of a material that is thoroughly soaked with a fluid such as water. (3) *See* SATURATED AIR. (4) *See* DEGREE OF SATURATION in soils.

saturation coefficient. The ratio between the volume of water absorbed by a material, such as brick, and the volume of its pore spaces.

saturation line. *See* WATER TABLE.

saw. A tool that cuts wood and other materials by means of a cutting edge of teeth set on a steel blade. *See also* BAND SAW, CIRCULAR SAW, CROSS-CUT SAW, HACKSAW, JIGSAW, PANEL SAW, RIP SAW, TENON SAW.

saw bench. A steel table in which the blade of a CIRCULAR SAW is set.

saw doctor. A person skilled in the care of SAWS and their blades.

sawdust concrete. A concrete made with a mix of cement and sawdust.

saw horse. A four-legged stand used to support timber while it is sawed.

sawn veneer. Timber that is QUARTER-

SAWN to produce thin VENEER. *Compare* ROTARY VENEER.

saw set. A tool that is used to give the correct SET to the teeth of a SAW.

saw tooth. A single cutting blade in the cutting edge of a SAW.

sax. *See* ZAX.

SB. *abbr.* SUB-BASEMENT.

SC. *abbr.* (1) STRENGTH CLASS. (2) SKY COMPONENT.

scabbing hammer. A hammer with a pick on one end of the head. It is used for the rough dressing of stone.

scaffold (scaffolding). A temporary framework of metal tubing or timber used in the construction or repair of a building. The scaffolding supports platforms that give easy access to the construction. *See also* STANDARD.

scaffold board. One of a number of timber boards laid as the platform in a SCAFFOLD. The boards are usually protected by metal bands at their ends.

scaffold pole. One of the poles that make up SCAFFOLDING. Modern scaffold poles are usually made from tubular metal.

scagliola. Plasterwork coloured and polished to look like marble.

scalar illumination. Lighting or that part of an illumination that arrives on a surface by indirect means such as reflections. *See also* VECTOR/SCALAR RATIO.

scale. (1) The ratio of the measurements on a drawing to the actual measurements of objects represented by the drawing. (2) *See* SCALE RULE. (3) A coating of iron oxide that forms on heated steel. (4) FURRING caused by HARD WATER.

scale rule (scale). A measuring RULE marked in various scale ratios. It allows final measurements to be read directly from a drawing.

scant cut timber. Timber that is cut under size.

scantling. Square-sawn framing timber of relatively small dimensions.

scarf joint. A joint in which the ends of boards or veneers are cut at an angle before they are glued together. *Compare* BUTT JOINT.

scarifying. The process of scratching and loosening a surface as a preparation for paint or other surface coatings.

SCET. *abbr.* (*UK*) Society of Civil Engineering Technicians.

sch. *abbr.* Schedule (used on drawings).

schedule of dilapidations. A list of repairs to be carried out by a tenant at the termination of a lease.

schedule of prices. An alternative to the BILL OF QUANTITIES for certain types of project.

Schmidt hammer. A spring-loaded hammer used in the NON-DESTRUCTIVE TESTING of concrete. The rebound of the hammer is used to predict the compressive strength of the concrete.

scissors truss. A simple roof TRUSS in which the bottom rafters are V-shaped inward rather than horizontal. *See* ROOF TRUSS diagram.

sclerometer. An instrument that uses a diamond point to determine the HARDNESS of a material.

score. *vb.* To scratch a surface to provide a better mechanical BOND for a coating such as plaster.

scoria. Loose pieces of porous volcanic rock.

Scotch bond. *See* ENGLISH GARDEN WALL BOND.

scotia. A MOULDING with an inward curve. *See* MOULDINGS diagram. *Compare* OVOLO.

scraper (scraper plane). A hand tool with a thin vertical blade that is used for removing marks from wood before sandpapering.

scratch awl. *See* AWL.

scratch coat. The first coat of plaster or stucco. It is roughened by scratching to provide a key for the next coat.

scratch work. *See* SGRAFFITO.

screed. (1) A straight strip of wood or metal that is moved over fresh concrete or plaster to give a flat, even surface. (2) A narrow band of concrete or other coating that is carefully laid to the correct level and acts as a guide to the thickness. (3) A layer of mortar laid on top of a concrete slab to provide a level surface. *See also* MONOLITHIC SCREED.

screed rail. A RULE used to level the surface of concrete or a screed.

screen. (1) A movable partition used to give protection or privacy. (2) A frame fitted with a wire mesh used to protect a ventilation inlet or to separate pebbles and sand.

screen analysis. *See* SIEVE ANALYSIS.

screw. A metal fastener that joins two objects by means of a thread on the shaft of the screw. When the head of the screw is turned the action of the thread pulls the screw into a material such as wood. *See also* COACH SCREW, CROSS-HEAD SCREW, RECESSED-HEAD SCREW, SELF-TAPPING SCREW, SLOTTED HEAD SCREW.

screw-down valve. A valve that cuts off a water by the action of screwing down a plate or disc.

screwdriver (turnscrew). A tool used to turn and drive a SCREW. The tip of the screwdriver shaft is shaped to fit the slot on the head of the screw.

screwed pipe. A type of water or gas pipe that is threaded at the ends for joining.

screw eye. A WOOD SCREW with a loop instead of a head. It can be used with a CABIN HOOK to fasten a window or door.

screw gauge. A series of numbers used to define the relative diameters of SCREWS.

screw nail. *See* DRIVE SCREW.

screw pile. A PILE that is turned rather than rammed into the ground.

screw plug. A temporary plug that expands and seals a drain for testing.

screw starter. *See* GIMLET.

scribe. (1) To scratch a line on the surface of a material. (2) The process of shaping a surface to fit an irregular surface. (3) *See* AWL.

scriber. A sharp-ended tool used to SCRIBE lines on a surface.

scrim. A coarsely-woven fabric of canvas, sacking or muslin that is used for SCRIMMING.

scrimming. The stretching of SCRIM over joins or irregularities in walls to make them smooth for plastering or papering.

scrub board. (*USA*) *See* SKIRTING BOARD.

scrub plane. A type of PLANE with a rounded cutting iron for making thick shavings.

scumble. A semi-transparent stain or paint through which GROUND COAT shows in patterns.

scumboard. A board dipped below the surface of a liquid to prevent surface scum flowing.

S-curve analysis. A method of predicting profits from a building project before the CONTRACT. If the value or cost of a project is plotted against time taken, the model assumes a straight line in the middle section with parabolic lead-in and lead-out sections, giving an overall S-shape.

scutch. *See* COMB HAMMER.

scutcheon. *See* ESCUTCHEON.

seal. (1) A fitting or material that prevents

the entry of air around a joint. (2) Water that remains in the TRAP of a drain and isolates the inlet from foul air in the drain.

sealant. Various materials used to make a SEAL in the long joints between components such as window frames and walls. Modern sealants are often MASTICS that remain flexible for a considerable time.

sealed system. A water-heating circuit from which air is excluded. It has no expansion tank like the OPEN-VENTED SYSTEM.

sealed unit. A glazing unit of two or more sheets of glass that are sealed together in the factory with a narrow gap of up to 20 mm between them. This cavity is filled with clean dry air, or another gas.

sealer. A liquid, sometimes clear, used to coat a porous surface before further treatment. The sealer closes the pores of the surface and may prevent base materials from bleeding through.

sea-level correction. For lengths surveyed at altitudes significantly above sea distance, a conversion to the equivalent distance at mean sea level.

sealing compound. (1) *See* SEALANT. (2) A BITUMEN applied cold to the joints between ROOFING FELT.

seam. A joint in FLEXIBLE-METAL roofing made by folding the edges of two sheets together.

seamer. A special pair of pliers used to form SEAMS in flexible-metal roofing.

seam roll. *See* HOLLOW ROLL.

seasoning. (1) The process of drying timber to a MOISTURE CONTENT suitable for the conditions where it will be used. *See also* AIR-DRIED, KILN DRIED. (2) The process of drying and hardening stone.

secondary beam. A beam that is supported by other beams rather than by columns or walls. *Compare* MAIN BEAM.

secondary circuit. A pipe circuit that delivers hot water or heat from a HEAT EXCHANGER, without being directly heated. *See* HEATING diagram. *See also* PRIMARY CIRCUIT, INDIRECT SYSTEM.

secondary filtration. In water treatment, a final stage of filtration such as SAND FILTERS or MICRO-STRAINERS. *Compare* PRIMARY FILTRATION.

secondary glazing. A system of DOUBLE GLAZING formed when a second layer of glass is added to an existing window. *Compare* SEALED UNIT.

secondary treatment. In the treatment of SEWAGE, the use of AEROBIC BACTERIA to treat the effluent produced by the PRIMARY TREATMENT.

second fixings. Joinery, such as skirting boards and cupboards, that is fitted after plastering. *Compare* FIRST FIXINGS.

second growth. New trees that grow in a forest area after the original trees have been removed.

second moment of area. *See* MOMENT OF INERTIA.

secret dovetail. *See* MITRE DOVETAIL.

secret fixing. The general fixing of joinery so that the fixing can not be seen at the surface. SECRET NAILING and secret screwing are common methods.

secret gutter (sunk gutter). A drainage GUTTER at the bottom of a roof valley that is almost completely covered by the overhang of tiles or slates. *Compare* OPEN VALLEY.

secret nailing (blind nailing). Nailing that is hidden from the surface. Tongued and grooved floor board, for example, may be slant nailed through the edges.

secret tack. A method of fixing large sheets of lead to steep roof surfaces. A strip of lead, which is soldered to the back of the sheet, passes through a slot in the ROOF BOARDS and is secured by screws.

secret tenon. A STUB TENON that does not pass right through the mortised wood.

secret wedging. The fixing of a TENON by the insertion of wedges into its end. As the tenon is driven into a MORTISE the wedges expand the tenon and form a grip.

section. (1) A drawing of a building or object seen as if a vertical plane passed through it. *See also* CROSS-SECTION. *Compare* PLAN. (2) A building material, such as a steel beam, with a fixed CROSS-SECTION but a continuous length. (3) A plot of land intended for housing.

sectional insulation. THERMAL INSULATION material that is shaped to fit around a pipe or other fitting.

security glazing. Various types of GLAZING designed to give protection against attack. Materials used include SAFETY GLASS, laminates, wired glass and transparent plastics.

sedimentary rocks. Rocks formed by the slow accumulation and consolidation of mineral fragments, such as in SANDSTONE. *Compare* IGNEOUS ROCKS, METAMORPHIC ROCKS.

sedimentation. The gradual sinking of small particles suspended in water. *Compare* FLOCCULATION.

seediness. A paint defect that appears as evenly distributed specks in the finish of GLOSS PAINT. It is usually caused by coarse pigment or old paint.

segregation. The separation of a concrete MIX into portions with different proportions of COARSE AGGREGATE and FINE AGGREGATE. Concrete can segregate when poured down a chute or dropped from a height.

seismology. The study of earthquakes.

self-cleansing velocity. That velocity of liquid in a pipe or channel, such as drain, that is high enough to carry solids and prevent them settling.

self-faced stone. Stone that splits cleanly and leaves a finished face.

self-finished roofing felt. A type of ROOFING FELT that has been saturated with BITUMEN, coated with bitumen on both sides and often dressed with talc.

self-levelling finish. *See* LEVELLING COMPOUND.

self-reading staff. A common type of LEVELLING STAFF graduated in a way that allows the observer to read off, through the telescope, the elevation at which the line of sight cuts the staff.

self-supporting. *adj* Describing a structure, such as a wall, that has no external load placed on it. *Compare* LOAD-BEARING.

self-tapping screw. A hardened steel SCREW that cuts its own matching thread as it enters a material, usually a metal.

semi-detached house (semi). A house that shares a PARTY WALL with a similar house. *Compare* DETACHED HOUSE, TERRACED HOUSE.

semi-engineering brick. A type of BRICK with slightly less crushing strength than an ENGINEERING BRICK but with higher strength and less absorption than a COMMON BRICK.

semi-hydraulic lime. A form of LIME that hardens mainly by drying, like HIGH-CALCIUM LIME, but which also has some properties like HYDRAULIC LIME.

sensible heat. The heat energy associated with a change of temperature. The quantity of sensible heat needed to raise the temperature of a body depends on the SPECIFIC HEAT CAPACITY of that material. The same quantity of sensible heat is given off when the body cools. *Compare* LATENT HEAT. *See also* ENTHALPY, SENSIBLE HEAT.

sensible horizon. The HORIZON that is seen.

separate application. A method of using synthetic resin adhesives in which the resin is spread on one side of the joint and the accelerator on the other side.

separate system. A system of drainage in which surface water and sewage are taken away in separate pipes. *Compare* COMBINED SYSTEM, PARTIALLY-SEPARATE SYSTEM.

septic tank. A tank for the local treatment of SEWAGE when a building cannot be connected to a sewerage system. Septic tanks contain several compartments such as a DIGESTION TANK where ANAEROBIC BACTERIA break down the solids, followed by a settling tank. The output of a septic tank is an effluent that can be safely drained away.

sequence of trades. The order in which various trades usually carry out their work in a new building. This order is also used for the listings in a specification.

serpentine wall (crinkle-crankle wall). A masonry wall that curves in and out along its length. The curves give some extra stability and a decorative effect. The method is sometimes used for boundary walls. *See also* TRAMMEL.

service cable (service conductor). An electrical power cable that connects a supply authority's SERVICE MAIN to a building.

service charges. In a shared building, such as a block of flats, a payment made for communal expenses such as heat and repairs. *See also* SINKING FUND.

service core. A central area of a building where services such as electrical supplies, gas supplies, water supplies and lifts can be grouped.

service drop. (*USA*) The overhead electric conductors that connect a building to the power company's distribution lines. *See also* DRIP LOOP.

service lift. A LIFT designed to carry goods rather than passengers.

service main. The cable or pipe that distributes electricity, water or gas to buildings in an area such a street.

service pipe. A pipe that brings water or gas from a supply authority's SERVICE MAIN into a building. *See also* SUPPLY PIPE.

service reservoir (distribution reservoir). A RESERVOIR that stores water at night and uses it to meet peak demand during the day. A service reservoir is usually protected by a cover.

service road. A road that connects a main road to nearby properties. A service road may be used for deliveries to shops.

services. The equipment in a building concerned with water supply, drainage and electricity

services engineer. A person trained in the design, installation and maintenance of various SERVICES in a building.

set. (1) The initial hardness of concrete, mortar or plaster. *See also* SETTING. (2) The slight bend given to the teeth of a SAW so that the cut is wider than the blade and the saw does not stick. (3) A finishing coat of plaster.

setback. The distance between the BUILDING LINE of a building and the boundary of the property.

set-head nail. *See* LOST-HEAD NAIL.

set screw. *See* GRUB SCREW.

set square. A flat piece of metal or plastic in the shape of a right-angled triangle. It is used to set out angles in technical drawing.

sett. A small block of stone or brick laid with others to form a ground surface. Setts have been used to pave roads and are used for decorative paving.

setting. (1) The first hardening of concrete, mortar or plaster. *See also* CURING, HARDENING, INITIAL SETTING. (2) The process of laying bricks, stones and beams into a wall. (3) The brickwork that surrounds a furnace or boiler.

setting block. *See* GLAZING BLOCK.

setting out. (1) The process of using pegs in the ground to show the position of a new structure. (2) The initial marking out of the position of new walls or other items. *See also* BUILDER'S SQUARE.

setting time. (1) The time taken for an ADHESIVE to harden. (2) *See* INITIAL SETTING.

setting up. The hardening of paint during storage.

settlement. The slight sinking of a new structure caused when the final weight of the structure compresses the soil beneath it. Settlement is not generally a problem unless it is DIFFERENTIAL SETTLEMENT. *Compare* SUBSIDENCE.

settlement tank. A chamber used in the early treatment of SEWAGE where solids are allowed to settle.

sett paving. *See* SETT.

sewage. Domestic waste matter that is carried away by water in system of SEWER drains. It may also include surface water and non-toxic industrial waste. *Compare* SEWERAGE.

sewage farm. A SEWAGE TREATMENT station that disposes of EFFLUENT into the ground.

sewage gas. Gas given off from the DIGESTION TANKS in SEWAGE TREATMENT. A high proportion of the gas is METHANE which can be used as a fuel.

sewage treatment. The separation of the pollutants from the water in SEWAGE. *See also* AEROBIC BACTERIA, ANAEROBIC BACTERIA, DIGESTION TANK, SEPTIC TANK, SETTLEMENT TANK.

sewer. A pipe or closed channel that carries SEWAGE.

sewerage. A network of SEWERS that disposes of SEWAGE from a community.

sewer brick. An ENGINEERING BRICK with low absorption that is especially suitable for the construction of sewers and drains.

sewer connection. The length of drain that joins the last MANHOLE of a private sewer to the public sewer.

sextant. An optical instrument that measures angles by sighting through a telescope and mirrors. It is used for navigation at sea by measuring the angle between the sun and the horizon.

sfb. *See* CI/SfB.

sgraffito (graffito, scratch work). A decoration produced on a soft plaster surface by scratching a pattern through the layer of plaster to reveal another colour beneath.

shade. (1) The process of darkening the colour of a paint. (2) The protection of a building from direct sunlight.

shaft. (1) The main part of a column. (2) A vertical passage through a building or a mine. It often contains services such as a lift. (3) A revolving rod that transmits power, such as from an engine.

shake. (1) A timber defect seen as the separation of wood fibres along the grain. *See also* RING SHAKE, SHELL SHAKE. (2) A type of hand-split wooden roofing tile.

shale. A type of laminated SEDIMENTARY ROCK formed by the compression of layers of clay or silt.

shallow well. A WELL that obtains water from near the surface, rather than from below the first impermeable layer. The classification of a well as deep or shallow depends on where it takes its water from and not on the depth of the bore.

shaping. The formation of cutting edge on the teeth of a SAW.

shaping machine. A stationary power tool with a table and rotating cutter head. It is used to cut edges grooves on timber. *See also* ROUTER.

sharp paint. A paint, such as a PRIMER, which dries quickly and contains a high proportion of PIGMENT.

sharp sand. Sand whose grains are angular rather than rounded. It is free from clay and suitable as an AGGREGATE in CONCRETE. *Compare* SOFT SAND.

shave hook. A hand SCRAPER.

shear. The tendency of two adjacent objects or portions of material to slide over one another when under load.

shear force. The value of that load that tends to shear a BEAM at the beam's support.

sheariness. A defect in the surface of a gloss paint film that appears oily rather than glossy.

shear legs (shears). A simple frame made of two poles set apart on the ground and tied together at the top. A pulley is suspended from them and used to lift loads.

shear modulus (modulus of rigidity). The ratio of SHEAR STRESS to SHEAR STRAIN. *Compare* ELASTIC MODULUS.

shear plate. A type of timber CONNECTOR.

shear resistance. The resistance of a material to the process of SHEAR.

shears. *See* SHEAR LEGS.

shear strain. The relative amount of deformation caused by SHEAR. *See also* SHEAR MODULUS.

shear strength. The maximum SHEAR FORCE that a material, such as a soil, can withstand before it fails. *See also* COHESIVE SOIL.

shear stress. A measure of SHEAR as the shear force divided by cross-section area. *See also* SHEAR MODULUS.

sheathing. (1) *See* BOARDING. (2) A protective CLADDING.

sheathing paper. BUILDING PAPER used with roof BOARDING.

sheen. A glossiness on a surface when viewed at certain angles.

sheet. A general term for layers of material, such as metals and plastics, when supplied in the form of wide measures.

sheet glass. Ordinary building glass whose surfaces are not perfectly flat and parallel. *Compare* FLOAT GLASS, PLATE GLASS.

sheet piles. A row of PILES that are set close together in the ground as a continuous barrier. They may be used around the edge of an excavation to keep out earth or water.

Sheetrock. (*USA*) Trade name for a form of GYPSUM PLASTERBOARD.

shelf life. The length of time that a paint or adhesive will keep in good condition when stored, sealed, in a shop or stockroom.

shell. A building with complete exterior cladding but an unfinished interior. *See also* TOPPING OUT.

shellac. A natural RESIN obtained from the secretions of tree insects. Shellac is soluble in alcohol and forms the basis of FRENCH POLISH and other types of natural VARNISH.

shell bedding. A method of laying hollow BLOCKS using thin strips of mortar along the front and back edges only. The method increases the resistance to moisture penetration.

shell construction. The use of thin curved plates, often made of reinforced concrete, to span large open spaces. *See also* HYPERBOLIC PARABOLOID.

shelling. FLAKING of paint.

shell shake. A RING SHAKE that is seen on the surface of sawn timber.

sheradizing. A form of zinc METAL COATING for iron and steel articles that are heated with zinc dust. The resulting coat is more durable than GALVANIZING.

shift. The displacement between the coordinates of two different survey grids whose information is to be combined into one.

shim. (1) A thin piece of wood with a flat or wedge shape used to level or secure an object during installation. (2) A thin metal slip used to fill a gap.

shingle. (1) A flat rectangular piece of timber or other material, used as cladding for roof or wall. It is usually tapered along the grain. (2) Smooth stones on a beach or river bed without much fine material between them. *Compare* GRAVEL.

shingle hatchet. A type of hammer with a HATCHET blade instead of a claw on its head.

It is often used for working with wooden roof SHINGLES.

shiplap. A form of WEATHERBOARDING in which the lower board fits behind a rebate in the upper board. *See* WALL diagram.

shoe. (1) A box-like metal sheath that fits around the end of a wooden post or rafter in order to protect it. (2) A bend at the bottom of a DOWNPIPE that directs water away the building.

shook. A set of timber parts ready for assembly.

shoot. The process of using a PLANE to smooth the edge of a board.

shooting board. A board used to keep another board steady during planing.

shop welded. *adj.* Describing welding done as SHOPWORK.

shopwork. Work that is completed in a factory rather than on site.

shore. A temporary support of timber or metal for a building while it is unsafe or being repaired. *See also* FLYING SHORE, RAKING SHORE.

shoring. The process or the materials used for the SHORE of a building.

short-bored piles. Light piles used to carry loads, such as a house, below the layer of clay affected by seasonal movement.

short circuit. A faulty connection in an electrical circuit. It allows an excessive current to flow and should cause a CIRCUIT BREAKER to operate.

short column. A column that could fail by crushing rather than bending.

short oil. A VARNISH that contains a relatively low proportion of oil and high proportion of RESIN. *Compare* LONG OIL.

shot blasting (blast cleaning). A method of cleaning the surface of steel with abrasive particles, such as steel shot, carried in a high velocity stream of air. *Compare* SAND BLASTING.

shotcrete. *See* GUNITE.

shoulder. The surface or ledge of wood at the root of a TENON and which lies outside the joint.

shouldering. (1) The diagonal cutting of corners from SINGLE-LAP TILES and some slates. (2) A thin bed of MORTAR laid beneath the head of slates in exposed areas in order to hold the tails down.

shower. A SANITARY APPLIANCE in which a person stands and washes beneath a spray of water delivered from a shower head. The shower water is usually contained by a waterproof cubicle or curtains and is drained away via a SHOWER TRAY. *See also* NEEDLE SHOWER.

shower tray. A shallow-sided sink that acts as the base of a SHOWER. It collects the water and has an outlet connected to a drain.

shrinkage. (1) A reduction in the dimensions of a material such as wood caused by a loss of MOISTURE CONTENT. (2) A permanent reduction in the size of a component caused by the ageing of a material such as concrete.

shrinkage joint. *See* CONTRACTION JOINT.

shrinkage limit. A water content of soil, below which the sample will decrease in volume.

shrink film. A thin transparent film of plastic that shrinks when heated. It can be used on the inside of window frames as a temporary STORM WINDOW or for SECONDARY DOUBLE GLAZING.

shutter. A protective cover that fits over a window to protect it, usually at night. Wooden or steel shutters often hinge outward and lie flat against the wall. Some decorative versions of these shutters do not close. Shutters may also be made from continuous strips of wood or metal that wind into a roll above the window for storage.

shutter bar. A hinged security bar used to keep shutters closed.

shuttered socket. An electrical SOCKET outlet that has SHUTTERS protecting the contacts. The shutters are lifted by the action of inserting a plug.

shutter hinge. *See* H-HINGE.

shuttering. FORMWORK, especially that part in contact with the concrete.

shutting post. The post against which a gate closes.

SI. *abbr.* Système International d'Unités. *See also* SI UNITS.

side cut. (1) *adj.* Describing timber that is cut so as to avoid the HEARTWOOD. (2) An angled saw cut made on a HIP or VALLEY RAFTER where it meets a ridge or common rafter.

side flight. One of the return flights of stairs in a DOUBLE-RETURN STAIR.

side gutter. A small GUTTER made at the junction of a roof slope and a vertical surface such as a chimney.

side lap. (1) For courses of roof tiles or slates, the overlap or horizontal distance between the vertical joint in one course and the adjacent vertical joint in the course above. (2) The amount of overlap between adjacent SINGLE-LAP TILES.

sidelight. A window panel set beside an external door. The frame usually shares the same structure as the door frame.

side rabbet plane. A small PLANE used to smooth a REBATE or groove.

side string. *See* OUTER STRING.

sidesway. A small sideways movement of a structural frame in its own plane.

siding. (*USA*) The exterior CLADDING on the walls of a framed building. *See also* WEATHERBOARDING.

sienna. A yellow-brown PIGMENT based on iron oxide. It was originally found near the town of Sienna, Italy.

sieve. A box or tray with a mesh or perforations in the bottom which pass particles of known size. It is used to separate soil and AGGREGATES into groups of different size. *See* SIEVE ANALYSIS, RIDDLE.

sieve analysis (screen analysis, sieve test). The classification of a soil or aggregate into a standard group according to the size and distribution of particles. A sample of material is passed through SIEVES of standard sizes and the quantities remaining are weighed.

sieve test. *See* SIEVE ANALYSIS.

sight line. (1) A line on a drawing that helps improve visibility, especially at road junctions. (2) A line of vision used to establish levels or edges during surveying or setting out.

sight rail. A horizontal board set at a fixed height and used to establish levels, or gradients for a drain, when viewed from other points.

sight size. The effective size of a window as measured between the frame edges.

SIL. *abbr.* (1) Sound intensity level. *See* SOUND LEVEL. (2) SPEECH INTERFERENCE LEVEL.

silica. Silicon dioxide, SiO_2. A hard solid material that is abundant on the earth's surface and found in sand, quartz, flint and SILICATES. It is the main raw material in the manufacture of glass. *Compare* ALUMINA, CLAY.

silica brick. A brick that is made with a high content of SILICA and can withstand high temperatures.

silica gel. A form of SILICA that can absorb large amounts of water. It can be used for drying air such as that within the cavity of DOUBLE GLAZING.

silicate paint. A water-based paint that contains sodium silicate. It has useful fire-retardant properties but is alkaline and cannot be easily used with normal oil paints.

silicates. A variety of salts based on the radical SiO_3. They are found in many types of minerals and rocks.

silicon. A non-metallic element (chemical symbol Si) comparable to CARBON in properties. Commonly found in SILICA.

silicon bronze. An ALLOY metal based on copper that also contains silicon and manganese. It is harder than copper and is used to make durable screws and nails.

silicon carbide (SiC). A hard compound in ABRASIVES such as CARBORUNDUM or SANDPAPER.

silk. *See* EGGSHELL.

sill (cill). In general, the lowest horizontal member of a frame. *See also* DOOR SILL. (1) (sole plate) The bottom member of a wooden wall frame. (2) The bottom of a window frame that projects from the wall and helps throw rainwater clear. *See* WINDOW diagram. (3) A timber fixed across the bottom of a trench.

sill anchor. A bolt set in a concrete or masonry base and used to hold down the frames of a timber structure.

sill bead. *See* DEEP BEAD.

sill board. *See* WINDOW BOARD.

silo. A tall cylindrical storage building, usually built of reinforced concrete and used for keeping grain.

silt. A deposit of fine particles, smaller than sand but larger than clay, usually found at the bottom of rivers and lakes. *See also* LOESS.

silver brazing. A method of joining lightweight copper pipes with a SOLDER made from copper, silver and phosphorus.

silver fir. A common type of FIR tree whose timber is used for general construction work.

silver grain. Pale flecks of RAY seen in some quarter-sawn timber such as oak.

silver solder. SOLDER used in SILVER BRAZING.

simple beam (simply-supported beam). A beam that simply rests on supports and is free to bend.

simple cell (voltaic cell). An electrical CELL that produces electrical energy when two metals are connected by a conducting solution. These combinations can occur in a structure and the current from this 'simple cell action' causes ELECTROLYTIC CORROSION.

simply-supported beam. *See* SIMPLE BEAM.

Simpson's rule. A formula used to calculate the area of an irregular figure by dividing it into an even number of parallel strips and measuring the length of the numbered ordinates or lines between the strips. The area is calculated by taking the sum of first and last ordinate plus twice the sum of the even-numbered ordinates plus four times the sum of the odd-numbered ordinates and multiplying by one third of the width of the strips. A related formula can be used to calculate volume. *See also* PRISMOIDAL FORMULA.

single bridging. A herringbone system of FLOOR STRUTTING.

single Flemish bond. Brickwork that shows a FLEMISH BOND on the face while the body of the wall may contain ENGLISH BOND. *Compare* DOUBLE FLEMISH BOND.

single-hung window. A form of SASH WINDOW in which only one sash moves.

single-lap tiles (interlocking tile). Various systems of roof tiles that only overlap one course beneath them, unlike plain tiles and slates that overlap two courses. *See also* DOUBLE ROMAN TILE, ITALIAN TILES, ROMAN TILE, PANTILE.

single-lock welt. *See* CROSS WELT.

single phase supply. The usual method of delivering ALTERNATING CURRENT to households by connection to just one of the three phases available in the distribution system. *Compare* THREE PHASE SUPPLY.

single-pitch roof. A MONO-PITCH ROOF.

single-point heater. A form of INSTANTANEOUS WATER HEATER that supplies one tap only. *Compare* MULTI-POINT WATER HEATER.

single roof. A roof structure made of COMMON RAFTERS without the support of PURLINS or TRUSSES. *See also* CLOSE-COUPLE ROOF.

single-sized aggregate. An AGGREGATE for concrete or road macadam in which all particles are within the same range of size. *See also* SIEVE ANALYSIS.

single-stack system. A form of ONE-PIPE SYSTEM of drainage that omits most of the ANTI-SYPHON PIPES and uses ANTI-SYPHON TRAPS.

sink. A SANITARY APPLIANCE in the form of a deep basin used for cooking and cleaning activities. A kitchen sink is usually made of stainless steel and set into a worktop. *See also* BELFAST SINK, BUCKET SINK.

sink bib. (*USA*) A tap installed above a kitchen SINK.

sinking. (1) An indentation in a surface, such as the recess for a hinge. (2) *See* SINKING IN.

sinking fund. Money invested to pay for the maintenance or other costs of a building. *See also* SERVICE CHARGES.

sinking in (sinking). A loss of gloss from the surface of a paint film caused when the undercoat absorbs the MEDIUM from the top coat.

sintered clay. *See* EXPANDED CLAY.

sintering. The process of heating small particles of materials so that they tend to fuse together.

siphon (siphonage). The movement of a liquid in a sealed pipe or tube system that rises upwards before descending to a lower level. The liquid moves because of differences in atmospheric pressure.

siphonic closet. A quiet type of WC with a double water TRAP. When water from the FLUSHING CISTERN flushes the second trap a suction action is set up which drains the first trap and the contents of the pan.

Sirapite plaster. *See* ANHYDROUS GYPSUM PLASTER.

site. The portion of land on which a construction is planned or takes place. *See also* LOT, PLOT, SECTION.

site investigation. A survey of the surface and subsoil features of a SITE to determine how they will affect the design and construction of a building, especially the foundations.

site lighting. A temporary system of electrical supply and light fitting used on a site during construction. The current is usually supplied via an isolating transformer which may also reduce the voltage.

SI units. The international system of metric units now used for all scientific and most technical purposes. The system is based on seven fundamental units: the metre, kilogram, second, ampere, kelvin, candela, mole; and two supplementary units: the radian and steradian. Other units can be made up from combinations of the fundamental units.

size. A liquid used to seal absorbent surfaces, such as plaster, before painting or papering. Size usually contains a solution of resin, glue or starch.

sizing. The process of sealing the pores of a surface with SIZE.

sk. *abbr.* Sketch (used on drawings).

skeleton. The basic frame of a building before it is enclosed with cladding.

skeleton core. The internal frame of a HOLLOW-CORE DOOR.

skew. (1) (skewed) *adj.* At an angle or 'out of square'. (2) A KNEELER in a gable roof.

skewback. A SPRINGER in an arch, or the upper surface of a springer that is sloped to carry the first arch stone.

skew corbel. *See* GABLE SPRINGER.

skew nailing (toe nailing, tusk nailing). Nails that are driven at an angle through one board and then penetrate an adjacent board.

skewsaw. A type of handsaw with a slight curve to its back rather than a straight edge.

skew table. A KNEELER in a GABLE coping.

skid. A length of timber, steel rail or pipe placed under heavy objects to help move it or prevent it sinking into the ground.

Skilsaw. (*USA*) Trade name for a hand-held CIRCULAR SAW.

skim (skimming coat). A finishing coat of plaster that may be applied on top of previous coats, or over plasterboard.

skimming coat. *See* SKIM.

skimming float. *See* HAND FLOAT.

skin. A VENEER or thin layer of plywood used to decorate or protect doors or other items.

skinning. The formation of a tough dry layer on the surface of paint or varnish while it is stored in the container. The skin is usually caused by OXIDATION of the drying oil.

skip. (1) A large open steel container used for the removal of building debris. It is usually delivered to the site empty and removed when full. (2) A HOLIDAY or break in the continuity of a paint film or in the smoothness of a planed timber surface.

skirting. (1) *See* SKIRTING BOARD. (2) An upstand in a roof material where it joins a parapet or similar boundary.

skirting block. An ARCHITRAVE BLOCK.

skirting board (skirting, *USA*: baseboard, mopboard, scrub board). A moulding or upstand of material installed on the base of a wall to protect the wall and cover the junction with the floor.

skutch. *See* COMB HAMMER.

sky component (SC). The quantity of daylight received at a particular point in a room, direct from the sky without reflection. It is the main DAYLIGHT FACTOR COMPONENT. *See also* DAYLIGHT FACTOR.

skylight. A ROOF LIGHT designed to admit natural light from the sky. *Compare* LANTERN LIGHT, MONITOR LIGHT.

sky-luminance distribution. *See* LUMINANCE, STANDARD SKY.

SL. *abbr.* Structural level.

slab. (1) A thin, flat piece of concrete or stone of regular shape. (2) A large flat area of concrete that forms a floor or base. (3) Timber cut from the edge of a log so that only one face is sawn.

slag. Waste mineral material left in a furnace during the recovery or refinement of metals. It is recovered and used for building materials such as aggregates. *See also* BLAST-FURNACE SLAG.

slagwool. A form of MINERAL WOOL.

slaked lime (hydrated lime, lime paste). Calcium hydroxide, $Ca(OH)_2$, which forms when water is added to LIME (calcium oxide). Slaked lime is the basis of HIGH-CALCIUM LIME.

slaking. The process of producing SLAKED LIME.

slamming stile. The upright in a door frame against which the door closes.

slap dash. *See* ROUGH CAST.

slash grain. The GRAIN of FLAT-SAWN TIMBER.

slat. A narrow, flat strip of wood or other material often used to form blinds and louvres. *See also* VENETIAN BLIND.

slate. (1) Various types of dark natural rock with a laminated structure which easily splits into thin flat sheets. The material can be used for roofing, cladding and floor slabs; it was formerly used for DAMP-PROOF COURSES. (2) Regular-sized sheets of slate used for roofing.

slate-and-a-half slate. A roof SLATE that is 50 per cent wider than the others and is used at valleys, hips and other edges.

slate axe. A slating axe or ZAX.

slate batten. One of a series of square timber BATTENS that are fixed across roof rafters at a spacing suitable for nailing roof SLATES.

slate cramp. A piece of SLATE used to join blocks of stone. Each end of the cramp is DOVETAILED and fits into matching MORTISES in the stones.

slate hanging (weather slating). A vertical wall cladding formed by nailing SLATES to BATTENS fixed on the walls.

slate nail. A type of ROOFING NAIL used to fix SLATES.

slate powder. A fine powder, mainly aluminium silicate, that is used as an EXTENDER in paints.

slate ridge (slate roll). A circular rod of SLATE used to cap the ridge of a roof. A V-shape cut into the underside of the rolls allows the slate to be fixed.

slate roll. *See* SLATE RIDGE.

sleeper. (1) A length of timber placed along the ground, or across piles, to act as a bearer for further construction. (2) One of a series of wooden or concrete beams to which the rails of a railway track are fixed. Each sleeper runs at right angles to the rails and holds them at the correct width or gauge.

sleeper clip. (*USA*) A steel FLOOR CLIP that is pushed into the surface of a concrete floor or screed before it sets. The wooden SLEEPERS of the floor structure are then attached to the sleeper clips.

sleeper plate. A WALL PLATE on a SLEEPER WALL.

sleeper wall (pony wall). A low brick wall built beneath a timber ground floor to give extra support for floor JOISTS. It is usually built as a HONEYCOMB WALL which allows ventilation.

sleepiness. A reduction in the surface shine of GLOSS paint as it dries.

sleeve. A tube or tubular case inserted in a structure to carry pipes or cables.

sleeve piece (thimble). A short reducing tube of brass, copper or plastic used to make a join between pipes of different materials.

slenderness ratio. For a simple column: the effective height of the column divided by its effective thickness. The ratio can be used as an indication of the tendency of vertical loads to cause bending. For brickwork and masonry walls a typical maximum ratio is 18, compared with 200 for steel.

sliced veneer. Timber VENEER that is cut by a long blade rather than by ROTARY CUTTING.

sliding door. A door that hangs on a top track and slides sideways along the track to open.

sliding sash. A window that slides opens horizontally. *Compare* SASH WINDOW.

slip. (1) A FIXING FILLET in brickwork. (2) A PARTING SLIP in a sash window. (3) *See* GLAZE. (4) A thin paint that is easy to brush.

slip form. (1) A section of FORMWORK that can be moved along the structure during the placing of concrete. (2) A small section of FORMWORK that can be removed first and allow the remaining formwork to be taken down.

slip joint. (1) A joint made between pipes or conduits by sliding the end of one section into a socket in the next. (2) A movement joint in a wall that allows a tongue in one section to move within a groove in the next section.

slipper. An open bend that feeds a branch drain into the main channel at the bottom of a MANHOLE.

slip sill. A window or door SILL that slides between the vertical frames and is not built into the wall. *Compare* LUG SILL.

slip stone. *See* OILSTONE.

slipway. A carriageway of concrete or stone that slopes down to the water's edge and is used for ship building or repair.

slope correction. When surveying land that slopes, an adjustment to the measured distance to give the true horizontal distance. The correction may be calculated from the rise in height or from the angle of the slope.

slope of grain. The difference in angle between the axis of a piece of timber and the direction of its GRAIN.

slop sink. A large SINK set at a low level convenient for emptying and filling a bucket.

slot mortise. *See* OPEN MORTISE.

slot screwing. A method of fixing boards using screws that pass through slots and allow slight movement of the material.

slotted-head screw. A traditional type of SCREW with a single slot running across the head.

slow sand filter. *See* SAND FILTER.

sludge. Waste material from SEWAGE that settles as a fine deposit at the bottom of a SEPTIC TANK.

slump test. A simple test for the stiffness and consistency of fresh concrete. The concrete is poured into a cone-shaped mould which is consolidated, turned over and carefully removed. The amount by which the cone of concrete settles or slumps is measured and used as an indication of the water content of the mix. *Compare* V-B CONSISTOMETER. *See also* WORKABILITY.

slurry. A liquid mixture of water and fine materials such as cement. *See also* CEMENT SLURRY.

slurrying. A method of protecting the surface of stonework with a temporary coat of water, lime and stone dust that can be washed off.

small-bore system. A heating system that circulates water by the pressure of a pump rather than by convection currents. The diameter of the pipework is typically 15 mm. *Compare* MICRO-BORE SYSTEM.

SMM. *abbr.* STANDARD METHOD OF MEASUREMENT.

smoke detector. An electronic device that reacts to the presence of smoke by monitoring the visibility or ionization of the air. It usually activates an alarm or FIRE VENT.

smoke outlet. *See* FIRE VENT.

smoke stop. *See* FIRE STOP.

smoke test. A type of AIR TEST for a drain that uses smoke to trace possible leaks.

smoke vent. *See* FIRE VENT.

smoothing plane. A short type of BENCH PLANE used to smooth wood after other types have removed the rougher surfaces. *See* WOODWORKING TOOLS diagram.

Sn. Chemical symbol for TIN.

snake. A long flexible spring used to unblock drains.

sneck. A small squared stone used to fill in between larger stones in a SNECKED RUBBLE wall.

snecked rubble. A wall built from squared stone of various sizes such as RISERS and SNECKS.

snow board. A low board fixed on the roof just above the eaves and designed to stop snow falling from the roof, especially near doors. *See also* SNOW GUARD.

snow guard. A low wire fence fixed along the edge of a roof, with the same purpose as a SNOW BOARD.

snowload. A LIVE LOAD used in the design of flat roofs to allow for possible snow.

soakaway (rummel, soakpit). A pit dug in the ground and filled with stones or rubble. It is used to dispose of drainage by soakage into the soil.

soaker. A small piece of FLASHING, cut at

various angles, that seals the junctions of a sloping roof at hips and valleys. *Compare* APRON FLASHING.

soakpit. *See* SOAKAWAY.

soapstone. A stone with a greasy feel to the surface.

socket. (1) A cavity, such as a MORTISE, into which something fits. (2) The end of a pipe that is enlarged to receive the plain end of a similar pipe. (3) An electrical SOCKET OUTLET. (4) (*USA*) The threaded part of a lamp fitting into which the electric lamp is screwed.

socket chisel. A large CHISEL with a handle that is strong enough to be hit with a MALLET.

socket head. A type of SCREW head in which a SOCKET with many sides replaces the normal slot. A special tool is needed to turn the screw.

socket outlet. An electrical fitting that is set in the surface of a wall or floor as an outlet for current. It usually contains slots into which the prongs of a PLUG are inserted. *See also* SHUTTERED SOCKET.

soda ash. Sodium carbonate, Na_2CO_3. A raw material for the manufacture of glass, where it lowers the melting temperature of SILICA.

soda-lime process. A method of WATER SOFTENING by the addition of SODA and SLAKED LIME. The chemicals form insoluble precipitates which can then be removed by sedimentation and filtration. *Compare* BASE EXCHANGE.

sodium. A silvery-white metal element (chemical symbol Na) that is very reactive and reacts violently with water. Sodium forms many important compounds such as sodium chloride (common salt).

sodium vapour lamp. A form of DISCHARGE LAMP that uses SODIUM vapour. Low-pressure (SOX) sodium lamps give a monochromatic yellow light, while high-pressure (SON) sodium lamps give a better colour rendering.

soffit. (1) The bottom face or surface of part of building such as a BEAM, CORNICE or STAIRS. (2) *See* SOFFIT BOARD.

soffit board (soffit). A board attached to the underside of the rafters of overhanging EAVES. *See* ROOF diagram.

soft-burnt. Brick and tiles made from clay that has been fired at low temperature. They have a relatively low compressive strength and high absorption of moisture.

soft sand. Sand with small, rounded grains rather than angular ones like SHARP SAND. Soft sand is used for making plaster, RENDERING and bricklaying MORTAR.

soft solder. A type of SOLDER that melts at relatively low temperature. *Compare* HARD SOLDER.

soft water. Water that easily forms a lather with soap. Soft water does not contain those dissolved salts that cause HARD WATER but it usually contains other chemicals. Natural waters tend to be soft if they are collected over hard insoluble rocks such as granite. Although soft water avoids the disadvantages of hard water, it can cause PLUMBOSOLVENCY. *See also* WATER SOFTENING.

softwood. Wood from trees that belong to the botanical group GYMNOSPERMS, most of which are conifers, are evergreen and have needle-like leaves. Not all softwood trees produce timber that is soft. Common softwood trees include FIRS and PINES. *Compare* HARDWOOD.

softwood lumber standards. (*USA*) A guide for the classifications, nomenclature, basic grades, sizes, and the marking of timber.

soil. (1) The layer of loose material on the ground down to the level of bedrock. (2) SEWAGE or contaminated drainage rather than surface water or other drainage.

soil drain. A DRAIN that carries SEWAGE or other contaminated drainage.

soil mechanics. The scientific study of the structure of soils and their engineering properties.

soil pipe. A pipe that carries SEWAGE, such as in a SOIL STACK.

soil stabilization. Various treatments of soil to increase its strength and prevent movement. Methods include compaction or binding with cement.

soil stack. A vertical drain pipe that carries SEWAGE from fittings in a building to the foul drain. The upper end of the stack usually projects through the roof and is left open as a VENT PIPE.

sol-air temperature. An ENVIRONMENTAL TEMPERATURE for outside a building used in the calculation of heat loss through the fabric. It takes account of the effect of solar radiation as well as of the outside air temperature.

solar collector. A device designed to capture and distribute the energy of SOLAR RADIATION.

solar constant. A measure of the radiant energy from the sun received by the earth, in the absence of atmosphere. The figure is approximately 1400 watts on a square metre of a surface placed at the mean distance of the Earth from the sun, at right angles to the radiation.

solar construction. *See* PASSIVE SOLAR CONSTRUCTION.

solar control. A material or device on a window that reduces heat and glare from the sun. *See also* SOLAR HEAT GAIN.

solar degradation. The deterioration of a material caused by sunlight, especially by the ULTRA-VIOLET component.

solar heat gain. Heat energy from the sun that enters a building by direct radiation through the windows and by transmission through the fabric of the roof and walls. *See also* HEAT GAINS, SOLAR CONTROL.

solar radiation. Energy from the sun that reaches Earth as electromagnetic radiation containing heat, light and ULTRA-VIOLET wavelengths.

solder. An ALLOY with a low melting point that is used to join metals. *See also* FINE SOLDER, HARD SOLDER, SOFT SOLDER, WIPED JOINT.

soldering iron. A heating tool used to melt and distribute SOLDER on the joint of metals.

soldier. (1) A brick set on its end, rather than laid flat. (2) An upright timber or beam set in the side of a trench.

soldier arch. A FLAT ARCH formed from bricks set on end.

soldier course. A COURSE of bricks set on end, sometimes used to form the COPING of a wall.

sole. The smooth surface on the bottom of a PLANE.

solenoid. (1) A coil that sets up a magnetic field when current flows in it. (2) An electromagnetic device that moves a mechanism when current flows in the solenoid coil. It may be used to operate an RELAY switch or a valve. (3) *See* SOLENOID VALVE.

solenoid valve (solenoid). A gas or water VALVE that is controlled by electric current in a SOLENOID.

sole plate (foot plate, ground sill). (1) A horizontal timber at the bottom of a wall frame, a DEAD SHORE or a RAKING SHORE. *See* WALL diagram. (2) *See* SILL.

solid angle. An angle that measures three-dimensional spread from a point. *See also* STERADIAN. *Compare* PLANE ANGLE.

solid bridging. A form of STRUTTING that reinforces a floor with solid 'strutting pieces' of timber that are as deep as the joists. *Compare* HERRING-BONE STRUTTING.

solid door (solid-core door). A FLUSH DOOR with the facing surfaces mounted on a solid material such as COREBOARD. *Compare* HOLLOW-CORE DOOR.

solid floor. (1) A floor SLAB built of solid concrete without HOLLOW BLOCKS. (2) A PLANK-ON-EDGE-FLOOR.

solid frame. A door frame in which the RE-

BATE and STOP is formed from the same length of wood as the STILE rather than added on.

solid-newel stair. A type of SPIRAL STAIR in which the inside end of each step narrows to a cylinder that interlocks with other cylinders to form a continuous NEWEL.

solid roll. A join between sheets of FLEXIBLE METAL roofing that are folded together around a WOOD ROLL.

solid stop. A DOOR STOP formed by cutting a REBATE in a SOLID FRAME.

solstice. Either the longest day or the shortest day.

solubility. A measure of the extent to which a SOLUTE will dissolve in a SOLVENT, at a given temperature.

solum. (1) The upper layer of soil on the ground. (2) The OVERSITE surface of the ground beneath a floor.

solute. A substance that is dissolved in a SOLVENT to form a solution. *Compare* DISPERSION.

solvency. The ability of a company to raise cash in order to meet debts.

solvent. (1) A substance, usually liquid, that will dissolve other substances called SOLUTES. (2) A liquid used in the manufacture of paint to dissolve and carry the other materials. The solvent evaporates from the paint film after application.

solvent weld. A joint formed between plastic objects, such as pipes, by applying a solvent to the plastic.

sone. A renumbering of the PHON scale or sound loudness level so that the sone values are directly proportional to the magnitude of the loudness. One sone is equivalent to 40 phons; 2 sones seem twice as loud as 1 sone.

SON lamp. *See* SODIUM VAPOUR LAMP.

soot door. A door set in the base of a chimney or flue through which soot can be removed.

sound. A variation in the pressure or density of fluids or solids caused by a vibrating object. A certain range of sound frequencies can be detected by human hearing.

sound absorption. A reduction in the sound energy reflected by the surfaces of a room. It is measured by the ABSORPTION COEFFICIENT and used to control the acoustic quality of sound rather than the transmission of sound. Sound absorption is provided by porous materials, PANEL ABSORBERS and CAVITY ABSORBERS. *Compare* SOUND INSULATION.

sound absorption coefficient. A measure of relative SOUND ABSORPTION. It is calculated as the sound energy absorbed (or not reflected) compared to the sound energy falling on the surface. Absorption tends to increase with the frequency of sound. The maximum value for a perfect absorber is 1.

sound boarding. Horizontal boards that are fitted between floor JOISTS to carry PUGGING for sound insulation.

sounding line. *See* LEAD LINE.

sound insulation. A reduction in the sound energy transmitted into an adjoining air space. The amount of sound reduction can be measured by the SOUND REDUCTION INDEX. *See also* AIRBORNE SOUND, IMPACT SOUND. *Compare* SOUND ABSORPTION.

sound intensity level (SIL). *See* SOUND LEVEL.

sound knot. A KNOT that is hard and firmly fixed in the surrounding wood. *Compare* UNSOUND KNOT.

sound level. A practical measurement of sound strength that uses the DECIBEL. Sound intensity or sound pressure is converted to a LOGARITHMIC SCALE of numbers that corresponds reasonably with the way that the ear judges sounds. The term 'level' in sound intensity level (SIL) or sound pressure level (SPL) implies that the THRESHOLD OF HEARING value is taken as the reference for comparison.

sound level meter. An electronic instrument designed to give constant and objective measurements of SOUND LEVEL. *See also* WEIGHTING NETWORK.

soundness. A desirable property of cement so that it does not significantly expand in volume when it sets. Unsoundness may be the result of excess amounts of free LIME or calcium sulphate. Soundness is tested by the LE CHATELIER METHOD.

sound pressure level (SPL). *See* SOUND LEVEL.

soundproofing. *See* SOUND INSULATION.

sound reduction index (SRI). A measure of the SOUND INSULATION provided by a partition against the direct transmission of AIRBORNE SOUND. It is calculated as 10 times the logarithm (base 10) of the reciprocal of the sound TRANSMISSION COEFFICIENT. The unit of SRI is the DECIBEL.

Southwater reds. A hard durable type of ENGINEERING BRICK with a red colour.

SOX lamp. *See* SODIUM VAPOUR LAMP.

space. On a SAW, the distance between the point of one tooth and the next.

space frame (space structure). A structural framework or construction that performs in three dimensions, unlike a plane frame that needs stiffening. Space structures are used to cover large spans; examples include decks, domes and vaults.

space heating. The equipment or process of heating the interior of a building, as distinct from heating the water supply or other needs. *See also* CENTRAL HEATING.

space structure. *See* SPACE FRAME.

spall. (1) A process that causes flakes of stone or concrete to break off a large block when subject to weathering or to knocks. (2) A flake of stone or concrete broken from a block by spalling.

span. The distance between two supports for an overhead structure, such a roof or bridge. *See also* CLEAR SPAN, EFFECTIVE SPAN.

spandrel. (1) The triangular area between the outer curve of an arch and the right angle that surrounds the arch. (2) The triangular area filled in beneath the STRING of a STAIR. (3) Part of a CURTAIN WALL or similar cladding that fills the space between the top of one window and the sill of the next window above.

spandrel beam. A horizontal beam concealed in a wall and used to support a CURTAIN WALL above it.

spandrel step. A solid stone or concrete step with a triangular cross-section that forms a straight underside or soffit to the stairs.

Spanish tiles. *See* ITALIAN TILES.

span piece. *See* COLLAR BEAM.

spar. (1) *See* COMMON RAFTER. (2) A thin piece of wood, such as willow, that is split and used to fasten the THATCH in a roof.

spar finish. A roof surface, such as white stone chippings, that reflects radiation from the sun and reduces its destructive effects.

sparge pipe. A water pipe drilled with a series of outlet holes, such as the supply to flush a URINAL.

spar varnish. A durable type of VARNISH with a good resistance to heat and weather.

spatterdash. A wet mix of cement and sand that is thrown on to smooth brick or concrete surfaces. When hard, the spatterdash provides a KEY for later finishes, such as plaster.

special. An object or fitting that is not of the standard size or shape. Bullnose bricks, corner bricks, tee fittings for pipes are examples of specials. *Compare* STANDARD SPECIAL.

specification. A document that defines the details of a construction project and supplements the drawings. It is usually written in the order chosen for the SEQUENCE OF TRADES. *See also* BILL OF QUANTITIES.

specific gravity. *See* RELATIVE DENSITY.

specific heat capacity. A measure of the THERMAL CAPACITY of a material under specified conditions. It is the quantity of heat energy needed to raise the temperature of 1 kilogram of the material by 1°C (or 1 K). *See also* SENSIBLE HEAT.

spectrophotometer. An instrument that produces or measures the qualities of a SPECTRUM.

spectrum. (1) A dispersion of white light into the individual frequencies present that are seen as separate colours. (2) *See* NOISE SPECTRUM.

specular. *adj.* Describing a quality of light that is not scattered, as after a direct reflection compared to a DIFFUSE REFLECTION.

speech interference level (SIL). A measure of that level of background noise that will interfere with speech in a particular situation, such as in an office.

spigot. (1) A cylindrical projection on a component, such as a pipe, that is designed to fit into a socket on another component. (2) A water TAP.

spigot-and-socket joint. A joint between two pipes in which a plain end fits into an enlarged end.

spindle. (1) A shaft that turns like an axle. (2) A circular length of wood, such as the BALUSTER of a stair.

spine wall. An internal load-bearing wall that runs along the length of a building.

spiral grain. A pattern of wood GRAIN caused by the spiral growth of fibres in a tree.

spiral stair. A stair that climbs in a tight circle around a central NEWEL post. All the treads are WINDERS.

spiriting off (spiriting out). A final stage in FRENCH POLISHING in which the film is hardened and polished.

spirit level (level). A STRAIGHT EDGE used to establish an accurate level or vertical for a surface. A bubble of air in a tube of liquid moves into the centre of tube to show the right position. *Compare* PLUMB LEVEL.

spirit stain. A dye dissolved in industrial alcohol and used to stain a wood surface.

spirit varnish. A type of VARNISH made from RESINS dissolved in alcohol which evaporates during the drying process.

spit. The depth of soil penetrated by one spade blade.

SPL. *abbr.* Sound pressure level. *See* SOUND LEVEL.

splashback. A vertical panel that protects the wall behind a worktop, particularly near taps.

splay. An angled edge or surface that runs the full width of an object, unlike a CHAMFER.

splay brick. A special brick made with a SPLAY on the header or the stretcher.

splayed heading joint. A BUTT join between the ends of two floorboards that are cut at an angle so that one end overlaps.

splayed skirting. A SKIRTING with a bevel along the top edge.

splay knot. A KNOT in timber that is cut and revealed along the direction of the knot. *Compare* ROUND KNOT.

splice. (1) A join between the ends of two timbers made by fish plates attached on either side of the join. (2) A connection made between two ropes, cables or wires.

spline. *See* FEATHER.

split. A timber defect seen as a crack that passes through the wood or a veneer. *Compare* CHECK.

split course. A COURSE of bricks or blocks that are less than normal thickness.

split-ring connector. A timber CONNECTOR in the form of a ring which fits in a circular groove cut in both timbers.

split shake. Wooden SHINGLES that have been split rather than sawn.

spoil. Material such as earth that is surplus after excavations for foundations and drains have been completed. *Compare* FILL.

spokeshave. A small PLANE with handles either side of the blade that is used to smooth curve surfaces.

spontaneous combustion. The IGNITION of a substance by internal oxidation rather than by an external source of heat.

spot board. A GAUGE BOARD used to work plaster before it is applied.

spot finishing (spotting in). The repair and refinishing of small areas of a coating, such as paintwork.

spot height. *See* SPOT LEVEL.

spotlamp. *See* CS LAMP, ISL LAMP, PAR LAMP.

spot level (spot height). The ELEVATION of a point on a site.

spotting in. *See* SPOT FINISHING.

spot welding. A WELD that joins two pieces at one spot. The weld is approximately the same area as the electrode.

spouting. *See* DOWNPIPE.

sprag. A length of timber or steel bar that is wedged against a wall or floor so as to hold a component in place.

sprayed concrete. *See* GUNITE.

spray gun. A device that uses compressed air to break up paint into a fine mist and eject it from a hand-held nozzle. A use of the principle of ATOMIZATION.

spray painting. The application of paint in the form of a stream of fine droplets from a SPRAY GUN.

spread (spread rate). The area that can be covered by a specified amount of paint or adhesive.

sprig. A small nail without a head. *See also* GLAZING SPRIG.

spring. (1) A WARP in a length of timber. (2) A pipe with a bend of less than 90°. (3) A source of SPRING WATER.

springer (skewback). The first stone at the base of an arch. It is based on the springing line; the upper surface or skewback is sloped.

springhouse. A shed that surrounds and protects a SPRING of water.

springing line. The level at which the base of an ARCH begins to curve away from the vertical supports. *See also* RISE.

spring strip. A strip of flexible metal that runs around the opening frame of a door or window to seal it.

spring water. A source of GROUNDWATER at places where geological conditions allow it to emerge naturally, such as at the sides of valleys.

sprinkler head. The outlet of a SPRINKLER SYSTEM.

sprinkler system. A FIRE EXTINGUISHER system consisting of sprinkler heads at ceiling level that are fed by water pipes. The system responds to temperature or fire detectors and controls the fire by the discharge of water through groups of heads. *Compare* EMULSIFIER SYSTEM.

sprocket (cocking piece). A length of wood nailed to each COMMON RAFTER at the EAVES to produce a flatter slope to the roof at the overhang.

sprung. *adj.* Describing timber that has developed a WARP.

sprung floor. A timber floor laid on battens that are supported by spring clips attached to the subfloor. The surface has a resilience which is suitable for dance floors.

spud. (1) A DOWEL that fixes the bottom of a door post to the floor. (2) A small outlet nozzle in a gas burner. (3) A nail used to hang a PLUMB BOB at a survey point.

spudding. The process of enlarging the diameter of the hole when driving a PILE.

spur. (1) A short concrete or metal post set in the ground to carry a wooden post. (2) A connection to an electric SOCKET OUTLET that is branched off a RING MAIN.

square. (1) An L-shaped hand tool used to set out right-angles. (2) A length of timber with a square section up to about 150 mm each side.

square chisel. *See* MORTISING MACHINE.

squared rubble. A stone wall made from squared stones. *Compare* RANDOM RUBBLE.

square-edged timber. Timber without WANE on any edges.

square joint. *See* BUTT JOINT.

square-sawn timber (converted timber). Timber that is cut to a square or rectangular cross-section, with or without WANE.

squaring up (squaring, working up). The calculation of areas and subtotals in a BILL OF QUANTITIES. *Compare* TAKING OFF.

squatting closet (eastern closet). A WC with a wide bowl that is installed at floor level and used in a squatting position.

squint. A brick with a special shape to make a SQUINT QUOIN.

squint quoin. A corner of a building that is not a right angle. *Compare* QUOIN.

SRI. *abbr.* SOUND REDUCTION INDEX.

SRPC. *abbr.* SULPHATE-RESISTING PORTLAND CEMENT.

SS. *abbr.* STAINLESS STEEL.

SSBC. *abbr.* (*USA*) Southern Standard Building Code.

SST. *abbr.* (*UK*) Society of Surveying Technicians.

stability. (1) The ability of a material to resist changes in properties while it is stored and in use. (2) The resistance of a structure to movement or collapse. (3) *See also* FIRE RESISTANCE.

stabilized soil. A foundation material that had been treated to increase its bearing capacity. *See also* COMPACTION, CONSOLIDATION.

stable door (*USA*: Dutch door). A door cut in half across the width and hinged so that each half opens independently.

stack. (1) A CHIMNEY stack or FLUE. (2) A vertical conduit in a building, such a SOIL STACK.

stack effect. The rise of warmed air within a building or up a chimney. *See also* CONVECTION CURRENT.

stadia hairs. Two short horizontal lines seen in the telescope of a THEODOLITE. They are spaced apart in a way that allows distances to be read optically through the telescope. *See also* INTERCEPT.

stadia staff. A LEVELLING STAFF that is graduated for measuring distance with the help of the STADIA HAIRS.

staff. *See* LEVELLING STAFF.

Staffordshire blues. A very durable variety of ENGINEERING BRICK with a bluish-gray colour.

stage payments. An agreed schedule of payments that are made to a builder parts of the construction are completed.

stagger. An arrangement of adjacent fastenings, such as nails, in alternate positions so that they do not line up and weaken the structure.

staggering (breaking joint). A staggered arrangement of components, such as the BONDING of brickwork, that avoids the alignment of adjacent joints.

staging. A working platform, such as on a scaffold.

stain. (1) A liquid colouring agent that changes the colour of a material such as wood by penetrating the surface rather than by hiding it. (2) *See* BLUE STAIN.

stainers (tinters). Solutions of coloured PIGMENT that are added in small quantities to alter the colour of paint.

stainless steel. An ALLOY made from low CARBON STEEL with the addition of up to 20 per cent chromium. The chromium compounds formed on the surface give the steel a high resistance to corrosion. *See also* FERRITIC STEEL.

stair. *See* STAIRS.

staircase (stairway). A complete set of STAIRS and associated construction.

stairs. A series of steps that connects one level of a structure with another. Stairs take various forms, sometimes with intermediate LANDINGS. The construction of stairs involve many components such as BALUSTER, BALUSTRADE, FLIER, HANDRAIL, GOING, NEWEL, NOSING, TREAD, RISER, SPIRAL STAIR, STRING, WINDER. *See* STAIRS diagram.

stairway. *See* STAIRCASE.

stairwell. An opening between the floors of a building in which a STAIRCASE is built.

stallboard. A SILL at the bottom of a shop window.

stallboard riser (stall riser). The vertical surface between the pavement and the STALLBOARD at the bottom of a shop window.

stanchion (staunchion). A vertical strut or COLUMN, usually made of steel.

stanchion base. A concrete base that distributes the load from a STANCHION.

standard. (1) An upright member of a SCAFFOLD. (2) *See* STANDARDS ORGANIZATIONS.

standard deviation. A statistical measure of the spread or dispersion about the MEAN value of a set of results. A relatively low standard deviation, on successive concrete strength tests for example, indicates a low variation in standards. Standard deviation is calculated as the square root of the mean of the squared deviations from the mean in a FREQUENCY DISTRIBUTION. *Compare* PROBABILITY.

standardization correction. In the accurate survey of lengths, an allowance for the slight difference between the length marked on a steel band or tape and the true length.

Standard Method of Measurement (SMM). (*UK*) A structured system of ESTIMATING.

standard mix. A method of specifying CONCRETE in which the quantities of ingredients for the mix are given in terms of dry weight per bag of cement. *Compare* DESIGNED MIX, NOMINAL MIX.

standard sky. A set of constant properties assumed about the quantity and distribution of light given by the sky. Because the sky varies with geography and weather a standard sky is needed for design purposes such as DAYLIGHT FACTORS. *See also* ARTIFICIAL SKY, CIE SKY, UNIFORM SKY.

standards organizations. Various national and international organizations that establish acceptable standards for the manufacture, design and installation of materials and equipment such as those used in construction. The British Standards Organization publish BRITISH STANDARDS (BS) and codes of practice (CP). In the USA standards are issued by the National Bureau of Standards (NBS), the American Society for Testing Materials (ASTM) and promoted by the American National Standards Institute (ANSI). Institutions in other countries include: AFNOR (Association Française de Normalisation) in France, DIN (Deutsche Industrie Normen) in Germany and ISO (International Organization for Standardization).

standard special. An item, such as a bullnosed brick or pipe bend, that is stocked but needs to be ordered.

standard wire gauge (SWG). A traditional but non-metric method of specifying the thickness of metal sheets, nails and wire.

standing leaf. In a double folding door, the

leaf that is usually kept closed. *Compare* OPENING LEAF.

standing seam. An upstanding join made between adjacent sheets FLEXIBLE METAL. The two ends are bent up then folded down over a TINGLE.

standing timber. Trees that are growing. *See also* TIMBER.

standing waves. Noticeable variations between sound levels at different positions in a room. They occur when repeated reflections between parallel surfaces cause reinforcement of the sound waves at certain frequencies.

standpipe. In general an outlet pipe that extends vertically from a supply. It may be used as a temporary street outlet from a water main.

Stanley knife. A brand name for a type of UTILITY KNIFE.

staple. (1) A U-shaped metal fastener with two sharp points that can be driven into wood or plaster. (2) The U-shaped part of a HASP AND STAPLE door lock.

star connection. A method of connecting electrical devices to a THREE-PHASE SUPPLY so that the three coils of the device share a common neutral at the centre of the 'star' shape. *Compare* DELTA CONNECTION.

star drill. (*USA*) A PLUGGING DRILL with a star-shaped face.

stat. *See* THERMOSTAT.

states of matter. The three forms in which a substance might exist: solid, liquid and gas.

statically determinate. In STRUCTURAL DESIGN, a structure that can be analysed by the use of the conditions of static equilibrium. *Compare* STATICALLY INDETERMINATE.

statically indeterminate. In STRUCTURAL DESIGN, a structure that cannot be correctly analysed by the use of the conditions of static equilibrium. *Compare* STATICALLY DETERMINATE.

static head. The component of PRESSURE in a pipe system caused by a difference in levels between two points. *See also* HEAD.

static load. A load imposed on a structure without any allowance for sudden loading. *Compare* IMPACT LOAD.

statics. The study of the behaviour of materials and structures under the action of forces that produce no motion.

station. A reference point on the ground, used for TRAVERSES and other measurements in land surveying. Permanent stations have markers set in the ground.

stator. The stationary frame of a large DYNAMO. *Compare* ROTOR.

staunchion. *See* STANCHION.

stave. A RUNG of a LADDER.

stay bar. (1) A horizontal bar that strengthens the MULLION of a window. (2) A TIE bar in a structure.

steady-state conditions. In heat calculations, the assumption that temperatures remain constant. The results of the assumption become more accurate over longer periods of time.

steam curing. The use of steam to accelerate the CURING of concrete.

steam trap. A device used to control the flow of steam and returning water in a steam-heating system.

steatite. A form of TALC.

steel. An alloy of IRON and carbon. *See also* CARBON STEEL, STAINLESS STEEL.

steel casement. A type of WINDOW frame made from galvanized steel.

steel section. *See* ROLLED-STEEL JOIST, UNIVERSAL BEAM, UNIVERSAL COLUMN.

steel square (framing square). An L-shaped instrument, similar to a SQUARE, used to set out the lengths of rafters and the angles of cuts.

steel wool. An abrasive material made from pad of fine steel threads. It is available in various grades that are used to smooth surfaces or to strip old paint.

steening (steining). The use of stones or bricks to line the inside of a WELL or SOAKAWAY.

step. (1) One unit of a STAIR with a RISER and a TREAD. *See also* FLIER, WINDER. (2) The change in level between different parts of a foundation.

step-down transformer. *See* TRANSFORMER.

Stephens and Bate formula. An acoustic formula that predicts an appropriate REVERBERATION TIME by taking into account the volume of the room and the type of the sound activity in it.

step iron. A strong metal RUNG built into the side walls of a deep MANHOLE to provide permanent access steps.

step joint. A notched join made between two timbers such as those in a roof frame.

step ladder. A LADDER with a folding prop that allows it to stand by itself, such as when decorating a ceiling. It has flat treads or steps rather than round RUNGS.

stepped flashing. A FLASHING that makes the join between a sloping roof and a wall. The flashing material is dressed into the wall in a series of steps.

stepped foundation. A FOUNDATION that changes level in a series of steps so as to follow the slope of the ground.

step turner. A slotted wooden tool used to shape the material for STEPPED FLASHINGS.

step-up transformer. *See* TRANSFORMER.

steradian. The SI unit of SOLID ANGLE. It is defined as the solid angle that encloses a surface on the sphere equal to the square of the radius. *See also* LUMEN.

stereography. The study of geometrical solids.

stereometric map. A map that gives an impression of the hills and valleys of the land.

stereoscopic. *adj.* Pertaining to the process of seeing objects in three dimensions.

sterilization. In the treatment of water supplies sterilization usually means the same as DISINFECTION.

sticker. (1) A strip of wood used to separate stacks or layers of timber to help drying. (2) A machine used to cut MOULDINGS.

stiffener. A metal angle or other piece welded or bolted to the WEB of a structural section to prevent buckling.

stiff frame. *See* REDUNDANT FRAME.

stiffness. *See* RIGIDITY.

stile. A vertical member in a framework such as a door. *See* DOOR diagram. *See also* CLOSING STILE, HANGING STILE, LOCKING STILE. *Compare* RAIL.

Stillson. *See* PIPE WRENCH.

stipple. (1) A textured paint surface produced by disturbing the wet paint film with a brush or other tool. (2) A paint surface broken up with spots of a different colour paint.

stirrup strap (beam hanger). A metal strap built into a wall and used to support the end of a structural timber. *Compare* JOIST HANGER.

stock. (1) *See* STOCK BRICK. (2) *See* CONVERTED TIMBER. (3) The body or handle of a PLANE or other tool.

stock brick. The form of BRICK most commonly available in a particular district. *See also* LONDON STOCK.

stock brush. A brush used to wet a wall before plaster is applied.

Stoke's law. An expression that uses the velocity at which particles of a soil sample settle to determine the particle diameters.

stone. A material used for walling, usually obtained from natural rocks.

stone saw. A SAW blade that cuts stone by the use of an abrasive rather than teeth.

stone slate. A thin slab of stone used for roofing.

stone tongs. *See* NIPPERS.

stoneware. A hard CERAMIC material used for some drainage pipes. *Compare* EARTHENWARE.

stool. A flat MOULDING that can be fixed over the frame of the window SILL.

stoop. A small platform or set of stairs at the entrance to a building.

stop. (1) A REBATE or strip of MOULDING around a frame against which the door or window closes. (2) A decorative end to a MOULDING. (3) *See* BENCH STOP. (4) *See* STOPCOCK.

stopcock (stop). A VALVE in a water or gas pipe that turns the flow on or off but does not regulate the flow like a COCK.

stopped chamfer. A CHAMFER that gradually merges into a sharp ARRIS.

stopping (stopper). A stiff paste used as a FILLER in holes and other surface defects that are to be painted.

stopping knife. A PUTTY KNIFE used to apply STOPPING.

storage heater. A heating system that uses the high thermal capacity of materials, like bricks, to store heat and to release that heat for use at other times. The storage system is usually heated by cheap-rate electricity by night or at certain times of the day.

storage tank. A water CISTERN in a roof.

storey. A floor or level of a building.

storey rod (storey pole). A strip of wood cut to the height of one STOREY and used to mark out the positions of window sills and other features.

storm clip. A metal clip fixed to the outside of a GLAZING BAR so as to hold the glass in place.

storm drain (storm sewer). A DRAIN (or SEWER) that only carries STORMWATER.

stormwater. Water that must be carried from a CATCHMENT AREA after heavy rain. *See* STORMWATER DRAIN.

stormwater overflow. A system that allows a SEWER to divert excess STORMWATER in heavy rains.

storm window. A temporary system of SECONDARY GLAZING that gives extra protection and insulation to a window during the winter. *See also* SHRINK FILM.

story. (*USA*) *See* STOREY.

stoving. The process of drying a paint film by the use of warm air or by INFRA-RED radiation.

straddle pole. A sloping pole that lies in the roof in a SADDLE SCAFFOLD.

straddle scaffold. *See* SADDLE SCAFFOLD.

straight arch. *See* FLAT ARCH.

straight edge. A length of wood or metal with a true straight edge that is used in SETTING OUT a building. *See also* SPIRIT LEVEL.

straight flight. A set of STAIRS that rises in a straight line without change of direction.

straight grain. Grain in wood that runs along the length of the piece. *Compare* CROSS GRAIN.

straight joint. (1) *See* BUTT JOINT. (2) An undesirable effect in brickwork BONDING where a vertical joint is immediately above another.

straight-run bitumen (blown bitumen). Solid BITUMEN that needs to be heated before use. It is obtained from the distillation of crude petroleum, assisted by 'blowing' air or steam into the fractionating column. Bitumen in this form must be heated before use in flooring, roofing and roads, but the rapid hardening is useful. *Compare* CUTBACK BITUMEN.

straight tongue. A TONGUE formed on the edge of a board by two REBATES formed on each side.

strain. The change in shape of an object caused by an imposed load. Linear strain is calculated as change in length divided by original length. *Compare* STRESS.

strain energy. The energy possessed by an elastic material or structure that is under stress.

strain gauge. An instrument used to measure small deflections in structures.

straining piece (strutting piece). A horizontal timber in a FLYING SHORE to which the inclined struts are attached.

S-trap. A type of water TRAP on the discharge of a sink, basin or WC pan. The bends in the trap are formed like the letter S. *Compare* P-TRAP. *See also* SEAL.

strap anchor. A steel plate used as a TIMBER CONNECTOR to join floor joists that butt together.

strap hinge. A HINGE with projecting bands that are usually screwed to a gate or post.

strapped wall. (*USA*) A form of wall construction with additional horizontal members fixed over the wall STUDS to allow extra insulation to be installed.

strapping. COMMON GROUNDS on a wall that form the base for a lining.

strategic stock. Stocks of materials that are held in case of shortage or price rise.

strawboard. *See* COMPRESSED STRAW SLAB.

streamline flow. *See* LAMINAR FLOW.

strength. The ability of a material or structure to resist force in a particular direction. It is usually measured by STRESS. *See also* COMPRESSION, TENSION.

strength class (SC). A system of marking STRESS-GRADED TIMBER according to species of tree and ELASTIC MODULUS values. For example, SC3 is often used for general structural purposes.

stress. A measure of the force applied to a material or structure. It is measured as the force divided by the area over which the force acts. *See also* COMPRESSIVE STRESS, TENSILE STRESS, ULTIMATE TENSILE STRESS. *Compare* STRAIN.

stressed-skin construction (geodetic construction). A form of construction in which a skin of plywood or metal stretched over a framework helps spread the load on the structure.

stress-graded timber. Timber for structural purposes that is tested and divided into classes according to its strength. Visual grading takes account of the slope of the GRAIN and the number of defects like KNOTS. Machine grading tests the stiffness of timber and relates it to minimum strength. There are various systems of marking stress-graded timber such as STRENGTH CLASSES.

stress-strain curve. A curve produced by plotting the STRESS applied to a material against the STRAIN that stress produces. The curve gives information about the STRENGTH, the ELASTICITY and DUCTILITY of the material. *See also* LOAD-EXTENSION CURVE.

stretcher. A brick or stone laid so that its length runs parallel to the face of the wall. *Compare* HEADER. *See also* RYBATE, STRETCHER BOND.

stretcher bond (running bond). A brickwork BOND in which all courses show only STRETCHERS. It is commonly used for CAVITY WALLS. *See* BRICKWORK diagram. *Compare* ENGLISH BOND, FLEMISH BOND.

stretcher face. The long face of a brick seen when it is laid in a course.

striated. *adj.* Describing a surface finish, such as on concrete, that is formed by parallel grooves or scratches.

strike (striking). (1) The removal of FORM-

WORK after concrete has hardened. (2) The removal of excess mortar in a newly-laid masonry wall.

striking plate. A metal plate fixed on a door frame so that catch or lock mechanism engages a hole in the plate. *See also* BOLT, LIP.

string (*USA*: stringer). The inclined board that runs up the side of a STAIR and supports the steps. *See also* CLOSED STRING, CUT STRING. *See* STAIRS diagram.

string course (sailing course). A horizontal COURSE of brick or stone that projects from a wall. It may serve as decoration or to throw water clear of the wall.

stringer. (*USA*) (1) *See* STRING. (2) A fence rail.

string piece. The horizontal tie beam in a BELFAST TRUSS.

strip. (1) Lengths of timber with relatively narrow cross-section such as 50 mm by 100 mm, or less. (2) Narrow sheets of metal such as aluminium or copper.

strip flooring. *See* PARQUET STRIP.

strip footing. *See* STRIP FOUNDATION.

strip foundation (strip footing). A FOUNDATION or footing for a wall that is built as a continuous ribbon of concrete and projects beneath each side of the wall. It may be built from brick or concrete. *See also* DEEP STRIP FOUNDATION.

stripping. (1) The process of clearing vegetation and topsoil from a site before construction starts. (2) The removal of old paint, wallpaper or other decoration from a surface. *See also* BLOW LAMP, PICKLING. (3) The STRIKING of FORMWORK from around concrete.

stripping knife. A hand tool with a wide flat blade that is used to lift old paint or paper when STRIPPING a surface.

str.S. *abbr.* Structural steel.

struck joint. *See* WEATHER-STRUCK JOINT.

structural. (1) A description of parts of a building, such as walls or beams, that carry loads in addition to their own weight. (2) (*USA*) General construction work that includes all buildings and civil engineering projects.

structural clay tile. A hollow or partially solid building BLOCK made from fired clay.

structural design. The design of structural members so that their size and position are correct for the loads that they might carry. *See also* BOW'S NOTATION, POLYGON OF FORCES, TRIANGLE OF FORCES.

structural engineer. A person who is professionally qualified in the area of STRUCTURAL DESIGN.

structural lumber. (*USA*) LUMBER intended for structural use where known strength is important. It is usually SQUARE-SAWN and STRESS-GRADED.

structural module. A space on a PLANNING GRID used for structural elements.

structural steelwork. A building framework made from ROLLED STEEL JOISTS or other members that are welded, bolted or riveted together.

structural timber. TIMBER intended for structural use where strength is important. It is usually STRESS-GRADED.

structure. (1) The parts of a building, such as walls or beams, that carry loads. *See also* STRUCTURAL DESIGN. (2) A building or other construction project.

structure-borne sound. Sound generated in a building as IMPACT SOUND and transmitted through the structure of the building rather. *Compare* AIRBORNE SOUND.

strut. A long structural member in COMPRESSION, like a COLUMN but not necessarily vertical. *Compare* TIE.

strutting. (1) The use of temporary supports for FORMWORK and SLABS. (2) *See* FLOOR STRUTTING.

strutting piece. (1) *See* STRAINING PIECE.

(2) A piece of timber used in SOLID BRIDGING or HERRING-BONE STRUTTING.

stub. The nib of a TILE.

stub tenon. A short TENON that fits into a BLIND MORTISE and does not penetrate completely through the wood.

stub wall. (*USA*) An internal wall that is not full height.

stuc. Plasterwork that is made to look like stone.

stucco. A smooth type of plaster or RENDER COAT used on outside walls. It was traditionally made from LIME and sand but now contains CEMENT. It may be coated over brickwork or on to wire netting.

stuck moulding. A MOULDING cut from solid material.

stud anchor (stud bolt, stud fixing). A metal pin or bolt inserted in surface as a fixing for further work, such as STUDDING. The bolts may be have male or female threads.

studding. *See* STUDS.

stud fixing. *See* STUD ANCHOR.

stud gun. A cartridge-fired gun used to shoot hardened STUD BOLTS or pins into concrete or brickwork.

stud partition wall. A type of PARTITION wall built from lining attached to a frame of STUDS. *See also* STUD HEAD.

studs (studding). The vertical members in a wall frame made of timber or light metal partition. *See also* NOGGING, TIMBER-FRAME CONSTRUCTION. *See* WALL diagram.

subbase. (*USA*) The lowest horizontal parts of a structural base. A SKIRTING BOARD.

sub-basement. A storey below the basement.

sub-contract. Part of a project, usually a specialist service, that is carried out by an individual or firm for the main CONTRACTOR rather than for the client.

sub-floor. A rough floor that carries the structural load but is covered by another floor finish, such as blocks.

sub-frame. A secondary frame that is built within the main structure and which supports part of the building.

subletting. The process where a main CONTRACTOR arranges for various SUB-CONTRACTORS to carry out parts of the work.

subsidence. The sinking of a ground level or foundation. *Compare* HEAVE, SETTLEMENT.

sub-sill. An extra SILL that projects from the window frame and throws water clear from the wall.

subsoil. The layer of earth immediately below the TOPSOIL.

substation. One of the installations of an ELECTRICITY SUPPLY system where the voltage is changed and supplies are routed to various destination. *See also* SWITCHGEAR.

substrate. A backing or surface on which another material or coating is applied.

substructure. The lower part of a building or structure that is below the walls.

subtractive colour. The reproduction of COLOURS by subtracting various components of colour from a source such as white light. The three subtractive primary colours are cyan (which absorbs red light), magenta (which absorbs green light) and yellow (which absorbs blue light). Most colour is seen as a result of the subtractive process: paint, colour photographs and printing all subtract colour from the light that illuminates them. *Compare* ADDITIVE COLOUR, MONOCHROMATIC LIGHT.

subway. (1) (*UK*) A walkway below the ground or beneath a building. (2) (*USA*) An underground railway.

suction. (1) In general, a movement or force that arises when the surrounding atmospheric pressure is greater than the local pressure. (2) The adhesion of wet mortar or plaster to a surface. (3) The tendency of a porous surface to absorb moisture from a material such as paint.

Suffolk latch. *See* THUMB LATCH.

sugar soap. An alkaline soap preparation used to wash down old paintwork before redecoration.

suite. (1) A set of different locks that can all be opened by the same master key. (2) A set of related rooms in a building, such as a hotel.

sulfate. (*USA*) *See* SULPHATE.

sulfur. (*USA*) *See* SULPHUR.

sullage. Waste water from a basin, sink or bath but not usually from a WC.

sulphate (*USA*: sulfate). A combination of the elements sulphur and oxygen that commonly occurs in a SALT. *See also* SULPHATE ATTACK.

sulphate attack. The chemical reaction of components in PORTLAND CEMENT with the sulphate SALTS found in some groundwater and materials. This reaction causes an expansion and disruption to concrete or mortar made from the cement.

sulphate-resisting Portland cement. A modified form of PORTLAND CEMENT that is less likely to have a disruptive reaction with sulphate SALTS. *See also* SULPHATE ATTACK.

sulphur (*USA*: sulfur). A non-metallic element (chemical symbol S) that exists as a yellow powder. *See also* SULPHATE.

sump (sump pit). A pit or hollow below the level of a basement or excavation in which unwanted water collects.

sump pump. A special pump that empties water when it collects in a SUMP. The pump may automatically operate at a certain water level.

sun control. *See* SOLAR CONTROL.

sunk gutter. *See* SECRET GUTTER.

sun path. The curved track that the sun makes across the sky on any one day. The daily tracks of the sun can be plotted on charts and used to predict the sunlight that will fall on a building face.

sunporch (sunspace). A room or addition to a building that has many windows and is exposed to maximum sunshine.

super (superficial). A term used in estimating to mean area measured in, for example, square metres.

super-glue. An adhesive that makes a very fast and strong bond on contact. *See also* ISOCYANATE.

superimposed load. *See* LIVE LOAD.

superimposition. A principle that allows the stresses in a structural member caused by one type of load to be added to the stresses due to another type of load. This assumption be used to find the forces acting in a REDUNDANT FRAME.

superstructure. The parts of a structure that are built above the foundations or ground level. *Compare* SUBSTRUCTURE.

supersulphated cement. A CEMENT that contains mainly BLAST-FURNACE SLAG with a some GYPSUM and some PORTLAND CEMENT. The cement has a high resistance to SULPHATES.

supply pipe. A water SERVICE PIPE that runs between the building and the boundary or stop valve.

surbase. (1) A MOULDING on the top of a DADO. (2) A MOULDING that runs along the top of a SKIRTING BOARD.

surcharge. Any load placed on the earth that is level with the top of a RETAINING WALL.

surface condensation. A form of CONDENSATION in buildings where a film or beads of water form on the surfaces of walls, windows, ceilings and floors. *Compare* INTERSTITIAL CONDENSATION.

surface dry. *adj.* Describing a paint film at the stage in drying when it is dry on top but still wet beneath.

surfaced timber. *See* SURFACE TIMBER.

surfacer (surface planer). A mechanical PLANE.

surface resistance. One of the types of THERMAL RESISTANCE that contribute to the U-VALUE of a structure. The resistance to heat flow offered by the inside and outside surfaces of an enclosure depend on the EMISSIVITY of the surfaces and the exposure of the building to climatic effects.

surface retardant. A liquid applied to the surface of FORMWORK so that the surface layer of concrete remains soft. This makes it easier to remove formwork and leaves a rough surface that gives a good key for plasterwork.

surface spread of flame. The tendency of a material in a fire to spread or to discourage the spread of a fire across the surface of the material. This particular property is important in the fire rating of lining materials.

surface tension. The ability of liquid surface to behave cohesively as if it had a skin.

surface timber (surfaced timber). Pieces of timber that have at least one side planed.

surface water. Rainwater that is collected and drained away from roofs and paved surfaces. *See also* COMBINED SYSTEM, SEPARATE SYSTEM.

surface water drain. A drain designed to carry only SURFACE WATER, rather than SOIL or WASTE.

Surform file. Trade name for a hand tool that contains a RASP for smoothing wood.

surround. A material, such as concrete around a drainpipe, that is placed around an object in order to protect or decorate it.

survey. In general, an examination and record of the land together with natural or man-made features such as buildings. *See also* CADASTRAL SURVEY, GEODESY, LAND SURVEYOR, PLANE SURVEYING, QUANTITY SURVEYOR.

surveyor. A person qualified in one of the areas of SURVEY work. *See also* CADASTRAL SURVEY, GEODESY, LAND SURVEYOR, PLANE SURVEYING, QUANTITY SURVEYOR, RICS.

suspended ceiling (drop ceiling, false ceiling, hung ceiling). A ceiling that is formed on a framework hanging from the floor above. The gap above the ceiling is often used for services, but can also form a fire risk.

suspended floor. A floor, such as on beams or joists, that is only supported at intervals. *Compare* SOLID FLOOR.

suspended scaffold. A CRADLE or PROJECTING SCAFFOLD.

suspension. (1) A liquid that contains fine solid particles distributed through the liquid but not dissolved in it. (2) An EMULSION paint in which fine droplets of one liquid are distributed through another liquid.

Sussex garden wall bond. *See* FLEMISH GARDEN WALL BOND.

SVP. *abbr.* SATURATED VAPOUR PRESSURE.

swage. A tool used to work hot or cold metal such as rivet heads. One half of the tool acts as a hammer against the other half.

swage-setting. A method of setting the teeth of circular saws so that they are most effective at ripping rather than cross-cutting.

swan neck (*USA*: goose neck). (1) A pipe in the form of an S-bend, such as the connection between a roof gutter and a downpipe. (2) A sudden change in level of a HANDRAIL.

swan-neck chisel. A long curved CHISEL that is used to lever the wood out of a mortise.

swatch. A sample piece of material such as a surface lining or paint colour. A number of such samples is often distributed together.

sweated joint. (*USA*) *See* CAPILLARY JOINT.

sweating. (1) The process of joining metal surfaces by heating them while SOLDER flows between them. *See also* CAPILLARY JOINT. (2) A paint defect caused when oil from the undercoat shows through the top coat. (3) *See* CONDENSATION.

sweat out. A PLASTER defect in which the surface remains soft and damp because the plaster continues to hold moisture.

swept valley. A roof VALLEY in which the tiles or slates are cut or tapered over the valley board so that a metal gutter is not needed. *Compare* LACED VALLEY.

SWG. *abbr.* STANDARD WIRE GAUGE.

swing door. *See* REVOLVING DOOR.

swinging post. The post on which a gate is hinged.

switch. An electrical device placed in a CIRCUIT to open or close the path of current flow in that circuit. *See also* FLOAT SWITCH, RELAY.

switchboard. (1) A group of hand-operated switches that control different circuits. (2) A small telephone exchange for a building.

switch-fuse. An electrical SWITCH that also contains a FUSE.

switchgear. (1) Equipment used to control the distribution of electricity, such as in the SUBSTATIONS of a power transmission system. (2) Equipment that controls and protects electrical machines.

synthetic resin. An artificial POLYMER, such as MELAMINE FORMALDEHYDE or BAKELITE. In the raw state they have some properties similar to natural RESINS.

synthetic stone. *See* CAST STONE.

systematic error. A constant form of error that can occur in a measurement such as surveying. The error always happens in the same instrument or type of operation.

system building. *See* INDUSTRIALIZED BUILDING.

T

tacheometer. *See* TACHEOMETRY.

tacheometry. An optical method of surveying distances or heights by means of the angle an object makes with the viewing instrument such as a THEODOLITE or tacheometer.

tack. (1) A short type of NAIL with a flat head. (2) The degree of stickiness of a wet paint film or glue.

tack coat. A coating applied to a surface, such as a road, to improve the ADHESION of the coating that follows.

tack rag. A cloth that is kept sticky by varnish and used to remove dust from a surface before the next coat.

tacky. *adj.* Indicating the sticky quality of a coat of paint or glue that is partially dry.

tafy joint. A joint between two lead pipes made by fitting the reduced end of one joint into the enlarged end of the other and filling the joint with SOLDER.

tag. A folded strip of copper used to wedge copper sheet into a masonry or brick joint.

tail. (1) The end of a RAFTER or other member that overhangs a wall. (2) The end of a stone step that is built into the surroundings. (3) The lower edge of a SLATE OF TILE.

tailing in (tailing down). The process of building-in the end of a step or other member that projects as a cantilever.

tailing iron. A steel support built into a wall above the TAIL of a cantilever.

tail joist. A JOIST that rests on a TAIL TRIMMER rather than on a wall.

tailpiece (*USA*). A drainpipe that connects a SINK strainer or similar appliance with the TRAP.

tail trimmer. A trimmer JOIST that supports the TAIL JOISTS close to a wall.

taking off. The systematic measurement and listing of items from a drawing as the first step in the preparation of a BILL OF QUANTITIES. *Compare* SQUARING.

talc (talcum). A MINERAL composed of hydrated compounds of MAGNESIUM and SILICATES.

tally slates. SLATES that are sold by number rather than by weight.

tambour. (1) A circular wall such as that which supports a dome. (2) The sliding cover of a roll-top desk.

tamp. *vb.* To pack down a material, such as soil, by repeated impact.

tanalized timber. Construction timber that has been treated with PRESERVATIVE.

T and G. *abbr.* Tongued and grooved. *See* TONGUED AND GROOVED JOINT.

tang. The pointed end of a tool blade, such as a chisel, that is inserted into the handle.

tangent. A line or surface that touches a curved line or surface at one point, without intersecting it.

tangential. A process that occurs in the same direction as a TANGENT.

tangential shrinkage. Timber shrinkage

that occurs parallel to the growth rings of a log. *Compare* RADIAL SHRINKAGE.

tank. (1) *See* CISTERN. (2) A water storage container that is fully enclosed and may therefore be pressurized, unlike a CISTERN which is open.

tanking. A continuous waterproof layer, such as ASPHALT or POLYTHENE sheet, installed in the walls and floor of a basement to keep the area dry.

tap. (1) A threaded tool used to cut internal threads in a pipe or hole. (2) A valve that controls the flow of water or gas at an outlet. *See also* BIB TAP, FAUCET, JUMPER, PILLAR TAP, WASHER.

tape. (1) A graduated ribbon of steel or other material that is used for measuring distances. It coils back into a case when not in use. (2) *See* JOINTING TAPE.

tape corrections. In surveying, adjustments that are sometimes made to TAPE readings to allow for say, STANDARDIZATION, temperature and other changes.

tapered-edge plasterboard. PLASTERBOARD with a bevel on one long edge which can be used to make invisible joints with the next board.

taper pipe. A pipe that changes in diameter to act as an INCREASER or REDUCER.

taper thread. A standard type of screw thread used to join water and gas pipes. *Compare* PARALLEL THREAD.

taping strip. A strip of BITUMEN FELT laid used to seal joints in DECKING before the roof layers are laid.

tar. A black viscous liquid obtained from the destructive distillation of coal. The tar is collected as a vapour while pitch is left as the residue. The adhesive properties of tar are used in ROAD TARS. *Compare* BITUMEN.

tare. The weight of a container or vehicle before it is loaded with goods or people.

target. An object that is set over a survey point so that it can easily be sighted by a THEODOLITE or by EDM equipment.

target rod (target staff). A type of LEVELLING STAFF with a TARGET that is raised or lowered on the rod while it is being sighted through a level or theodolite. *Compare* SELF-READING STAFF.

Tarmac. Trade name for a form of TARMACADAM.

tarmacadam (coated macadam). A paving material or road surface formed from graded stones with a binder of TAR or BITUMEN. *See also* MACADAM.

tarpaulin. A large sheet of tough waterproof material used for temporary protection of building works and materials. The traditional canvas tarpaulin may be replaced by synthetic materials such as polythene.

task lighting. Lighting, such as a reading lamp, that illuminates a particular work area rather than general lighting that illuminates all parts of an area.

T-beam (tee-beam). (1) A beam built into a reinforced concrete floor so that part of the floor slab forms a FLANGE. (2) A ROLLED-STEEL SECTION in the shape of a T.

tcd. *abbr.* Traced (used on drawings).

teak. A HARDWOOD tree with a dark, strong and durable wood that is especially suitable for doors, windows and garden furniture.

tee. A T-shaped device for connecting pipes or ducts. The stem of the T often connects a branch pipe to the main pipework of the horizontal part of the T.

tee-beam. *See* T-BEAM.

tee-hinge. *See* T-HINGE.

tee-square. *See* T-SQUARE.

teflon. A brand name for POLYTETRAFLUOROETHYLENE.

tegula. *See* CHANNEL TILE.

telemeter. A type of THEODOLITE that measures distance to a staff by using a glass prism at the lens and comparing two different light paths.

telescopic centering. Floor FORMWORK made from steel sections that slide into one another and can be adjusted to fit the span.

telltale. A piece of glass or other suitable indicator that is firmly fixed across a crack in a structure and watched to see if the crack is widening.

temper (tempering). (1) A process that increases the toughness of hardened STEEL by heating and QUENCHING. (2) Mechanical or heat treatment that increases the toughness of non-FERROUS metals. (3) The mixing of mortar or plaster to give correct properties.

temperature. The degree of 'hotness' of a body, a property that is related to the energy of the atoms. *Compare* HEAT. *See also* ABSOLUTE TEMPERATURE SCALE, CELSIUS TEMPERATURE, FAHRENHEIT TEMPERATURE.

temperature gradient. A measure of change of temperature with length. Temperature gradients can be plotted on to section drawings of walls and roofs and used to determine the likelihood of INTERSTITIAL CONDENSATION and the position of VAPOUR BARRIERS. *See also* BOUNDARY TEMPERATURE.

temperature steel. Reinforcement that is placed in concrete to help minimize cracks caused by temperature changes. The temperature steel is not needed if reinforcement for strength is used.

temperature stress. A STRESS produced in a structural member by a change in temperature. A restrained member will be compressed as it expands with rise in temperature and will be placed in tension as it contracts with a fall in temperature.

tempering. *See* TEMPER.

template. A full-sized pattern of an object made from a sheet of plastic, wood or other material. It is used for marking or cutting out copies of that object.

temporary benchmark. A BENCH MARK that is established on a site as a reference for the building works. *See also* ORDNANCE BENCH MARK.

temporary hardness. A form of HARD WATER in which the hardness can be removed by boiling, unlike PERMANENT HARDNESS. Temporary hardness is caused by the presence of dissolved salts such as calcium carbonate (limestone) and magnesium carbonate. The removal of temporary hardness by boiling produces the scale or FUR found inside kettles.

temporary threshold shift (TTS). A temporary loss of hearing that recovers in 1–2 days after exposure to noise. *Compare* PERMANENT THRESHOLD SHIFT. *See also* EQUIVALENT CONTINUOUS SOUND LEVEL.

tender. An offer from a contractor to carry out specified work for a stated sum of money.

tendering. The process of preparing and submitting a TENDER.

tendon. A rod or cable used in PRESTRESSED CONCRETE. *See also* RELAXATION.

tenon. A length of timber with a reduced end that fits into a MORTISE slot. *See also* BAREFACED TENON, HAUNCHED TENON, MORTISE-AND-TENON JOINT, SECRET TENON, TUSK TENON.

tenon saw (mitre saw). A hand SAW that is stiffened by a fold of metal along the top of the blade and used for fine work. *See* WOODWORKING TOOLS diagram.

tensile stress (tensile strength). A measurement of the force per unit area that produces TENSION in a material. Metric units in practical use include N/mm^2 and MN/m^2. *Compare* COMPRESSIVE STRESS.

tension. The effect of a pulling force that tends to stretch a structure. The opposite of COMPRESSION.

tension correction. A type of TAPE CORRECTION made to a surveying measurement to allow for a non-standard tension in the tape.

tension flange. The side or FLANGE of a beam that is in TENSION.

tensometer. An instrument used to measure TENSILE STRESSES in a sample of material. Information may be given in the form of a LOAD-EXTENSION CURVE.

tera-. A standard prefix that can be placed in front of any SI unit to multiply the unit by 1,000,000,000,000. The abbreviated form is 'T' as in TW for terawatt. *See also* GIGA-, KILO-, MEGA-, MICRO-, MILLI-.

terminal. (1) A point on electrical equipment for the connection of electrical cable. (2) A connecting point on an item of services equipment, such as a FLUE. (3) An outlet for ventilation systems. *See also* REGISTER.

termite. An ant-like insect found in tropical parts of the world and some parts of the USA. The insects live in colonies; their attack on wood and other building materials can soon damage a structure.

termite shield. A shield or lip of sheet metal fixed around foundation walls, pipes and other parts of the substructure to prevent the passage of TERMITES.

terne metal. An ALLOY of lead with some tin and antimony. It may be coated on to steel. *Compare* GALVANIZING.

terotechnology. A branch of technology that specializes in the cost-efficient installation and operation of machinery and equipment.

terrace. (1) A platform of land that is raised above the surrounding ground level. (2) A flat area of roof with an access doorway and other features that allow it to be used.

terrace house. A house in a row of three or more similar houses that are joined. A PARTY WALL separates one property from another.

terra cotta. A durable burnt clay product with a finer texture than BRICK. It has been used to make decorative blocks and moulded features. Glazed terra cotta may also be known as FAIENCE.

terrazzo. A floor or wall surface produced by a layer of stone chips laid in cement mortar. When set, the surface is ground and polished smooth.

tertiary treatment. A final treatment of SEWAGE effluent sometimes given after the SECONDARY TREATMENT.

tessellated. A surface, such as a MOSAIC floor, composed of small blocks of material laid in patterns.

tessera. One of many small blocks of glass, marble or other stone used in TESSELATED finishes such as a MOSAIC.

test cube. A standard-sized cube of concrete or mortar that is used to test concrete or cement, such as in the MORTAR-CUBE TEST.

test pit. *See* TRIAL HOLE.

tetrapod. A regular shape that has four arms radiating from a central point, each arm being at an equal angle to the others. Large concrete tetrapods are used to break the force of waves against BREAKWATERS.

textured finish. A variety of finishes that give a roughened surface to plaster, render or paint.

TFB. *abbr.* Taper flange beam (used in structural steelwork).

TFL. *abbr.* Tubular fluorescent lamp. *See* FLUORESCENT LAMP.

TG. *abbr.* Tongued and grooved. *See* TONGUED AND GROOVED JOINT.

thatch. A traditional ROOF CLADDING made of reed or straw. *See also* LIGGER, WITHY.

theodolite. A surveying instrument used to measure angles by means of a telescope that rotates on a horizontal axis. *See also* OPTICAL-READING THEODOLITE.

therm. *See* BRITISH THERMAL UNIT.

thermal. *adj.* Relating to heat. *See also* entries under HEAT.

thermal bridge. *See* COLD BRIDGE.

thermal capacity. The ability of a material to store heat energy. Thermal capacity is measured by SPECIFIC HEAT CAPACITY and is different from THERMAL INSULATION. Heavyweight structures such as masonry have a high thermal capacity and take longer to heat up or cool down. *See also* SENSIBLE HEAT, THERMAL STORAGE WALL.

thermal comfort. Those thermal variables that produce and affect satisfactory environmental conditions for humans. They include air TEMPERATURE, RADIANT TEMPERATURE, HUMIDITY and air movement. Individual thermal comfort also varies with activity, age, clothing and gender.

thermal conductance. The reciprocal of THERMAL RESISTANCE.

thermal conductivity. *See* COEFFICIENT OF THERMAL CONDUCTIVITY.

thermal insulation (thermal insulator). A material or method of construction that restricts the flow of heat. Good thermal insulators are materials with low COEFFICIENTS OF THERMAL CONDUCTIVITY to oppose CONDUCTION or low EMISSIVITY to oppose RADIATION. *See also* BAT, EXPANDED PLASTIC, LIGHTWEIGHT CONCRETE, MINERAL WOOL, U-VALUE.

thermal lance. A burning steel tube used to bore holes into concrete by means of high temperature. The tube is packed with steel wool and the flame is fed by a stream of oxygen down the tube.

thermal power station. A POWER STATION that generates electricity by burning fuels such as oil, gas and coal.

thermal resistance (R-value). A measure of the opposition to heat flow given by a particular component in a building element. The total thermal resistance of all the layers in a structure, including surfaces, is used to calculate the U-VALUE. *See also* THERMAL CONDUCTANCE.

thermal resistivity. An alternative measure of CONDUCTION in materials. It is calculated as the reciprocal of the COEFFICIENT OF THERMAL CONDUCTIVITY.

thermal response. The speed at which the temperature and other thermal properties of a building respond to changes. *See also* THERMAL CAPACITY.

thermal storage wall (Trombe wall). A masonry wall whose high THERMAL CAPACITY stores heat energy from the sun. The wall is usually set behind glass and absorbs radiant energy that is later passed to the rest of the building. *See also* PASSIVE SOLAR CONSTRUCTION.

thermal transmittance. *See* U-VALUE.

thermal wheel. *See* HEAT WHEEL.

thermal zoning. *See* ZONE CONTROLS.

thermistor. A semiconductor device that changes in electrical properties with changes in TEMPERATURE.

thermocouple. A device that uses the electrical properties of a metal junction to measure TEMPERATURE.

thermodynamic temperature. *See* ABSOLUTE TEMPERATURE.

thermograph. A device that measures and makes a continuous record of changing temperature.

thermometer. An instrument that measures TEMPERATURE. *See also* PYROMETER.

thermoplastic. A material that becomes soft when heated and hard when cooled. PLASTICS such as POLYETHYLENE, POLYPROPYLENE, POLYSTYRENE are thermoplastics. *Compare* THERMOSETTING.

thermoplastic tile. A floor tile made from a THERMOPLASTIC resin that binds other materials such as fibre.

thermosetting. *adj.* Describing a material that becomes hard when heated. PLASTICS such as MELAMINE FORMALDEHYDE and POLYURETHANE are thermosetting. *Compare* THERMOPLASTIC.

thermostat (stat). A device that reacts to changes in temperature and automatically controls heating or cooling equipment. Traditional thermostats use a BI-METALLIC strip. Modern thermostats often use a heat-sensitive resistor or other device in an electrical circuit.

thickening. *See* FATTENING.

thicknessing machine. A type of PLANING MACHINE used to reduce a piece of wood to a set thickness.

thimble. *See* SLEEVE PIECE.

T-hinge (tee-hinge). A HINGE with a fixing band that projects from the turning pin. It is commonly used on BOARDED DOORS. *See* DOOR diagram.

thinner. A volatile liquid, such as WHITE SPIRIT or TURPENTINE, added to PAINTS and VARNISHES to make them more fluid and increase their penetration.

thinning ratio. The proportion of THINNER added to PAINT for a particular purpose.

thistle plaster. A trade name for a form of RETARDED HEMIHYDRATE PLASTER.

thixotropic. A property of certain liquids that decrease in viscosity or 'body' when disturbed. For example, thixotropic or 'non-drip' paints have a jelly-like structure which temporarily flows under the pressure of the paint brush.

thread. A system of continuous ridges and grooves cut or moulded on to a cylinder or pipe. Components are joined when the thread on one is screwed into another. *See also* PARALLEL THREAD, TAPER THREAD.

three-coat work. (1) A high quality plaster or render finish formed by three separate coats: a rendering coat, a smoothing coat and a FINISHING COAT. *Compare* TWO-COAT WORK. (2) A paint coat built up from three layers of PRIMING COAT, UNDERCOAT, and top coat.

three-hinged arch. An arch that is hinged at each support and at the crown.

three-hinged frame (three-pinned frame). A PORTAL FRAME that is hinged at the crown and at the bottom of each column.

three-phase supply. An electrical supply that generates and delivers three alternating voltages having the same frequency but overlapping one another by 120°. The system allows the more efficient operation of large electric motors. *See also* DELTA CONNECTION, SINGLE PHASE SUPPLY, STAR CONNECTION.

three-pinned frame. *See* THREE-HINGED FRAME.

three-pin plug (three-prong plug). An electrical plug with three pins that insert into a SOCKET OUTLET. Two pins carry the current and the third is an EARTH connection. *See also* SHUTTERED SOCKET.

three-ply. *adj.* (1) Describing a construction, such as a roof, built up from three layers. (2) Describing a common form of PLYWOOD constructed from three layers.

three-port valve. A type of MOTORIZED VALVE.

three-prong plug. *See* THREE-PIN PLUG.

three-quarter bat. A BRICK cut straight across to reduce its length to three-quarters. *Compare* KING CLOSER.

three-way strap. A T-shaped steel plate used to help fix the junction of three members of a wooden frame, such as a TRUSS.

threshold. A strip of trim used to cover the joint between two types of flooring at a doorway.

threshold of hearing (threshold of audibility). The weakest sound that the average human ear can detect. It corresponds to a SOUND LEVEL of 0 DECIBELS.

threshold of pain. The strongest sound that the average human ear can tolerate. It corresponds to a SOUND LEVEL of approximately 140 DECIBELS.

throat (throating). (1) *See* DRIP. (2) The

GATHERING or lower end of a chimney flue where it narrows above the FIREPLACE.

through bonder. *See* BOND STONE.

through lintel. A LINTEL that extends the full thickness of the wall in which it is set.

throughstone. *See* BOND STONE.

through tenon. A TENON that protrudes through the mortised member.

thrust. (1) A COMPRESSION force that tends to produce motion. (2) A force that pushes outward, such as from the base of an arch or from behind a retaining wall. (3) *See* LINE OF THRUST.

thumb latch (Norfolk latch, Suffolk latch). A simple door LATCH in which a fall bar is raised by a lever passing through the door. The lever is lifted by pressing with the thumb.

thumb screw. A screw with a flattened end that can be turned by hand.

tie. (1) A structural member with TENSION. *Compare* STRUT. (2) A fixing clip for sheets of flexible roofing material.

tie beam. A horizontal TIE between the feet of a pair of rafters in a roof structure. *See* ROOF TRUSS diagram. *Compare* COLLAR BEAM.

tie point. The terminal point of a survey chain line that is also a point on another line.

tier. A vertical leaf of brickwork or stonework.

tie rod. A TIE that is usually made of steel rod with threaded ends.

tight. *adj.* Indicating a dimension that is exact or slightly undersized.

tight knot. A KNOT that is firmly set in the surrounding timber. *Compare* LOOSE KNOT.

tight size. In windows, the true dimensions between the REBATES. The size of the glass is a few millimetres less than the tight size.

tile. A thin covering unit made from burnt clay, concrete, ceramics, flexible plastics and other materials. Roof tiles are laid in overlapping rows to make the roof weathertight. *See also* ITALIAN TILE, PANTILE, PLAIN TILE, ROMAN TILE, TILE BATTEN. Floor and wall tiles are laid in regular patterns on a surface using an adhesive.

tile-and-a-half-tile. An extra-wide PLAIN TILE that is used for the VALLEYS and VERGES of a roof.

tile batten. One of a series of square timber BATTENS that are fixed across roof rafters at a spacing suitable for fixing roof TILES.

tile creasing. *See* CREASING.

tiled valley. A roof VALLEY that is lined with specially-shaped VALLEY TILES rather than built with a SWEPT VALLEY or VALLEY GUTTER.

tile fillet (tile listing). TILES that are cut and set in mortar at an angle between a roof and a wall or parapet. The fillet is used instead of a FLASHING. *Compare* CEMENT FILLET.

tile hanging. A method of covering an external wall with roof TILES by fixing them to battens on the wall and overlapping them like a roof.

tile listing. *See* TILE FILLET.

tilting fillet (eaves board). A horizontal board that is nailed along the bottom of the rafters or roof boards just before the DOUBLE EAVES COURSE of tiles or slates at the edge of the roof. The fillet, which may be triangular in cross-section, makes the slope less steep and tightens the tiles. *Compare* SPROCKET.

tilting level. A form of surveyor's LEVEL in which the telescope is not rigidly fixed to the levelling head. It is more advanced that the DUMPY LEVEL and quicker to use.

tilting mixer. A common type of concrete mixer that is emptied by tilting the complete mixing drum.

timber. (1) Wood from trees that has been sawn or processed ready for use in construction. *See also* CONVERSION. (2) (*USA*) The larger sizes of LUMBER. (3) Standing trees grown for use in construction.

timber connector. *See* CONNECTOR.

timber-frame construction. A building in which the timber frame of the walls carries a load, usually the roof. The frame may be faced with veneer of brickwork or other CLADDINGS. *See* WALL diagram. *See also* BALLOON FRAME, NOGGING, PLATFORM FRAME, STUD.

timbering. Timber used as temporary support during work such as excavations, CENTERING, FORMWORK, SHORING.

timber joints. *See* TIMBER JOINTS diagram.

timbre. *See* OVERTONE.

tin. A pure metal element (chemical symbol Sn) that is soft and ductile. It is used in many ALLOYS and SOLDER.

tingle. A strip of lead, or other metal, that is used to fix an edge or a join in FLEXIBLE METAL roofing. *See also* STANDING SEAM.

tin snips. Strong shears or scissors used to cut thin sheets of metal.

tint. A PAINT colour that is produced by mixing white paint and coloured paint.

tinters. *See* STAINERS.

tinting. A adjustment to the colour of PAINT by the addition of STAINERS.

titanium dioxide (titanium oxide, TiO_2). A white PIGMENT commonly used in paint, plastics and other materials. It is highly opaque and chemically stable.

toe. (1) The base of the SHUTTING STILE of a door. (2) The base of a dam or retaining wall on the free side of the structure, away from the retained material.

toe board (guard board). A board fixed around the edge of a working platform, such as a SCAFFOLD, to prevent material falling off.

toe nailing. *See* SKEW NAILING.

togglebolt. A special type of bolt used to make a firm fixing to a thin panel of material such as plasterboard. When the bolt is pushed through the hole in the togglebolt it expands inside the cavity and is tightened against the inside of the panel.

toilet. *See* WC.

tolerance. Acceptable variations in the dimensions of a component that allow for unavoidable variations in manufacture or working practice. The tolerance is usually indicated by a plus or minus figure for the dimension.

tommy bar. A bar that is inserted through holes in a box spanner to help turn the spanner.

ton. Unit of mass. A 'long' ton (UK) = 2240 pounds; a 'short' ton (USA) = 2000 pounds. *See also* TONNE.

toner. A pure form of DYE, without EXTENDER.

tongue. A long thin strip of wood or metal that fits in a groove to strengthen a joint. *See also* STRAIGHT TONGUE.

tongued and grooved joint (tongue and groove). A joint made between the edges of two boards by means of a TONGUE in one board that fits into a slot in the other board.

tonne (metric ton). Unit of mass. 1 tonne = 1000 kilograms (approximately 2205 pounds). *See also* TON.

tool. In general, a device used to shape and process timber, metals and other materials.

tooling. (1) *See* BATTING. (2) (*USA*) The process of trimming and smoothing the mortar joints in a masonry wall.

toothed connector. *See* CONNECTOR.

toothed plate. A type of timber CONNECTOR that uses a serrated metal plate to grip the timber.

toothing (indenting). A construction pattern at the end of a brick or stone wall in which alternate courses project to allow a good BOND with another wall.

top beam. *See* COLLAR BEAM.

top course tile. A short roof TILE that is used for the top course of tiles next to the roof RIDGE.

top-hat section. A section like a LINTEL that has a tall rectangular shape, like a top hat.

top-hung window. A WINDOW that is hinged at the top and opens outward.

top lighting. The lighting of an area in a building from overhead sources, such as ROOFLIGHTS.

topographical survey. A survey of the features on the surface of the earth. It is the work of a LAND SURVEYOR. *Compare* CADASTRAL SURVEY, GEODESY.

topography. The physical features of the surface of the land including elevations, depressions and rivers.

topping. (1) *See* SCREED. (2) The WEARING COURSE of a road surface.

topping out. The completion of the waterproof SHELL of a new building, before the internal finishes and services are completed.

top plate. A horizontal timber at the top of a wall frame. *See* WALL diagram. *See also* WALL PLATE.

top rail. The upper horizontal member of a frame or of a PANEL door. *See* DOOR diagram and WINDOW diagram.

topsoil. The uppermost layer of earth. *Compare* SUBSOIL.

torching. An outdated practice of using mortar or plaster on the undersides of roof tiles or slates.

torque. A measure of the rotating effect of a system, such as a turning shaft. It is calculated as the MOMENT of the force about the axis of rotation. *See also* TORSION.

torsion. The twisting effect of a turning force or TORQUE.

total float. In CRITICAL PATH ANALYSIS, the total spare time between the time available for an activity and the time actually needed to perform it. *See also* FLOAT TIME.

touch dry. *adj.* Indicating a stage in the drying of a PAINT film when no marks are left if the film is touched.

toughened glass. A form of GLASS that is made by rapidly cooling the surface during manufacture. This process leaves compressive stresses in the glass that increase its strength and allow the glass to break into small pieces as a SAFETY GLASS.

toughness. The general ability of a material to resist sudden cracking. Its opposite is BRITTLENESS. A tough material, such as putty, need not have high STRENGTH. Toughness in metals gives the related properties of DUCTILITY and MALLEABILITY.

tower bolt. A large BARREL BOLT.

town gas. Gas that is manufactured, formerly from coal, for distribution to homes and industry. *Compare* NATURAL GAS.

town planning (city planning). In general, the preparation and administration of overall plans for the buildings and other features of a town or city.

trabeated. *adj.* Describing a structure that uses horizontal BEAMS rather than ARCHES.

tracing. A copy of a drawing made by laying transparent paper or cloth over the original drawing and following or 'tracing' all the features with a pen or pencil.

track. (1) A metal section set on a ceiling or above an opening. It may be used to suspend various types of sliding doors, screens and lighting. (2) A continuous system of rails, sleepers and ballast on which a train or crane can run.

TRADA. *abbr.* (*UK*) Timber Research And Development Association.

trade (craft). An occupation such as carpentry or bricklaying that needs manual skill and training. *See also* APPRENTICE, JOURNEYMAN, WET TRADES.

traffic noise level (L_{10}). An assessment of the annoyance caused by noise from busy roads and motorways. The variations in SOUND LEVELS are measured for a period of time, such as 18 hours, and then analysed by statistics. L_{10} is that sound level, in dB(A), that is exceeded for 10 per cent of that time.

trammel. (1) A device for drawing ellipses or large circles on a drawing board. The feet of a special beam compass slide in two grooves that meet at right angles. (2) A length of timber used to set out circular brickwork shapes. One end of the trammel pivots on a 'striking point' at the centre of the arc.

transducer. A device that changes one form of energy into another. For example, changes in stress or temperature may be converted into electrical signals and used to control machines.

transfer. In PRESTRESSED CONCRETE, the moment when the concrete first takes some of the load from the steel.

transformer. A system of coils that uses the principle of electromagnetic induction to change the voltage or to isolate a supply of ALTERNATING CURRENT electricity.

transit. A form of surveyor's LEVEL on which the telescope can also pivot up and down so that the instrument can be used to establish vertical lines.

transitional flow. For fluids moving in pipes, the complex transitional state between LAMINAR FLOW and TURBULENT FLOW. *See also* CRITICAL VELOCITY.

translucent glass. *See* OBSCURED GLASS.

transmission coefficient. A measure of the sound energy passed through a partition. It is calculated as the transmitted sound energy divided by the incident sound energy. *See also* SOUND REDUCTION INDEX.

transmittance. (1) A measure of the ability of a material, such as glass, to pass heat or light RADIATION. It is calculated as the ratio of the transmitted radiation to the incident radiation. (2) *See* U-VALUE.

transom (transome). (1) A horizontal member in a window frame that divides the window into separate LIGHTS. *Compare* MULLION. (2) An intermediate horizontal member of a door frame, or similar structure. (3) (*USA*) A small window LIGHT set above a door, usually as part of the door frame. *See also* FANLIGHT.

transverse. A structural member that is set horizontally between end supports, such as a BEAM suspended between two walls.

trap. A bend or similar device in a pipe that remains full of water and so prevents the escape of gases and smells from the drain system. *See also* ANTI-SYPHON TRAP, GULLEY, INTERCEPTING TRAP, P-TRAP, SEAL, S-TRAP, VENT PIPE.

trap door. A small hinged or removable door that covers the access opening to an attic or similar concealed area.

traverse. In land surveying, a connected series of straight lines whose lengths and BEARINGS are known. *Compare* CLOSED TRAVERSE.

tread. The horizontal surface of a STEP. *See* STAIRS diagram. *See also* WINDER.

treenail. *See* TRENAIL.

trellis. A timber or metal frame in which thin strips of material cross one another to form an open pattern or lattice.

tremie (tremmie). A delivery chute and pipe used for placing concrete under water. The pipe is kept filled with concrete and the foot of the pipe is kept immersed in the delivered concrete.

trenail (treenail, trunnel). A strong hardwood or synthetic dowel used to fasten timbers. *Compare* DRAWBORE PIN.

trench. (1) A long narrow excavation in the ground, such as for a FOUNDATION. (2) A narrow groove cut in a piece of wood to receive another component.

trestle. A low stand made from a cross beam set between a pair of slanting legs. A pair of trestles is often used to support a low SCAFFOLD for decorating work.

trial hole (trial pit, test pit). A hole or shaft made on site to investigate the nature of the soil and rock. *See also* SITE INVESTIGATION.

triangle. (1) A closed geometrical figure with three sides. (2) (*USA*) A flat piece of metal or plastic in the shape of a right-angled triangle. It is used to set out angles in technical drawing.

triangle of forces. In STRUCTURAL DESIGN, the principle that if three forces acting at a point are in equilibrium then they can be graphically represented by a closed triangle.

triangulation. A method of land surveying in which a large area is subdivided into a network of triangles starting from a measured BASE LINE. The angles of the triangles are measured and used to calculate other lengths and areas. *Compare* TRILATERATION.

triaxial compression test. A laboratory test that surrounds a sample of soil with liquid under pressure and records the changes in the sample.

trig station. A permanent STATION used in the TRIANGULATION of the land.

trilateration. A method of land surveying in which a large area is subdivided into a network of triangles of which all the side lengths are measured. *Compare* TRIANGULATION.

trim. The MOULDINGS and other finishes applied around the edge of doors, windows, floors, ceilings and other junctions.

trimetric projection. A drawing PROJECTION in which three axes are at arbitrary angles and may use different scales.

trimmed joist. A common JOIST that is shorter than the other joists. It is supported at one end by a TRIMMER JOIST that frames an opening in a floor or roof structure.

trimmer. (1) *See* TRIMMER JOIST. (2) (*USA*) A beam or joist to which a HEADER JOIST is fixed.

trimmer arch. A relatively flat brick arch made under the floor between the chimney and TRIMMER JOIST. It usually carries the hearth.

trimmer joist (trimmer, *USA*: header joist). A timber that is fixed at right angles to the common JOISTS of a floor or roof. It is supported by the TRIMMING joists and forms part of the frame around an opening such as for stairs. The trimmer joist may support TRIMMED JOISTS.

trimming joist. A JOIST that is thicker than the common joists and runs parallel to them. Trimming joists form part of the frame around an opening in a floor or roof structure and also support the TRIMMER joist.

Trombe wall. *See* THERMAL STORAGE WALL.

trough gutter. *See* BOX GUTTER.

trowel. A hand tool with a flat metal blade used to spread and finish CONCRETE, MORTAR and PLASTER. *See* MASONRY TOOLS diagram.

trowelled face. The surface of concrete or plaster that has been finished with a TROWEL.

true bearing. A BEARING angle that is measured clockwise from true north to the survey line.

trunking. An enclosed DUCT of metal or plastic that carries service pipes or cables through a building.

trunnel. *See* TRENAIL.

trunnion axis. The horizontal axis about which the telescope of a THEODOLITE can rotate.

truss. A two-dimensional rigid framework used to support loads such as a roof or bridge. The members of the framework usually form a series of triangles. *See* ROOF TRUSS diagram. *See also* FAN TRUSS, FINK TRUSS, SCISSOR TRUSS, TRUSSED RAFTER, WARREN TRUSS.

truss clip. (*USA*) A TOOTHED PLATE used to assemble TRUSSES or other timbers.

trussed beam. A BEAM that is stiffened by an external TIE rod.

trussed rafter. One of a series of TRUSSES used to provide all or most of the RAFTERS for a pitched roof. The trusses are often prefabricated at a factory.

try plane (trying plane). A long form of BENCH PLANE.

try square. An L-shaped tool used to lay out and check right angles.

T-square (tee-square). A T-shaped drawing ruler used on a drawing board to draw horizontal lines and to support SET SQUARES.

TTS. *abbr.* TEMPORARY THRESHOLD SHIFT.

tub. *See* BATH.

tube. A length of steel or alloy pipe used in construction.

tubular lock. A type of CYLINDER lock.

tubular saw. *See* HOLE SAW.

tubular scaffolding. A SCAFFOLD made of metal TUBES connected by standard fittings.

tuck. A groove made in a horizontal MORTAR joint for TUCK POINTING.

tuck pointing. The use of fresh mortar to fill a TUCK groove in a horizontal mortar joint. The line of new mortar may project and form a decorative line.

tufa. A porous rock composed of CALCIUM CARBONATE deposited by spring-water.

tumbler. The mechanism inside a LOCK that holds back the bolt until the key is turned.

tumbler switch. An electrical SWITCH in which a lever rotates the contacts.

tung oil (China wood oil). A DRYING OIL obtained from the leaves and seed of the tropical tung tree. The oil is used as a BINDER in traditional oil paints.

tungsten. A hard metal element (chemical symbol W). It is used in FILAMENT LAMPS and as a steel alloy for cutting tools.

tungsten filament lamp. *See* FILAMENT LAMP.

tungsten-halogen lamp (quartz-halogen lamp). A form of tungsten FILAMENT LAMP that contains a halogen gas, such as bromine or iodine, so that efficiency is increased. The filament is contained within a quartz glass bulb and runs at a higher temperature than a regular tungsten filament lamp.

turbine. A motor in which a shaft is steadily rotated by a current of liquid or gas directed against the blades of a wheel. *See also* PELTON WHEEL, PENSTOCK.

turbulent flow. A form of fluid flow in pipes and channels where the particles move at random in irregular paths and collide with one another. In practical situations most flow is turbulent. *Compare* CRITICAL VELOCITY, LAMINAR FLOW.

turn button. A simple catch in which a screw passes through the door and turns a length of wood or metal.

turning. A cylindrical object, such as a spindle or baluster, that is produced on a LATHE. The process of making such objects.

turnkey contract. *See* PACKAGE DEAL.

turnscrew. *See* SCREWDRIVER.

turn tread. *See* WINDER.

turpentine. A paint SOLVENT made from distilling the OLEO-RESIN found in pine trees. *See also* THINNER.

turpentine substitute. A liquid with similar properties to natural TURPENTINE but lower in cost. WHITE SPIRIT is often used for thinning paint and cleaning brushes.

turret step. One of the steps in a stone SPIRAL STAIR.

tusk. One of the projecting elements in the TOOTHING of a wall.

tusk nailing. *See* SKEW NAILING.

tusk tenon. A small wedged TENON traditionally used to fix a TRIMMER JOIST to a TRIMMING JOIST. *Compare* JOIST HANGER.

twin cable. An electric CABLE that consists of two separate conductors within the same outer insulation.

twist. A timber defect that gives a spiral distortion along the length. *See also* WARP.

twist bit (twist drill). A BIT (or DRILL) with hardened steel helical cutting edges used to drill wood or metal.

twist gimlet. A type of GIMLET with a thread that removes the wood cuttings.

two-coat work. (1) A plaster finish applied in two separate coats. *Compare* THREE-COAT WORK. (2) A paint or other surface coating applied in two separate coats.

two-handed saw. A long SAW that is worked by two people, one on each side of the timber being cut. It was formerly used for felling trees and cutting logs.

two-pack (two-part). *adj.* Describing a material supplied in two separate containers the contents of which are to be mixed together before use. One of the packs usually contains a curing or setting agent. The use of two-pack materials gives a high quality product that must be used promptly after mixing. Examples include EPOXY and ISOCYANATE adhesives and POLYURETHANE paints.

two-pipe system. (1) A system of drainage from a building in which the SOIL output is kept in separate pipes from the waste water. *Compare* ONE-PIPE SYSTEM. (2) A heating circuit in which each radiator is connected to a separate flow pipe and a return pipe. *Compare* ONE-PIPE SYSTEM.

Tyrolean finish. A rough, textured wall finish made with a machine that projects and splatters a coating material such as RENDER or PLASTER.

U

UB. *abbr.* Universal beam (used in structural steelwork).

UBC. *abbr.* (*USA*) UNIFORM BUILDING CODE.

U-bolt. A steel rod shaped like a letter U, with screw threads and nuts on each end. It is used to fasten a round object to a flat surface.

UC. *abbr.* Universal column (used in structural steelwork).

udl. *abbr.* UNIFORMLY DISTRIBUTED LOAD.

UF. *abbr.* UREA FORMALDEHYDE.

U/G. *abbr.* Underground.

U-gauge (water gauge). A U-shaped glass tube that is half-filled with water and temporarily connected to a drain or pipe system. Observation of the water levels shows whether the pipes are air-tight. *Compare* MANOMETER.

UL. *abbr.* (*USA*) Underwriters' Laboratory.

ULOR. *abbr.* Upward light output ratio. *See* LIGHT OUTPUT RATIO.

ultimate bearing capacity. A measure of the maximum load that can be applied to the ground beneath a foundation without significant movement.

ultimate tensile stress (ultimate tensile strength). The maximum tensile STRESS that a material can withstand before it breaks.

ultrasonic. *adj.* Describing sound with frequencies well above the human hearing range. Ultrasound is used for NON-DESTRUCTIVE TESTS by analysing the velocities and reflections of ultrasonic waves within concrete and other materials.

ultra-violet radiation (UV). A form of electromagnetic radiation with wavelengths slightly less than those of blue light. It is a non-visible part of the RADIATION from the sun and can cause the deterioration of materials such as plastics. Ultra-violet radiation can also be generated by DISCHARGE LAMPS. *See also* FLUORESCENT LAMP. *Compare* INFRA-RED RADIATION.

umber. A natural brown PIGMENT made from clays that contain iron oxides.

uncoursed. *adj.* Describing stone walling where the stones are of different sizes and can not be laid in regular courses. *See also* SNECKED RUBBLE.

undercloak. (1) At a FLEXIBLE-METAL ROOFING seam, the part of a lower sheet that is overlapped by the upper sheet (the OVERCLOAK). (2) A course of tiles or slates laid beneath the tiles or slates at EAVES or VERGES. *See also* DOUBLE-EAVES COURSE.

undercoat. (1) A coat of paint or varnish applied after preparation of the surface and before the final coat. The undercoat is used to build up the surface and give opacity. *Compare* PRIMER. *See also* THREE-COAT WORK. (2) A coat of plaster applied before the FINISHING COAT.

undercuring. The effect when a GLUE does not fully harden, often because of low temperature or insufficient time.

undercut. The formation of an overhang

and recess on the underside of a surface, such as a window frame.

undercut tenon. A TENON that has the shoulder cut slightly off square to ensure that the edges bear perfectly on the mortise.

under-eaves course. A course of short tiles or slates laid beneath the final course to give the double thickness at the EAVES. *See also* EAVES COURSE.

underfloor heating. A system of heating a building by electric cables or hot water pipes sunk in the upper screed of a concrete floor.

underlay. A layer of material, such as plastic sheeting, laid beneath a roof or floor covering.

underpinning. The process of supporting a building while the foundations are rebuilt or during adjacent excavations. *See also* DRY MIX, PIN.

under-ridge tile. A short tile in the top course of a roof, just beneath the RIDGE TILES.

undertile. A tile that is shorter than normal so that it can be used beneath a finishing course such as the UNDER-EAVES COURSE. *See also* UNDER-RIDGE TILE.

undressed timber. *See* UNWROUGHT TIMBER.

uneven grain. A type of wood GRAIN in which a difference between spring and summer growth can be seen.

unfixed materials. Building materials that have been delivered to the site but are not yet used.

unframed door. *See* BATTEN DOOR.

ungauged lime plaster. A PLASTER made from sand, LIME and water without the addition of Portland cement or gypsum plaster.

Unibond. Trade name of a type of PVA ADHESIVE.

unified thread. A non-metric standardized series of THREADS for screws and bolts.

Uniform Building Code (UBC). (*USA*) A publication that details standardized requirements for materials and methods used in construction. It is used as the basis of most BUILDING CODES in the USA.

uniformly distributed load (UDL). In STRUCTURAL DESIGN, a load that is spread over the complete length of a beam or area of a slab. *Compare* POINT LOAD.

uniform sky. A type of STANDARD SKY that is taken to have the same LUMINANCE from every point of view. *Compare* CIE SKY.

union. A threaded fitting used to join lengths of pipe.

union bend. A pipe bend with a UNION fitting at one end.

unit heater. SPACE HEATING equipment that uses an electric fan to pass air over the heating elements.

units. *See* SI UNITS.

universal beam (universal column, universal section). A structural steel girder with an I-shaped cross-section and parallel FLANGES. It is produced by HOT ROLLING in a range of standard sizes that are larger than ROLLED-STEEL JOISTS.

universal plane. A hand-operated PLANE that contains more than one PLANE IRON for cutting different grooves, mouldings and shapes.

universal section. *See* UNIVERSAL BEAM.

unplasticized polyvinyl chloride (uPVC). A relatively stiff form of POLYVINYL CHLORIDE used for drain pipes.

unsound knot (rotten knot). A KNOT that is softer than the surrounding wood. *Compare* SOUND KNOT.

unsoundness. *See* SOUNDNESS.

untrimmed floor. A floor carried completely on COMMON JOISTS, without TRIMMED JOISTS for fireplaces or staircases.

unwrought timber. Timber that has been sawn but not planed.

up-and-over door. *See* OVERHEAD DOOR.

UPC. *abbr.* (*USA*) Uniform Plumbing Code.

upflush. A pumped drainage system that lifts waste from appliances in a building that are below the sewer level.

uplift. An upward force on earth, such as in a dam, caused by the entry of water under pressure.

uplighter (uplight). A light fitting or LUMINAIRE that is designed to direct most of its light towards the ceiling.

upset. A fracture in wood that runs across the grain.

upside-down roof. *See* INVERTED ROOF.

upstand. (1) The edge of a roof covering, such as FELT or FLEXIBLE METAL, that is turned up against a vertical surface. The upstand is then protected by a FLASHING. (2) The edge of a beam or other object that projects through a floor or flat roof.

uPVC. *abbr.* UNPLASTICIZED POLYVINYL CHLORIDE.

urea formaldehyde (UF). A SYNTHETIC RESIN with THERMOSETTING properties. It is used as an adhesive and, when foamed, as a thermal insulator.

urea resin. A type of SYNTHETIC RESIN with THERMOSETTING PROPERTIES. Urea resins are used in the manufacture of durable ENAMELS.

urethane. *See* POLYURETHANE.

urinal. A SANITARY APPLIANCE designed to receive and flush away urine. It may take the form of a continuous slab or a bowl on the wall.

utile. A HARDWOOD timber that may be used as a substitute for TEAK.

utilities. (*USA*) The equipment in a building concerned with the supplies of electricity, heat, mechanical ventilation, water, and gas. Communications, lifts, fire detection and security systems may also be included. *See also* SERVICES.

utility knife. A general-purpose knife with a replaceable razor-like blade.

utilization factor (utilization coefficient). In lighting calculations such as the LUMEN METHOD, an allowance for the loss of light caused by the lamp fitting and the room surfaces.

UTS. *abbr.* ULTIMATE TENSILE STRESS.

U-tube. *See* U-GAUGE.

UV. *abbr.* ULTRA-VIOLET RADIATION.

U-value (air-to-air transmission coefficient, thermal transmittance, transmittance). A measure of the overall rate at which heat is transmitted through a particular type of wall, roof or floor. The U-value is measured air-to-air and includes allowances for the THERMAL RESISTANCES of surfaces and cavities. U-values are often used by BUILDING CODES to define the THERMAL INSULATION required for the shell of a building.

V

V. *abbr.* VOLT.

vacuum breaker. *See* AIR-ADMITTANCE VALVE.

vacuum concrete. CONCRETE poured into FORMWORK that is faced with a screen of metal and fabric. A partial vacuum applied to a cavity behind the screen removes excess air and water and produces a dense concrete with early strength.

vacuum forming. A method of forming objects from sheet materials, such as plastics, by evacuating the space between the sheet and the mould. Atmospheric pressure then presses the sheet into shape.

vacuum return system. (*USA*) A steam heating circuit that uses a vacuum pump to help the steam circulate through the pipes.

valley. The inside angle at the intersection of two sloping roofs. *See* ROOF diagram. *Compare* HIP.

valley board. A board used to support the tiles in the corner of a LACED VALLEY or SWEPT VALLEY.

valley gutter. A GUTTER fixed under the edge of the tiles or slates in the VALLEY of a roof. *Compare* SWEPT VALLEY.

valley jack. (*USA*) A JACK RAFTER fixed between the VALLEY RAFTER and the ridge of a roof.

valley rafter. The RAFTER that runs beneath the line of a roof VALLEY and to which the JACK RAFTERS are attached. *See also* VALLEY BOARD.

valley tile. A large TILE shaped to turn the corner of a roof VALLEY without the need for a SWEPT VALLEY or VALLEY GUTTER. *See also* TILED VALLEY.

valve. A device that controls or cuts off the flow of liquid or gas in a pipe. *See also* COCK, STOP COCK.

vanadium. A pure metal element that may be added to ALLOYS for hardness.

vapor. (*USA*) *See* VAPOUR.

vaporization. The change of state from liquid to gas when LATENT HEAT of vaporization is supplied.

vapour (*USA*: vapor). A substance that is in the gas state and can also be liquified by compression. *See also* CRITICAL TEMPERATURE.

vapour barrier. A layer of building material that has a high resistance to the passage of WATER VAPOUR. Examples include aluminium-foil board, polythene sheet and bituminous solutions. Vapour barriers are important in the prevention of CONDENSATION. *See* WALL diagram.

vapour check. A part of a construction, such as a wall, that includes a VAPOUR BARRIER.

vapour diffusivity. A measure of the rate at which water vapour passes through a material for particular conditions of VAPOUR PRESSURE.

vapour pressure. The pressure exerted by the molecules of a VAPOUR. It is a useful measure of HUMIDITY for some purposes. *See also* SATURATED VAPOUR PRESSURE.

vapour resistance. A measure of the opposition to the passage of water VAPOUR provided by a particular thickness of material.

vapour resistivity. A measure of the opposition to the passage of water VAPOUR provided by a particular material, without taking account of its thickness. *Compare* VAPOUR RESISTANCE.

variance. (1) In statistics, a measure of DISPERSION calculated as the mean of the deviations of all observed values. (2) (*USA*) Permission to differ from a requirement of a BUILDING CODE or zoning law.

variation order. Written permission for a contractor to vary the details shown in the contract documents.

varnish. A transparent coating material used to give a glossy protective film. A varnish can be made like a PAINT without pigment. *Compare* LACQUER.

vault. (1) An arched roof or ceiling. (2) A strong secure room used to store valuables.

V-B consistometer (VeBe consistometer). A British Standard test for the WORKABILITY of fresh concrete. A weighted disc is lowered on to a sample of vibrated concrete and the time taken to cover the disc is recorded. *Compare* SLUMP TEST.

VDU. *abbr.* Visual display unit. *See* CAD.

VeBe consistometer. *See* V-B CONSISTOMETER.

vector/scalar ratio. A measure of the DIRECT LIGHT at a particular point. A high value of the ratio corresponds to lighting with strong directional qualities, such as from a SPOTLAMP.

vee joint. A small CHAMFER on each edge of MATCHED BOARDS so that a V-shaped depression is made at the join of the boards.

vegetable glue. A type of GLUE based on vegetable ingredients such as starch or protein.

vehicle. A general term for the liquid BINDER and THINNER in paint. This vehicle 'carries' the PIGMENT. *Compare* MEDIUM.

veiling reflection. A reflection of light from a surface that makes it difficult to see details on the surface, such as printing on shiny paper.

Velux. Trade name for a form of ROOF LIGHT.

veneer. (1) A thin uniform layer of wood that is used as a facing on another wood or in PLYWOOD. *See also* INLAY, ROTARY CUTTING. (2) A relatively thin layer of stone or other material used as a facing to hide a less attractive material.

veneered wall. (*USA*) A wall with a brick or stone facing that is attached to the structural wall. *Compare* CAVITY WALL.

Venetian blind. A type of window BLIND made up of horizontal SLATS whose angle can be adjusted to deflect light.

Venetian red. A reddish PIGMENT originally based on IRON OXIDE.

vent. (1) An outlet that allows air or water vapour to escape from a building or part of a structure such as roof. (2) An outlet from a pipe system.

ventilating brick. *See* AIR BRICK.

ventilation. The supply and circulation of air in enclosed spaces such as a rooms or buildings. *See also* AIR CHANGE, INFILTRATION, NATURAL VENTILATION.

ventilation pipe. *See* VENT PIPE.

ventilator. An inlet or other device that aids VENTILATION.

venting. The process of releasing liquid or gas from an opening. *See also* FIRE VENTING.

vent pipe (vent stack, ventilation pipe). A pipe on a SOIL STACK that extends through the roof and is left open as a VENT to release gases and to keep equal pressures on each side of any water TRAPS. *See also* BALLOON.

Venturi effect. A reduction in pressure that

accompanies an increase in the velocity of a liquid or gas. Examples of the effect include the tendency of a roof deck to lift when wind blows across it.

Venturi meter. A device that measures the FLOW RATE in a pipe by comparing pressures at different diameters. *See also* BERNOULLI'S PRINCIPLE, VENTURI EFFECT.

verdigris. A protective film or PATINA that slowly forms on the surface of COPPER exposed to the air.

verge. The edge or a roof at a GABLE. *See* ROOF diagram. *Compare* EAVES.

verge board. *See* BARGE BOARD.

verge fillet. A length of BATTEN fixed on a GABLE WALL as a neat finish to the roof battens.

verge tile. An extra-wide PLAIN TILE that is used for the VERGE of a roof. *See also* TILE-AND-A-HALF-TILE.

vermiculated. *adj.* Describing a pattern of random grooves used to decorate ASHLAR stone.

vermiculite. A natural mineral of the MICA group. *See* EXFOLIATED VERMICULITE.

vermillion. A bright red PIGMENT based on compounds of mercury.

vernier. A device that measures subdivisions of the scale on an instrument such as a THEODOLITE. It uses a small movable scale that runs parallel to the main scale.

vertical grain. *See* EDGE GRAIN.

vestibule. A space just inside the entrance to a building. *Compare* LOBBY.

vibrating roller. A roller that produces COMPACTION of soil by a mechanical vibration as well as weight.

vibrator. A vibrating mechanical tool used to compact fresh concrete. It may be inserted into the wet concrete (a POKER VIBRATOR) or applied to the FORMWORK.

Vicat apparatus. Equipment that tests the INITIAL SETTING time of concrete by the application of a needle to a sample.

vice. A tool with adjustable clamps used to hold an object on a workbench while it is worked on.

vinyl. A HYDROCARBON RADICAL that readily undergoes POLYMERIZATION and is used to make various POLYMERS. *See also* POLYVINYL CHLORIDE.

vinyl flooring. A sheet or tile of flexible flooring material based on POLYVINYL CHLORIDE.

vinyl resin. One of a large group of SYNTHETIC RESINS with THERMOPLASTIC properties. They are used for plastic products and as the basis of most EMULSION PAINTS.

viscosity. The natural resistance of a liquid or gas to flow. For example, honey has a higher viscosity than water. *See also* ABSOLUTE VISCOSITY.

visible radiation. Electromagnetic radiation that can be detected by human vision. *See also* INFRA-RED RADIATION, ULTRAVIOLET RADIATION.

visual field. The total extent of space that can be seen by the human eyes when looking in a fixed direction.

vitreous (vitrified). *adj.* Describing the hard, glassy and waterproof properties of CERAMIC materials, such as china, produced by exposure to high temperature. The surface of some clayware is vitrified by the addition of GLAZE.

vitreous enamel. *See* ENAMEL.

vitrified. *See* VITREOUS.

vitrified clayware. Clay drain pipes and other fittings that have been made VITREOUS by firing at high temperatures. *Compare* SALT GLAZE.

voids. The small spaces between separate particles of soil, sand or gravel. The voids may be occupied by air or water and their volume may be found from a 'voids ratio' that compares the total space to the total solid material in a sample.

voids ratio. In a sample of soil or aggregate, the ratio of the volume of voids to the volume of solids.

volt (V). The unit used to measure electric potential. *See* ELECTROMOTIVE FORCE, POTENTIAL DIFFERENCE.

voltage. A value of ELECTROMOTIVE FORCE or POTENTIAL DIFFERENCE expressed in VOLTS.

voltage drop. A difference in POTENTIAL DIFFERENCE between two point in an electric circuit. *See also* LINE DROP.

voltaic cell. *See* SIMPLE CELL.

voltmeter. An electrical instrument that measures POTENTIAL DIFFERENCE in VOLTS.

volume yield. The volume of CONCRETE of a particular mix produced from a specified weight of CEMENT.

vortex. A spiralling motion of liquid or gas, such as may occur with waste water in a vertical soil pipe.

voussoir. *See* ARCH BRICK.

W

W. (1) *abbr.* WATT. (2) Chemical symbol for TUNGSTEN.

wadding. The use of hemp and plaster to fix PLASTERBOARD to joists.

waffle floor. A reinforced concrete floor slab with deep recesses on the underside.

wainscot (wainscoting). An interior wall covering of wooden boards or panels usually attached to the bottom portion of a wall.

waisted bolt. A type of BOLT where the diameter of the shank is made less than the diameter of the thread.

Waldram diagram. A specially-marked grid diagram of the sky that can be used, with the outline of the windows, to predict DAYLIGHT FACTORS. *Compare* DAYLIGHT FACTOR PROTRACTOR.

wale. *See* WALING.

waling (wale). (1) A horizontal member used for the support of timbers in an excavation or of concrete FORMWORK. *See also* POLING BOARDS. (2) *See* LEDGE.

walking line. A line fixed at constant distance from the HANDRAIL and used to set out the STAIRS.

wall. A vertical structure that forms one of the sides of a building, or divides an internal space. *See* WALL diagram. *See also* CAVITY WALL, CURTAIN WALL, FACING WALL, HONEYCOMB WALL, LOAD-BEARING, SELF-SUPPORTING WALL, SLEEPER WALL, VENEER WALL.

wall anchor (joist anchor). A steel strap built into the brickwork of a wall and supporting a common JOIST. The anchor helps the joist give lateral support to the wall.

wall board. Panels of material used to line the interior surfaces of walls and ceilings. GYPSUM PLASTERBOARD is commonly used, also sheets of PLYWOOD and LAMINATED PLASTIC.

wall column. A load-bearing COLUMN contained within a wall.

wall hanger. *See* JOIST HANGER.

wall hook. A spike or nail driven into the mortar joint of a masonry wall and used to carry a pipe or timber.

wall panel. A section of non-load-bearing wall used to fill between the frames of a PANEL WALL. The wall panel is often prefabricated of concrete or timber.

wallpaper. A strong paper pasted on walls and ceilings as a decorative surface. It can be used as a LINING PAPER to cover defects in a surface. *See also* PAPERHANGING, ROLLER.

wall plate (top plate, raising plate). (1) A horizontal timber along the top of a wall on which the RAFTERS or JOISTS rest. (2) A vertical timber on which a RAKING SHORE bears.

wall plug. *See* PLUG.

wall string. The STRING on the side of a stair that is next to the wall. *See* STAIRS diagram. *Compare* OUTER STRING.

wall tie. A non-corrosive metal strap or wire bedded into the mortar joints of masonry walls to tie components together. The two leafs of a CAVITY WALL are held together by wall ties that have a twist to discourage the passage of water. *See* WALL diagram. *See also* BUTTERFLY TIE.

wall tile. A tile, usually made of CERAMIC or plastic material, that is set on a wall to give a decorative and waterproof surface in a kitchen or bathroom

wane (*adj.* waney). The presence of BARK or its undersurface on the edge of a piece of sawn timber.

warm air heating. A form of SPACE HEATING in which heated air is blown into the room. The air may be heated by a local heater or it may be distributed by ducts from a central heater.

warm roof. In general, a roof where the layer of THERMAL INSULATION is installed near the outside rather than near the ceiling. The air in the roof space is therefore similar to the warmer temperatures of the building interior and the roof is less likely to suffer from CONDENSATION than a COLD ROOF. *See also* INVERTED ROOF.

warning pipe. An overflow pipe from a water storage tank such as those in a roof. The pipe usually protrudes through a wall or roof and simply directs water away from the building. It is intended to draw attention to a problem rather than act as a true OVERFLOW PIPE.

warp (spring). A timber defect that shows as a distortion of straight or flat surfaces, such as by a twist. Warping in timber is caused by changes in MOISTURE CONTENT.

Warren truss (Warren girder). A TRUSS made up of horizontal top and bottom members connected by sloping members without verticals.

wash. (1) (*USA*) An upper surface that is sloped to shed water. *See also* WEATHERING. (2) *See* WASH COAT.

washbasin (basin, lavatory basin). A common SANITARY APPLIANCE used to wash hands and face. *See also* PEDESTAL WASHBASIN.

wash coat. A thin coat of paint or PRIMER used as preparation for other treatments.

washdown closet. A common type of WC in which the pan is cleared by the flush of water, without the help of suction as in the SIPHONIC CLOSET.

washer. (1) A flat metal ring that is placed under a bolt head or nut. When tightened, the washer spreads the load and helps the grip. (2) A flat ring or disc made of rubber or plastic material, used to make a watertight joint, such as inside a TAP OR FAUCET. *See also* JUMPER.

waste. (1) (waste water) Water discharged into the drains from a bath, basin, sink, shower or bidet, but not from a WC. *See also* SULLAGE. (2) Building rubbish. (3) An allowance made in quantities for material that must be ordered but can not be used.

waste pipe. A pipe that carries WASTE water away from a SANITARY APPLIANCE, such as a washbasin, to a drain or SOIL STACK. *See also* TRAP.

waste water. *See* WASTE.

water bar (weather bar). (1) A flat steel bar formerly used to make a seal between a stone or concrete SILL and a wooden sill. It is bedded, on edge, into grooves in the sills. (2) A strip of flexible waterproof material that is set in construction joints and expansion joints.

water bond. A type of brickwork BOND sometimes used for MANHOLES. Within the one-brick thickness of the wall, the courses are laid at split levels that helps make the bond waterproof.

water-cement ratio. In a mix of CONCRETE or MORTAR, the total weight of water present divided by the weight of cement. A low ratio indicates a strong or 'rich' mix. *See also* WORKABILITY.

water channel. *See* CONDENSATION GROOVE.

water closet. More commonly known by its abbreviation, WC.

water gauge. *See* U-GAUGE.

water hammer. A shock wave, and noise,

that travels through water pipes when the flow is suddenly interrupted. Causes of water hammer include the rapid operation of a tap or pump.

water-jetting. A method of driving PILES using a pressurized jet of water to displace the earth at the point of the pile.

water joint. *See* SADDLE JOINT.

water main. *See* MAIN.

water of crystallization. A fixed proportion of water that is chemically combined with certain substances, such as GYPSUM, when they are crystals.

water paint. Any PAINT in which the volatile part of the VEHICLE is water. The paint can be thinned by water. *See also* EMULSION PAINT.

waterproofing. A process or coating that makes a surface or structure impervious to liquid water. Layers based on bitumen, cement, rubber and plastic are commonly used. *See also* INTEGRAL WATERPROOFING.

water-repellent. *adj.* Describing a material or surface treatment that resists the passage of water for a time but is not totally WATERPROOF.

water seal. Water that remains in a TRAP and prevents air or gases passing through it. *See also* VENT PIPE.

water seasoning. A method of SEASONING timber by soaking it in water and then air drying it.

watershed. (1) A boundary, such as a mountain range, between CATCHMENT AREAS. (2) (*USA*) *See* CATCHMENT AREA.

water softening. The general process of making SOFT WATER by removing or chemically changing the dissolved salts that cause HARD WATER. *See* BASE EXCHANGE, SODA-LIME PROCESS. Water softening is not usually part of the treatment of public water supplies as hard water tends to be good for health although it is often undesirable for industrial purposes.

water spotting. A defect on a PAINT film seen as pale spots caused by water drops.

water stain. (1) A type of decorative STAIN made of colouring matter dissolved in water. (2) A discolouration caused when timber is wetted.

water table (plane of saturation, saturation line). The level of ground below which there is GROUNDWATER. *See also* ARTESIAN WELL, ZONE OF AERATION, ZONE OF SATURATION.

water test. *See* HYDRAULIC TEST.

watt (W). The SI unit of POWER. 1 watt = 1 joule per second. *Compare* HORSEPOWER, KILOWATT HOUR.

wavy grain. A wood GRAIN with a curved appearance. It is often seen in wood from birch and mahogany.

WC. *abbr.* Water closet. A SANITARY APPLIANCE designed to receive and flush away urine and solid waste from the human body. It may be installed in a bathroom or in a separate room called a toilet or lavatory. *See also* FLUSHING CISTERN, SIPHONIC CLOSET, SQUATTING CLOSET, WASHDOWN CLOSET.

wearing course. The top surface layer of a roadway on which traffic runs. *See also* FINE COLD ASPHALT, MASTIC ASPHALT.

weather. The distance along the roof slope by which one SHINGLE overlaps the one below.

weather bar. (1) A device that seals the edges of a CASEMENT window against water and draughts. (2) *See* WATER BAR.

weatherboard. (1) A sloped moulding that is fixed to the bottom of a door so as to throw rainwater clear of the threshold. (2) A board used in WEATHERBOARDING.

weatherboarding (bungalow siding, siding). A form of external timber CLADDING where horizontal boards are nailed to the building. Each board overlaps the board beneath to allow the rain to run off. In shiplap boarding the upper board has a rebate to receive the lower board. CLAPBOARD has no rebate. *See* WALL diagram.

weathering. (1) A change in the colour or texture of a material after exposure to rain and sunlight. (2) An upper surface that is sloped to shed water.

weather joint. *See* WEATHER-STRUCK JOINT.

weather slating. *See* SLATE HANGING.

weatherstrip (weatherstripping). Narrow strips of metal, plastic and other materials used around the edges of door and windows to prevent the passage of air or water through the joints.

weather-struck joint (struck joint, weather joint). A mortar joint sloped by the trowel so that water is led outward. *See* BRICKWORK diagram. *See also* POINTING.

web. (1) The vertical connecting part of a ROLLED-STEEL JOIST or UNIVERSAL BEAM. *Compare* FLANGE. (2) In a TRUSS, the structural members that connect the top and bottom CHORDS. (3) The cross members of a hollow block or wall.

web stiffener. *See* STIFFENER.

wedge. A tapered piece of wood or metal that tightens a joint when forced into a gap.

weep hole. One of the small openings at the base of a masonry wall or RETAINING WALL that allow water to escape from behind the structure.

weight box (weight pocket). The space in the casing of a SASH WINDOW where the sash weights travel.

weighting network. In a SOUND LEVEL METER, an electronic filter that responds to frequency in a specified manner. The A-SCALE is the most used weighting.

weight pocket. *See* WEIGHT BOX.

weir. A low-level dam over which a river flows. It may be used to control and measure the flowrate. *See* HYDRAULIC JUMP.

weld. A method of joining pieces of metal or plastic by melting the materials at the edges. *See also* SPOT WELDING.

well. (1) A hole sunk in the ground to obtain water from underground supplies. *See also* ARTESIAN WELL, DEEP WELL, SHALLOW WELL. (2) A vertical shaft through a building that encloses a stair or lift. *See also* LIGHT WELL.

Welsh arch. A small FLAT ARCH built into a masonry wall.

welt. A SEAM in FLEXIBLE-METAL roofing.

welted drip. ROOFING FELT that is folded at the EAVES of a roof to make a DRIP for rain water.

welting strip. *See* TINGLE.

wetback. A type of HEAT EXCHANGER that is fitted behind a fireplace or stove and usually connected to the hot water supply.

wet-bulb thermometer. A thermometer that has its bulb surrounded by a permanently damp cloth. Evaporation of moisture from the cloth cools the wet bulb to an extent that depends on the dryness of the air. *Compare* DRY BULB. *See also* WHIRLING SLING.

wet-on-wet. A system of special PAINTS that can be applied while the first coat is still wet.

wet rot. A decay in timber structures caused by a FUNGUS that grows in alternating wet and dry conditions. *Coniophora cerebella* and *Coniophora puteana* are common forms of fungus that give wet rot. *Compare* DRY ROT.

wetting agent. A substance, such as detergent, that allows a liquid to spread more readily by its surface tension.

wet trades. The techniques of building and the TRADES that use CONCRETE, MORTAR or PLASTER.

wheel barrow. A simple vehicle, pushed by one person, used to move small loads on a building site. It usually has a single wheel at the front and two legs at the rear. *See also* BARROW RUN.

wheeling step. *See* WINDER.

whet. *vb.* To sharpen and finish the cutting edge of a tool by rubbing it on a HONE.

whetstone. *See* HONE.

whirling sling. A popular hand-held form of WET-AND-DRY BULB HYGROMETER. The DRY-BULB THERMOMETER and WET-BULB THERMOMETER are rotated in the air to get a rapid measurement of HUMIDITY.

white cement. A type of PORTLAND CEMENT with a low proportion of iron compounds and others that give the grey-green colour to ordinary Portland cement. A white pigment may also be added.

white lead. An opaque white PIGMENT based on compounds of LEAD. Its use in paints is now limited by its possible harm to health.

white lime. *See* HIGH-CALCIUM LIME.

white spirit (mineral spirit, mineral turpentine). A colourless liquid obtained from petroleum that is commonly used as a cheaper substitute for natural TURPENTINE.

whitewash. Various forms of white coating based on water and a cheap white pigment such as LIMEWASH or WHITING.

whitewood. A soft, general-purpose construction timber from various trees such as PINE and SILVER FIR.

whiting. A white pigment made from crushed CHALK. It is used as an EXTENDER in paints and as the main ingredient of GLAZIER'S PUTTY.

whole-brick wall. A wall with a thickness equal to the length of one brick. *Compare* CAVITY WALL, HALF-BRICK WALL.

whole circle bearing. A BEARING that quotes an angle as measured clockwise from true north.

WI. *abbr.* Wrought iron.

wide-grained timber. A timber in which the ANNUAL RINGS are spaced wide apart to give a coarse grain.

wiggle nail. *See* CORRUGATED FASTENER.

wind (winding). A WARP in timber where the timber twists about its long axis.

wind beam (wind brace). A reinforcing beam inserted in a structure to help resist WIND LOAD.

winder (wheeling step). A tapered TREAD where a stair changes direction. *See* STAIRS diagram. *See also* KITE WINDER. *Compare* DANCING STEP, FLIER.

winding. *See* WIND.

wind load. A force on the side of a structure caused by wind pressure.

window. An opening in a wall that is usually GLAZED with glass to allow light into the building. A CASEMENT in the window may open to allow ventilation. *See* WINDOW diagram. *See also* BAY WINDOW, CLERESTORY, FENESTRATION, JALOUSIE, LIGHT, MULLION, SASH WINDOW, SILL, TRANSOM.

window bar. *See* GLAZING BAR.

window board (sill board). A horizontal board fixed at sill level on the inside of a WINDOW.

window frame. The outer part of a window that surrounds the CASEMENTS or SASHES.

window furniture. Various fittings used to move and secure an opening part of a window. *See also* CASEMENT STAY, SASH FASTENER.

window ledge, window sill. *See* SILL.

windowstool. (*USA*) A horizontal board fixed at sill level on the inside of a WINDOW. *See also* WINDOW BOARD.

wing. A secondary section of a building that extends from the main building.

wing nut. A type of NUT with two finger

tags attached to that the nut can be tightened without a spanner.

wiped joint. A SOLDER joint between pipes made with the help of a special cloth pad to distribute the soft solder around the joint.

wiping cloth. *See* WIPED JOINT.

wirecomb. *See* DRAG.

wirecut brick. A brick shaped by the extrusion of a long bar of clay that is then cut to length by a set of wires. *Compare* PRESSED BRICK.

wired glass. *See* WIRE GLASS.

wire frame. In computer-aided design (*see* CAD), an image on screen formed by a series of lines that define all the edges of an object. This can give a three-dimensional effect, especially if the HIDDEN LINES are removed.

wire gauge. A system of numbers used to specify the diameter of wire. *See also* STANDARD WIRE GAUGE.

wire glass (Georgian glass, wired glass). A type of SAFETY GLASS that contains a wire mesh to keep the glass together if it is broken.

wire nail. A NAIL made by cutting and shaping a length of steel wire.

wire tie. Lengths of wire used to secure REINFORCEMENT steel.

withdrawal load (pull-out strength). The load that can be applied to a fixing such as a NAIL before it pulls out.

withe (wythe). A leaf of brickwork such as a HALF-BRICK WALL, MID-FEATHER or TIER.

withy (osier). A flexible stick that is twisted together with others and used to tie down THATCH on a roof.

wood block floor. A decorative pattern of flat wood blocks that are fixed on to a subfloor. *See also* PARQUET.

woodchip wallpaper. A thick form of WALLPAPER that covers small cracks and is used as a base for painting.

wood flour. A fine type of sawdust that may be used as an EXTENDER in GLUES.

wood preservative. *See* PRESERVATIVE.

wood roll. A round-topped length of wood used to make a join in FLEXIBLE-METAL roofing. *See also* SOLID ROLL.

wood slip. *See* FIXING FILLET.

woodwool slab. A board made from wood shavings bound together by cement. It provides reasonable thermal INSULATION and been used for wall linings and roof deckings.

woodworking tools. *See* WOODWORKING TOOLS diagram.

woodworm. *See* FURNITURE BEETLE.

workability. The ease with which a batch of concrete or mortar can be handled and finished. Factors that affect workability include the WATER-CEMENT RATIO and the type of AGGREGATE. It can be measured by the SLUMP TEST or the V-B CONSISTOMETER.

working. The movement of timber as it swells or shrinks with changes in the humidity of the surrounding air. *See also* MOISTURE CONTENT.

working drawing (production drawing). An up-to-date scale DRAWING from which measurements are taken during construction. *Compare* DETAIL DRAWING.

working plane. In lighting calculations, the height above the floor at which the ILLUMINATION is calculated.

working stress. *See* PERMISSIBLE LOAD, STRESS.

working up. *See* SQUARING UP.

work study. The examination of the time taken and methods used for a particular job.

worktop. A surface in kitchen used for the preparation of food. It is often made of LAMINATED PLASTIC.

worm hole. A hole left in timber by the action of BEETLES or other insects.

woven fencing. *See* INTERLACED FENCING.

wreath. On the bend of a STAIR, the part of the HANDRAIL that curves and rises at the same time.

wrecking bar (pinch bar). A steel bar used as a tool or LEVER. One end is usually hooked and has a claw while the other end is flattened. *See also* CROW BAR and JEMMY.

wrench. A large type of spanner that can usually be adjusted. *See also* FOOTPRINTS, PIPE WRENCH.

wrinkle. (1) A defect in VENEER where is in not glued. (2) A surface defect in a coat of paint.

wrinkle finish. *See* RIPPLE FINISH.

wrot. *See* WROUGHT TIMBER.

wrought iron. A pure form of IRON with a low carbon content. It is very MALLEABLE and can be used for chains, hooks and decorative ironwork. *Compare* STEEL.

wrought timber (wrot). Timber that has been planed on one or more sides. *See also* ABATEMENT.

W-truss. *See* FINK TRUSS.

WWPA. *abbr.* (*USA*) Western Wood Products Association.

wye branch (Y-branch). A plumbing or drainage pipe with a branch leading off at 45°.

wythe. *See* WITHE.

X

xylem. The plant tissue that forms the WOOD of trees.

xylol. A SOLVENT obtained from tar.

Y

Yale lock. A common brand of CYLINDER LOCK.

Yankee gutter. (*USA* *See* ARRIS GUTTER.

yard. An area of ground enclosed by buildings or fences. It may be used for the storage of building materials.

yard gulley. *See* GULLEY.

yard lumber. (*USA*) LUMBER intended for light framing and finish work.

yard trap. *See* GULLEY.

Y-branch. *See* WYE BRANCH.

year ring. *See* ANNUAL RING.

YG. *abbr.* Yard gulley. *See* GULLEY.

yield point (yield stress). The STRESS at which a component under tensile load begins to elongate without increase in load.

yoke. Timber or steel members fixed around the FORMWORK for a column or beam.

Yorkshire bond. *See* MONK BOND.

Young's modulus. *See* ELASTIC MODULUS.

Y-value. *See* ADMITTANCE.

Z

zax (sax). A bladed roofing tool with a point for punching holes in SLATES.

Z-bar (USA: zee-bar). A length of metal channel or FLASHING used at the join of wall or ceiling panels. The cross-section makes a Z-shape.

zenith. The highest point, such as when the sun is directly overhead.

zeolite. Materials used in ION EXCHANGE and the BASE EXCHANGE method of water softening. Natural zeolites are obtained from processed sands and synthetic zeolites are made from organic resins such as those of POLYSTYRENE.

zigzag rule. A metal or wooden measuring rule with pivoted sections that fold together.

zinc. A hard, bluish-white metal element (chemical symbol Zn) with a good resistance to corrosion. *See* GALVANIZING. Various ALLOYS of zinc are used in FLEXIBLE-METAL roofing.

zinc chrome. A yellow PIGMENT based on compounds of ZINC and chrome that is used in paints.

zinc oxide (zinc white). A fine white powder used as a PIGMENT in paints.

Zn. Chemical symbol for ZINC.

zone controls. A control system for central heating or air conditioning that can maintain different conditions in different areas.

zone of aeration. The ground that is above the WATER TABLE and therefore contains air rather than water.

zone of saturation. Ground that is below the WATER TABLE and therefore contains GROUNDWATER.

zoning. Planning schemes and regulations that control the type and use of buildings in various parts of a town.

Diagrams

BRICKWORK

Bonding

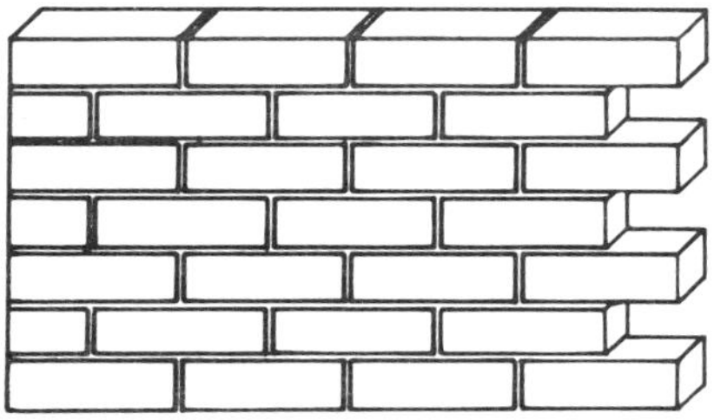

Stretcher bond

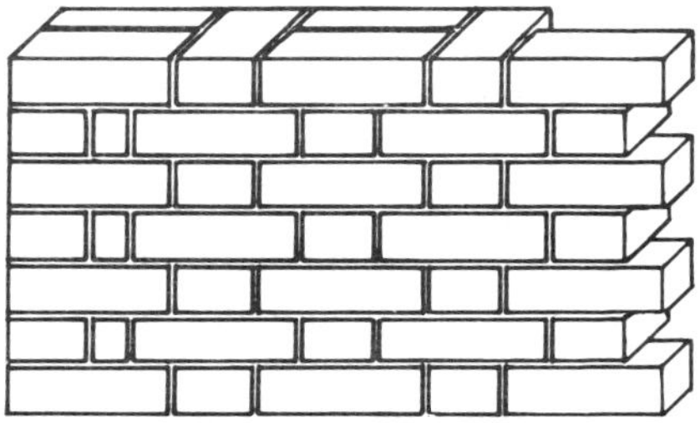

Flemish bond

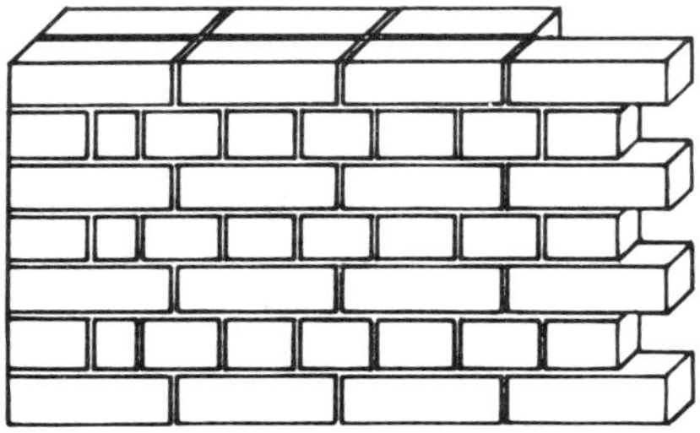

English bond

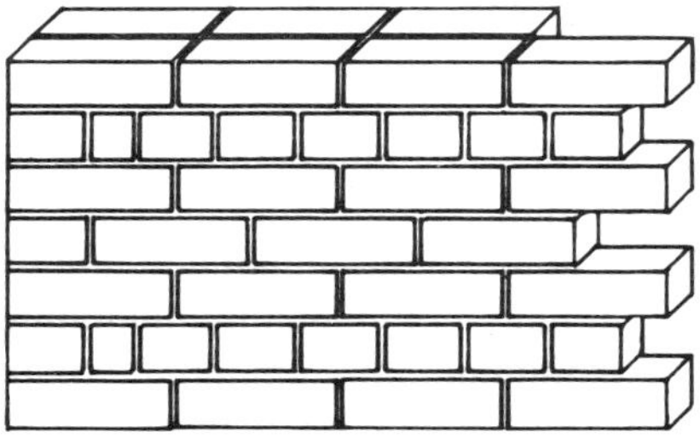

English garden wall bond

Mortar joints

Flush

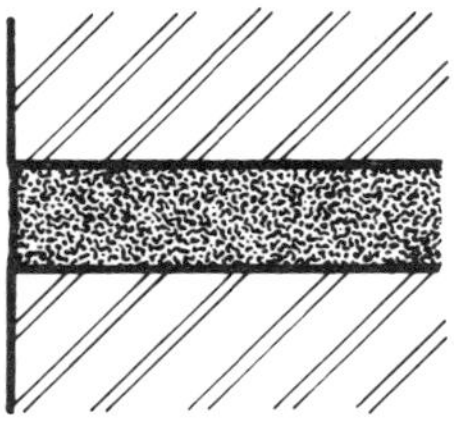

Recessed

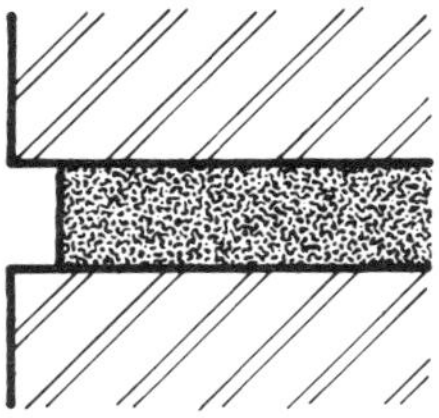

Keyed

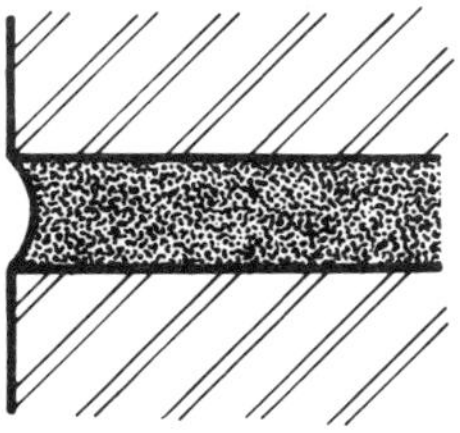

Weather-struck

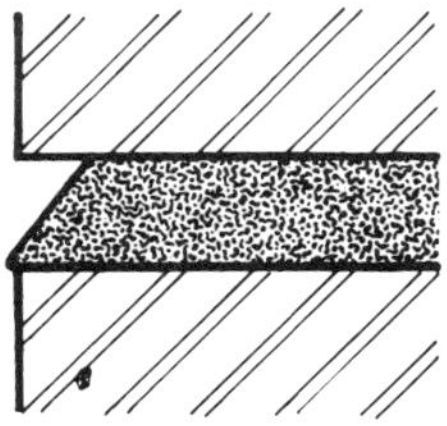

DOOR

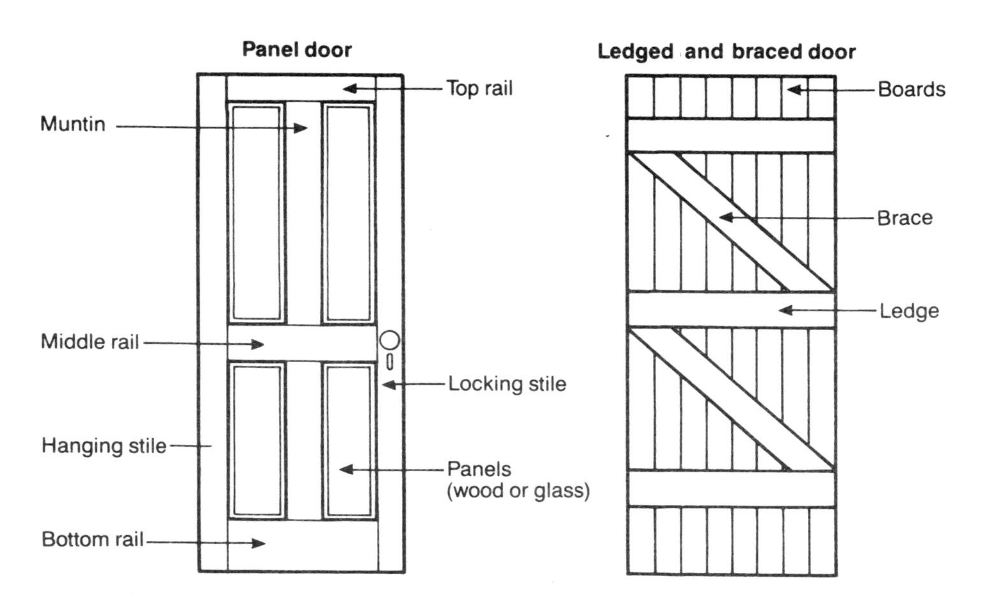
Panel door
Top rail
Muntin
Middle rail
Locking stile
Hanging stile
Panels
(wood or glass)
Bottom rail
Ledged and braced door
Boards
Brace
Ledge

Hinges

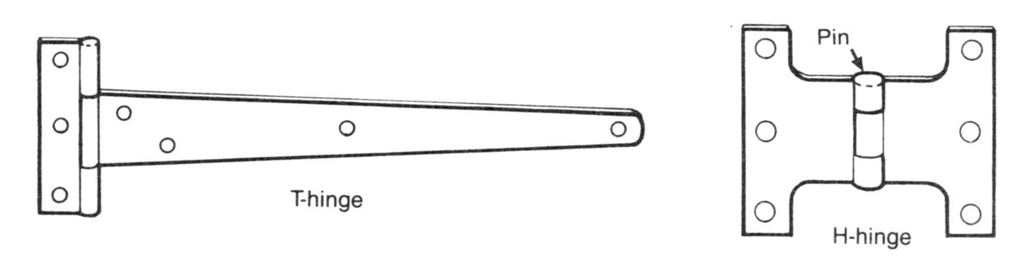
T-hinge
Pin
H-hinge

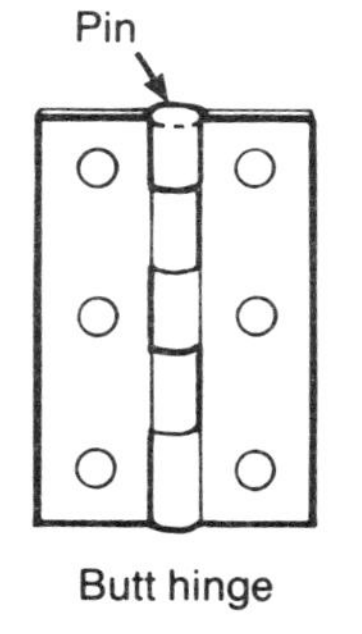
Pin
Butt hinge

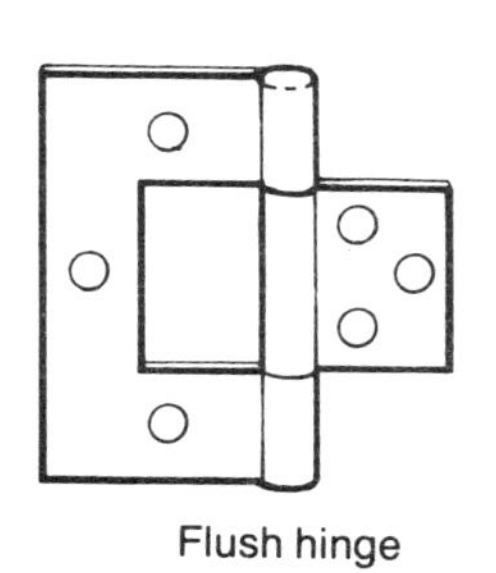
Flush hinge

HEATING

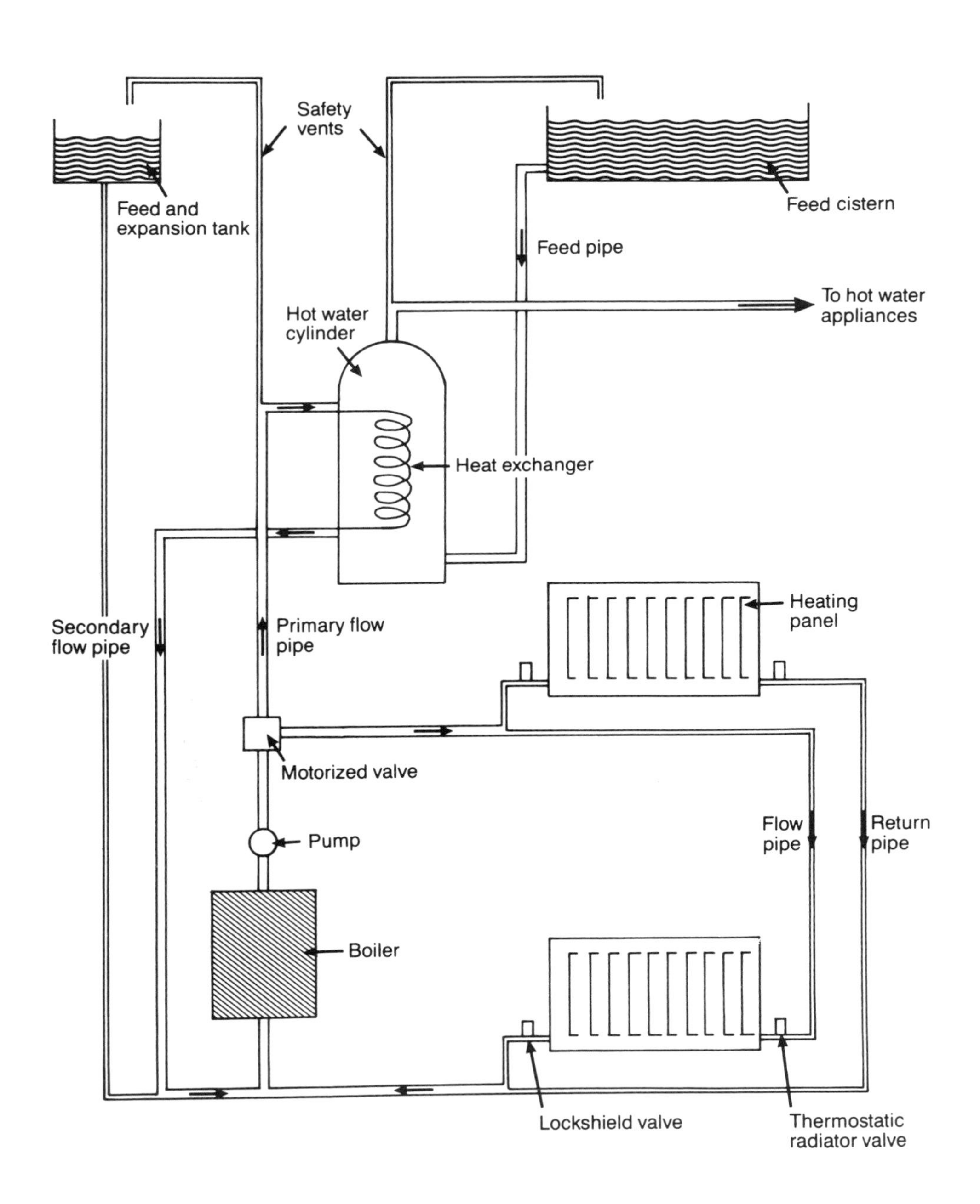

Vented hot water and heating system (schematic diagram)

MASONRY TOOLS

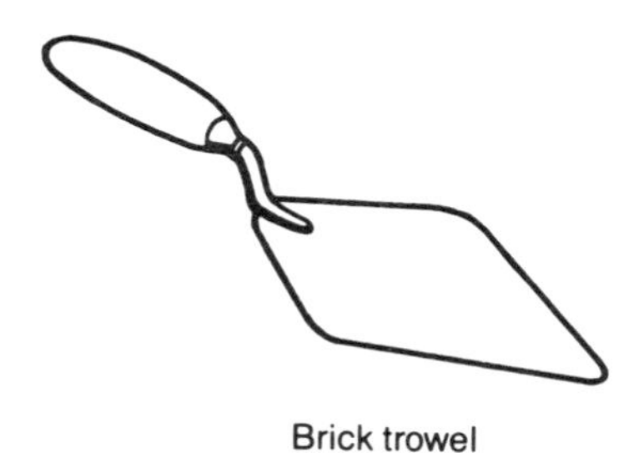

Brick trowel

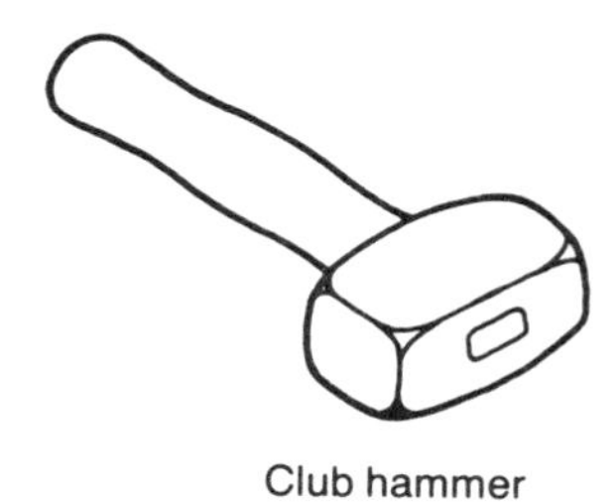

Club hammer

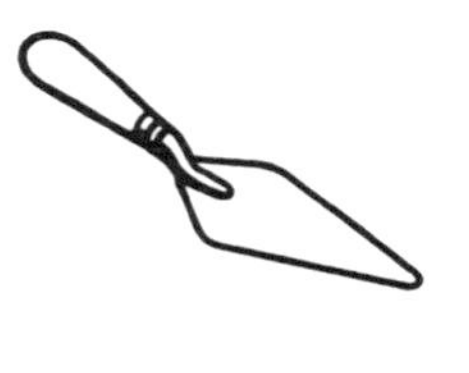

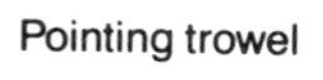

Pointing trowel

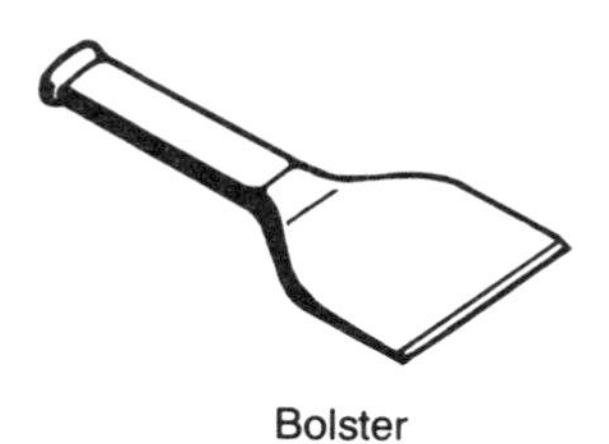

Bolster

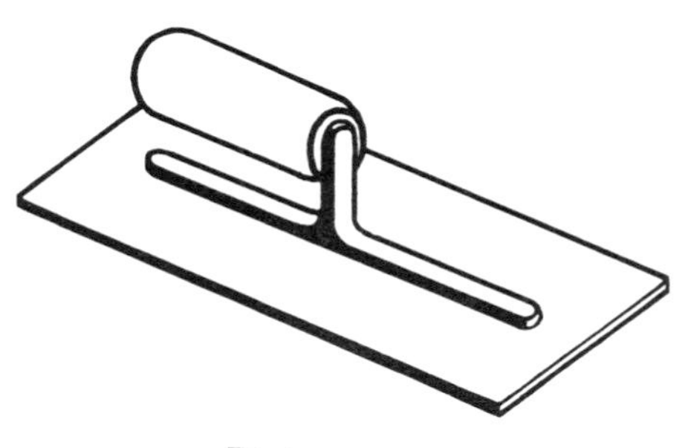

Float

Cold chisel

MOULDINGS

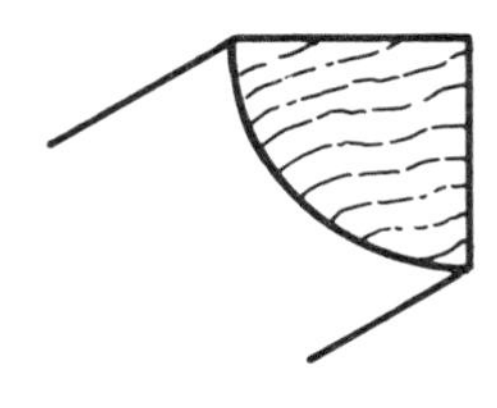

Quarter round

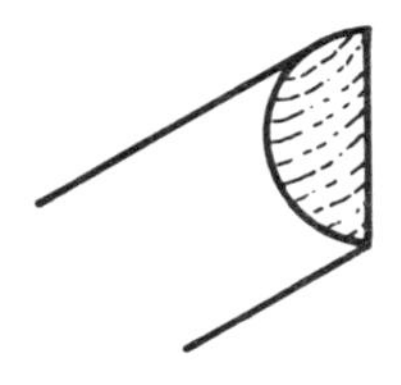

Half round

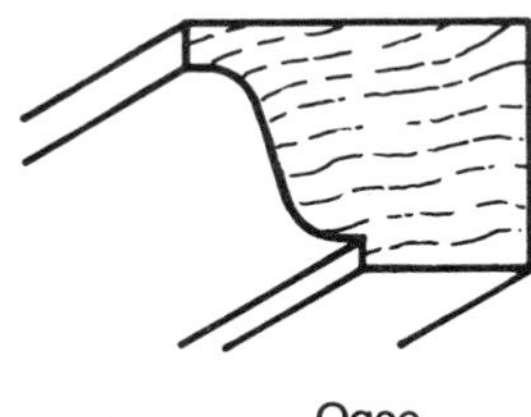

Ogee

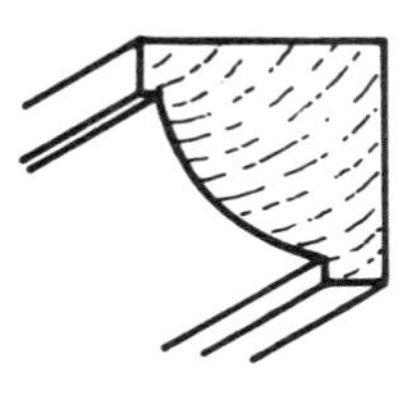

Ovolo

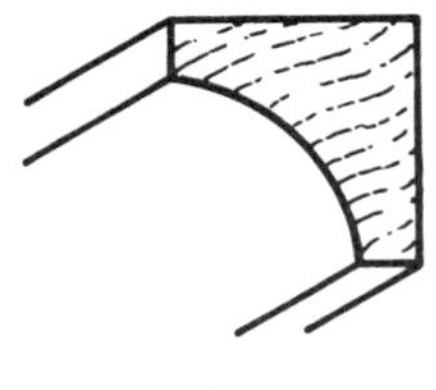

Scotia

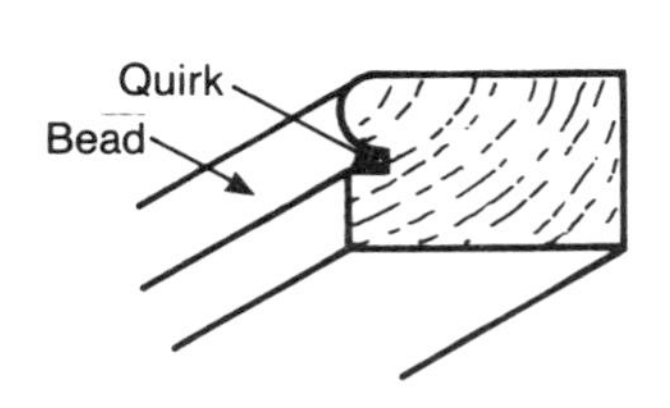

Quirk bead

ROOF

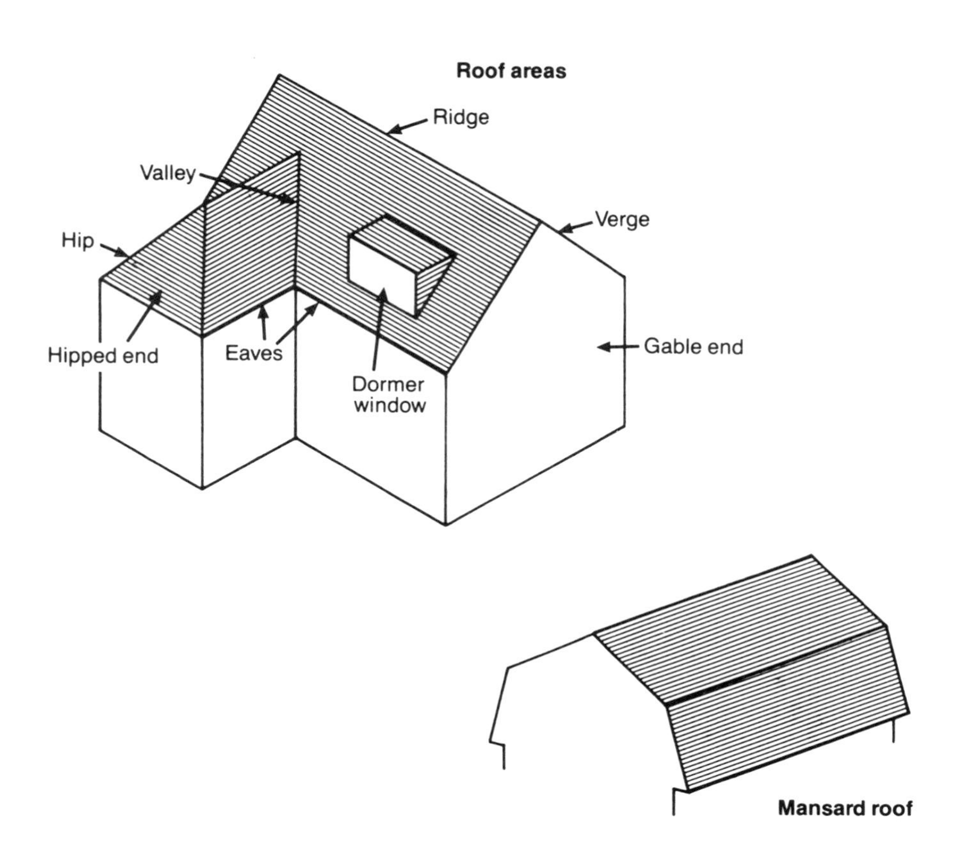
Roof areas
Ridge
Valley
Verge
Hip
Hipped end
Eaves
Dormer
window
Gable end
Mansard roof

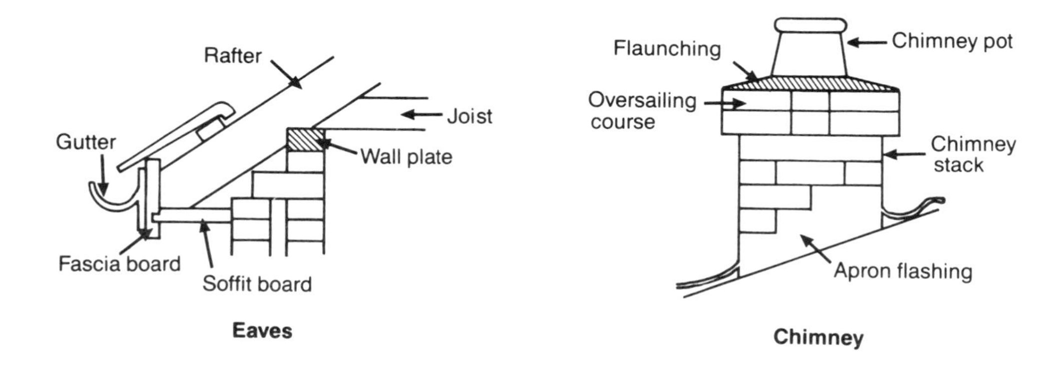
Rafter
Joist
Gutter
Wall plate
Fascia board
Soffit board
Eaves
Flaunching
Chimney pot
Oversailing
course
Chimney
stack
Apron flashing
Chimney

ROOF TRUSS

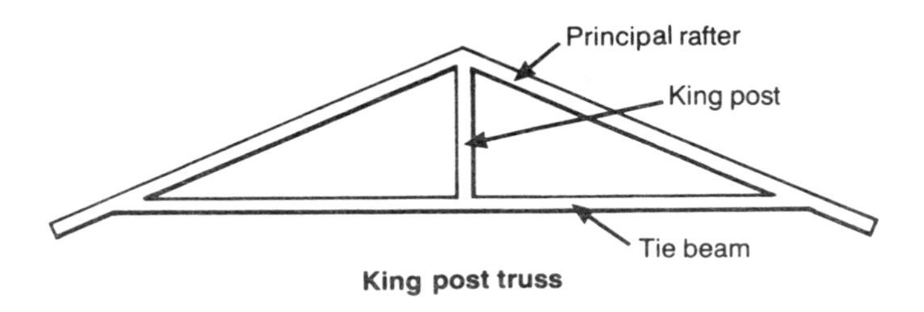

King post truss

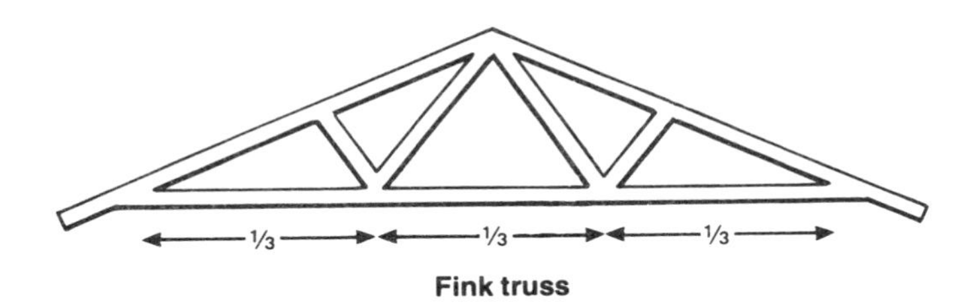

Fink truss

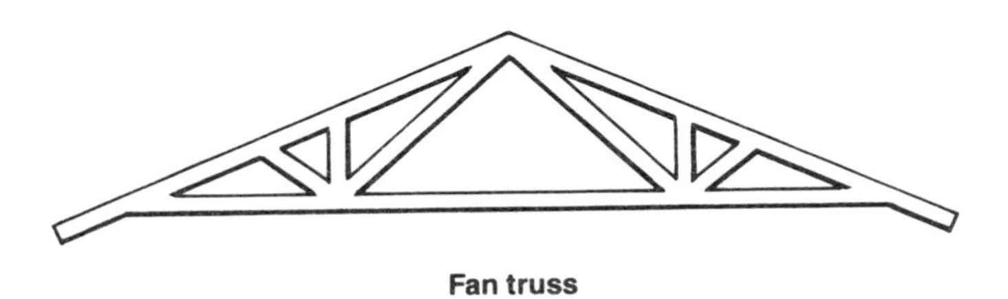

Fan truss

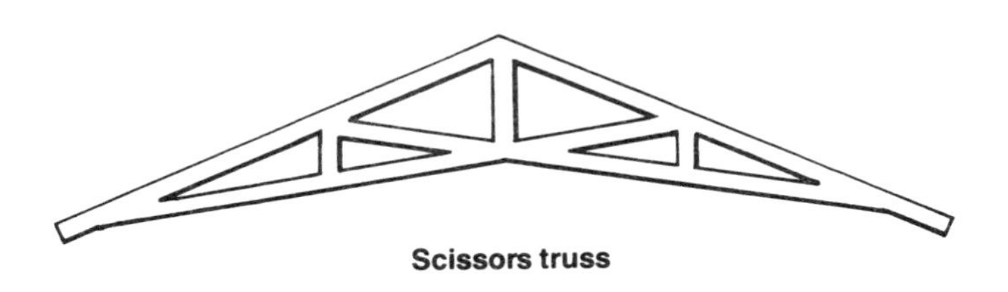

Scissors truss

STAIRS

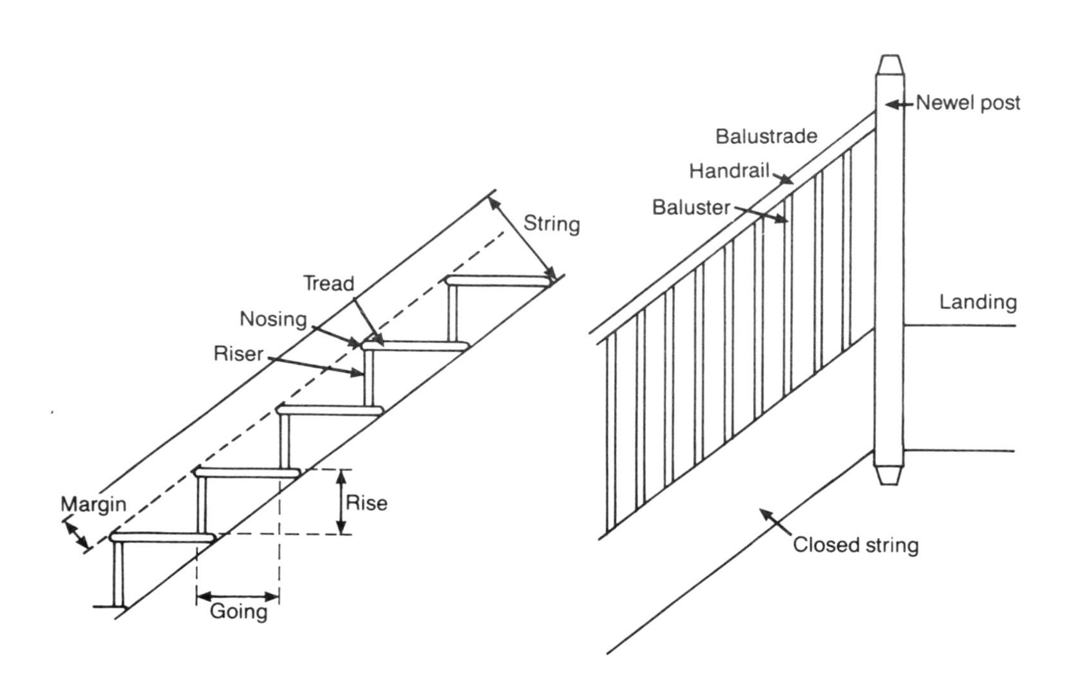
Newel post
Balustrade
Handrail
Baluster
String
Tread
Nosing
Riser
Landing
Margin
Rise
Going
Closed string

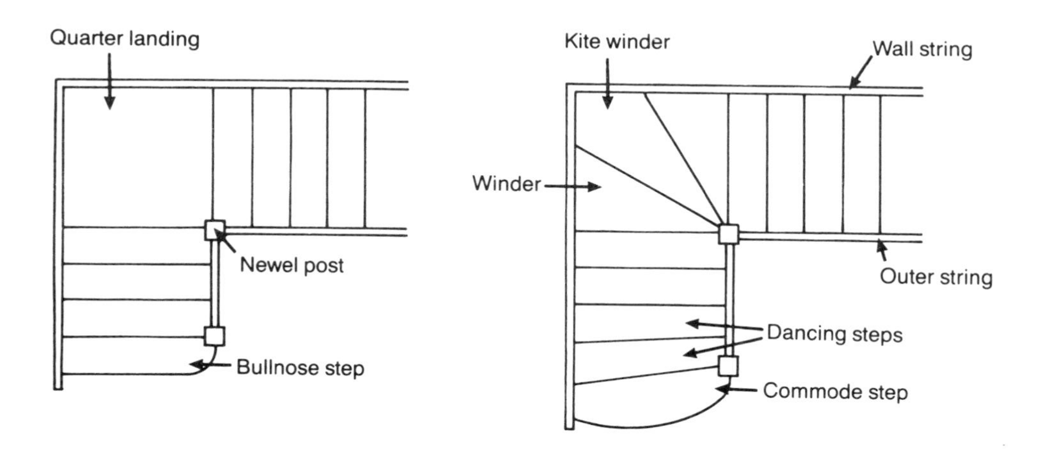
Quarter landing
Newel post
Bullnose step
Kite winder
Wall string
Winder
Outer string
Dancing steps
Commode step

TIMBER JOINTS

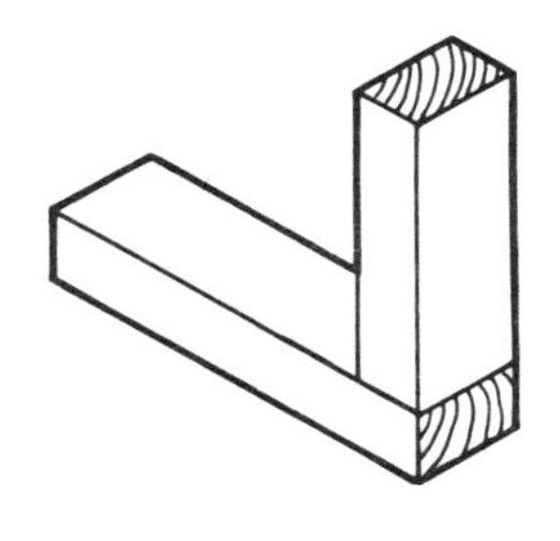

Butt joint

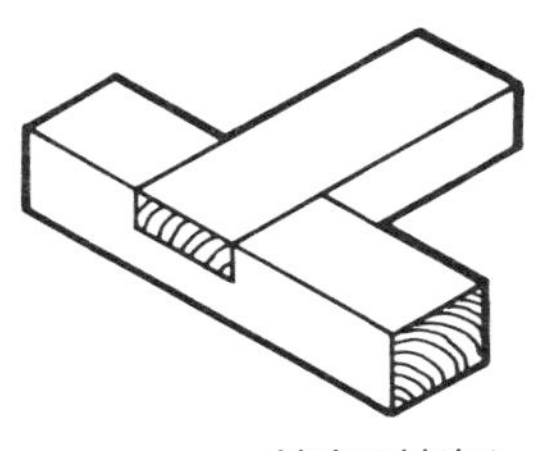

Halved joint

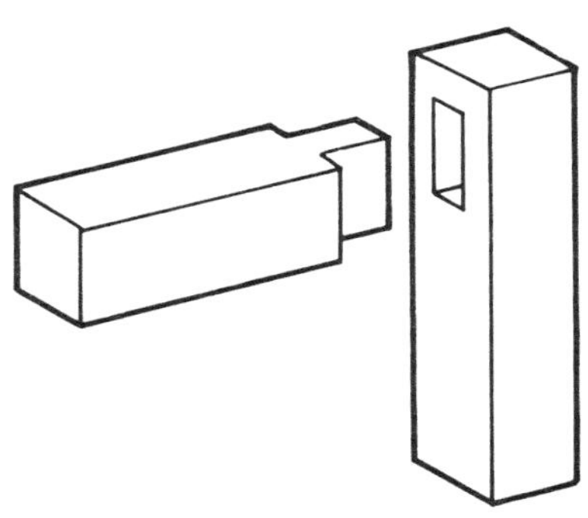

Mortise and tenon

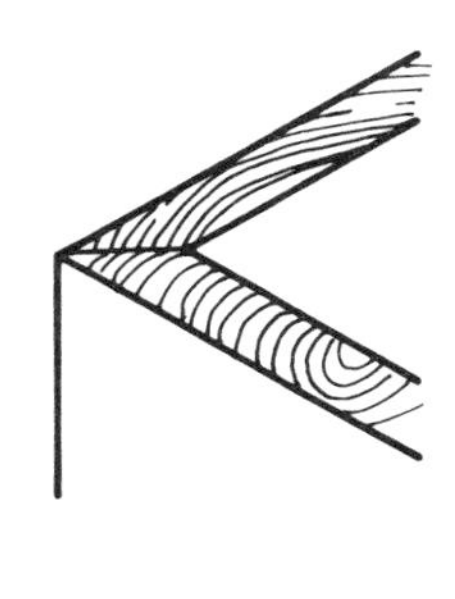

Mitre

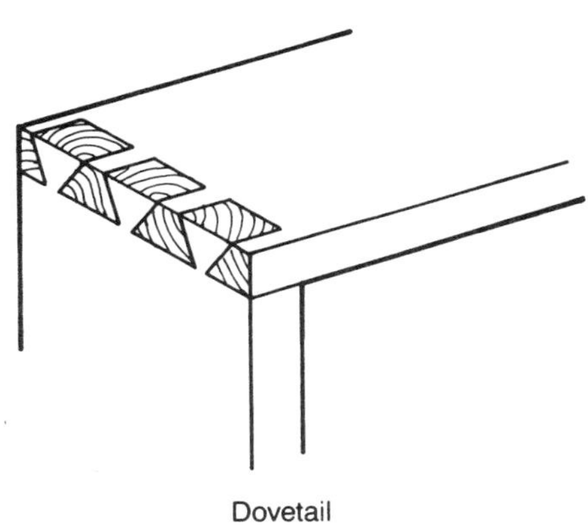

Dovetail

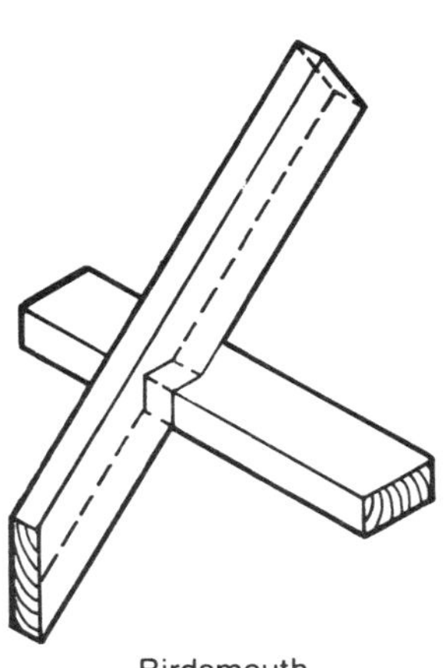

Birdsmouth

WALL

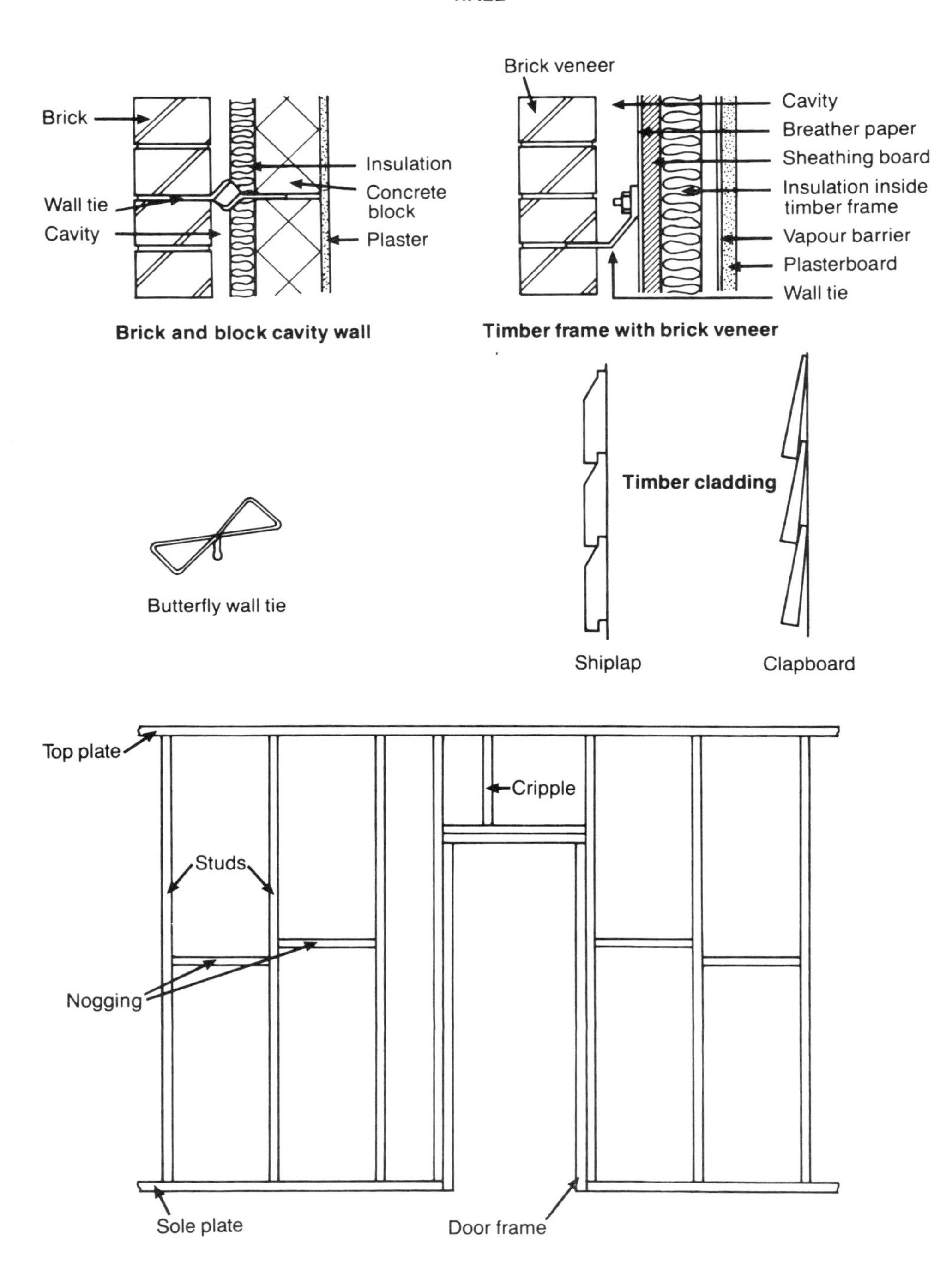

Timber-frame wall

WINDOW

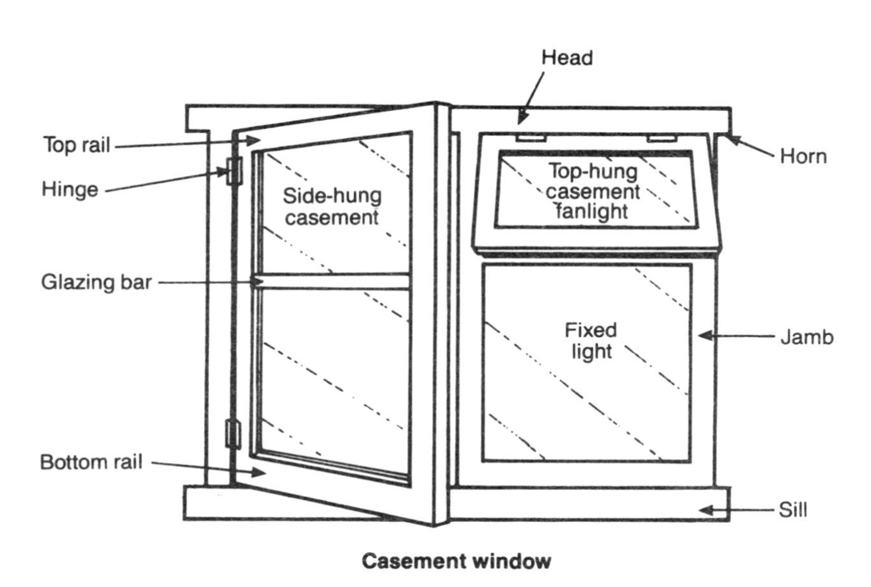

Casement window

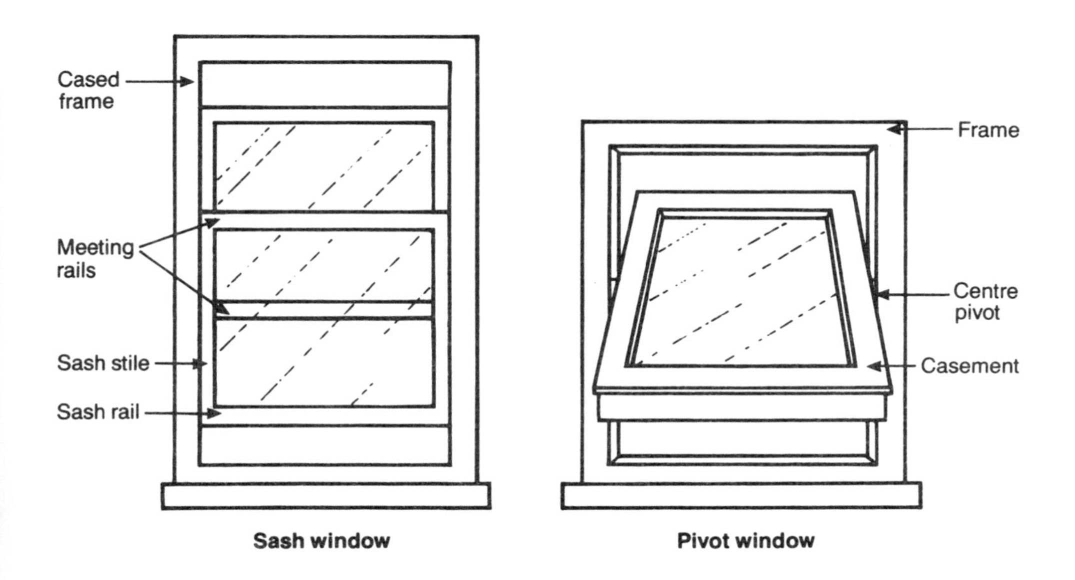

Sash window **Pivot window**

WOODWORKING TOOLS

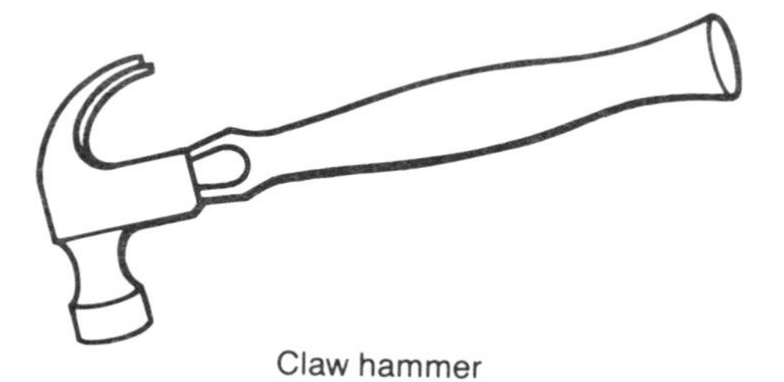

Claw hammer

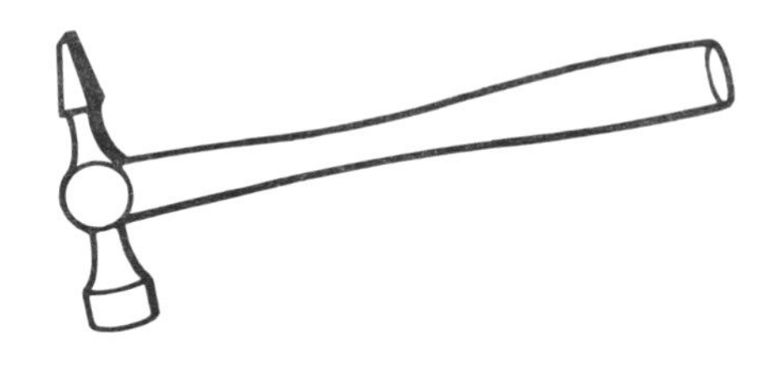

Cross-peen hammer

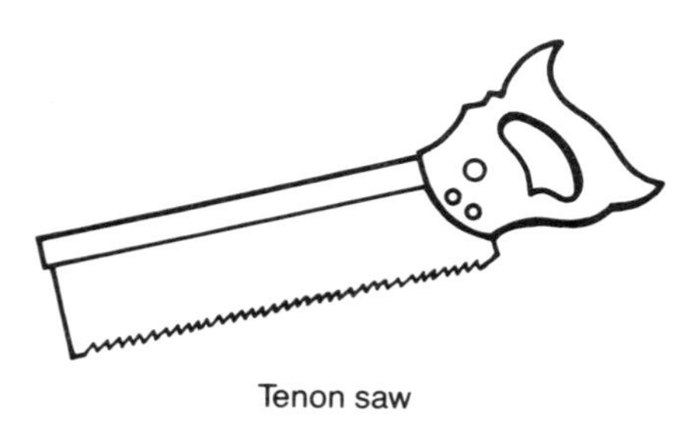

Tenon saw

Pad saw

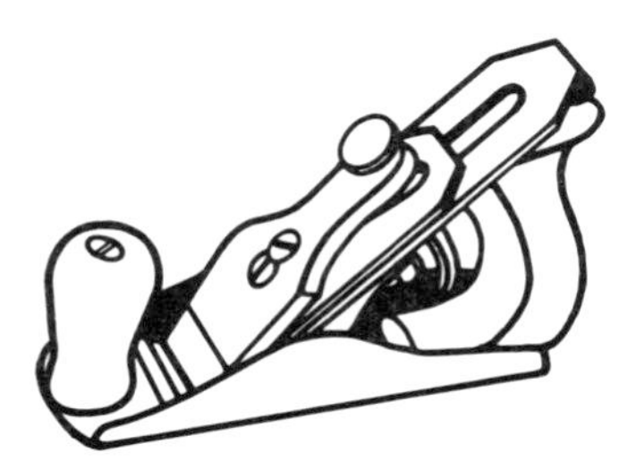

Smoothing plane

Bradawl